AXIOMATIC SET THEORY

PROCEEDINGS OF SYMPOSIA
IN PURE MATHEMATICS

VOLUME XIII, PART II

AXIOMATIC SET THEORY

AMERICAN MATHEMATICAL SOCIETY
PROVIDENCE, RHODE ISLAND
1974

PROCEEDINGS OF THE SYMPOSIUM IN PURE MATHEMATICS
OF THE AMERICAN MATHEMATICAL SOCIETY

HELD AT THE UNIVERSITY OF CALIFORNIA
LOS ANGELES, CALIFORNIA
JULY 10–AUGUST 5, 1967

EDITED BY
THOMAS J. JECH

Prepared by the American Mathematical Society
with the partial support of National Science Foundation Grant GP–6698

Library of Congress Cataloging in Publication Data CIP

Symposium in Pure Mathematics, University of California, Los Angeles, 1967.
Axiomatic set theory.
The papers in pt. 1 of the proceedings represent revised and generally more detailed versions of the lectures.
Pt. 2 edited by T. J. Jech.
Includes bibliographical references.
1. Axiomatic set theory–Congresses. I. Scott, Dana S. ed. II. Jech, Thomas J., ed. III. Title.
IV. Series.
QA248.S95 1967 511'.3 78-125172
ISBN 0-8218-0246-1 (v. 2)

AMS (MOS) subject classifications (1970). Primary 02K99; Secondary 04-00

Printed in the United States of America

CONTENTS

FOREWORD

This volume is the second (and last) part of the Proceedings of the Summer Institute on Axiomatic Set Theory held at U.C.L.A., July 10—August 5, 1967.

Many of the lectures delivered during the Institute have been published in the first volume of these PROCEEDINGS, edited by Dana S. Scott. Although we were unable to obtain all the remaining manuscripts, this volume contains most of them. A small number of the contributions was meanwhile published elsewhere; the complete list of lectures is provided at the end of this volume.

For several reasons, the publication of this volume was slightly delayed. I wish to thank the authors of the papers for their patience.

THOMAS J. JECH

Proceedings of Symposia in Pure Mathematics
Volume 13, Part II, 1974

CURRENT PROBLEMS IN DESCRIPTIVE SET THEORY

J. W. ADDISON

At the suggestion of the Program Coordinator I am speaking to you today[1] on recent progress in and the current state of descriptive set theory, with an emphasis on the problems that remain open. Rather than attempting to give a cursory survey of the whole field, I have thought it would be more useful if I concentrate on certain parts of the theory that I regard as especially fundamental and in which the time seems ripe for further progress.

Descriptive set theory historically has had as its purpose the study of the properties of the continuum. It was founded in the years 1898–1906 by the three eminent French analysts Borel, Lebesgue, and Baire, and germinated in the context of the great debates associated with the discovery of the paradoxes and with the introduction of the axiom of choice. It was the period of the celebrated *cinq lettres* between these three and Hadamard (who also made contributions to the theory and to whom may be traced the origin of its famous uniformization problem).

Always implicit and occasionally explicit in the foundations of the theory was the concept of the *nameable* or *representable* or, as we would more likely say today, the *definable*. Although the languages were not precisely given and perhaps only at best vaguely conceived it would seem fair to say that this was the real beginning of the study of definability in logic. Certainly from our present day standpoint we can regard descriptive set theory as constituting one of the best developed and most fascinating chapters of that branch of logic nowadays often referred to as *the theory of definability*.

The fundamental problem of descriptive set theory has always been to *characterize* and *classify* those sets of real numbers definable in a given grammatical form

AMS (*MOS*) *subject classifications* (1970). Primary 02K30, 04A15, 02F35, 54H05; Secondary 28A05, 26A21.

[1] Monday, July 17, 1967.

and to study the relationships among such classes. Although there have been times in its development when it has gotten wrapped up in interesting sidetracks, it has always returned to these basic problems. These central problems are, of course, the central problems of the general theory of definability.

Each chapter of descriptive set theory has usually begun with the consideration of particular sets and the "test problems" of finding what kinds of conditions are needed to define them. More often than not it is reasonably easy to find various definitions of a given set which turn out to be the simplest possible according to corresponding criteria of simplicity—the challenge then comes in showing that these definitions actually are the simplest, or in other words in proving *undefinability*. For the solution of this problem one is forced to develop conditions characterizing the sets in a given class. From there the theory normally moves on into general structural questions.

From almost the beginning it has been recognized that the natural setting for descriptive set theory is the set ${}^{\omega}\omega$ of functions from ω into ω (sometimes disguised as the set of irrational numbers in their continued fraction representation)—or alternatively the set ${}^{\omega}2$ of functions from ω into $2\ (= \{0, 1\})$ (sometimes disguised as the set of real numbers in the Cantor set in their decimal representation)—rather than the set $\boldsymbol{R}$ of real numbers itself. Not only is the theory simpler and more elegant there—chiefly because of the "simply definable" maps between ${}^{\omega}\omega$ and its countable direct powers—but also there is no real loss of generality because of the "simply definable" maps between ${}^{\omega}\omega$, ${}^{\omega}2$, and $\boldsymbol{R}$. Throughout this talk we regard ${}^{\omega}\omega$ as the basic universe the subsets of which we are considering.

Classical descriptive set theory is now generally considered as being essentially unified with the study of hierarchies in recursive function theory. This unification has brought a simplification of the theory and many new insights into it. Just as Rogers has viewed the theory of general recursive functions as a kind of idealized, simpler, asymptotic limit of more practical and complicated theories of computation, so classical descriptive set theory may be viewed, to a certain extent, as a similar limit of more complicated theories arising in the study of hierarchies in recursive function theory. In this idealized theory we pretend as known and understood each particular element of the continuum and concentrate our attention on the subsets of the continuum. In the present talk, both to emphasize the set-theoretic aspects over the computational aspects and also to focus attention on certain facets of the problems by stripping away secondary details, we restrict our attention to classical descriptive set theory. But the knowledgeable listener will be able in most cases to read effective analogs of the results and problems into what is said.

The first example of a classification in descriptive set theory was the Borel hierarchy, first introduced in 1898 although it was not proven at that time that it was really a properly increasing "hierarchy." It appeared again in later years in many different essentially equivalent formulations associated with the names Baire, de la Vallée Poussin, Hausdorff, and others.

In one formulation we can define the Borel hierarchy generated by a class $\mathscr{C}$ of subsets of ${}^{\omega}\omega$ recursively as follows. For any ordinal ν let

$$\mathrm{Bo}_\nu(\mathscr{C}) = \mathscr{C} \qquad \text{if } \nu = 0,$$
$$= \{\bigcup \mathrm{Rn}\, A : A \in {}^{\omega}\mathrm{Bo}^{-}_{(\nu)}(\mathscr{C})\} \quad \text{otherwise},$$

where, for any function $\mathfrak{H}$ from ordinals to subsets of ${}^{\omega}\omega$, $\mathfrak{H}_{(\nu)} = \bigcup\{\mathfrak{H}_\rho : \rho < \nu\}$ and, for any class $\mathscr{C}$ of subsets of ${}^{\omega}\omega$, $\mathscr{C}^- = \{X : \sim X \in \mathscr{C}\}$.

In particular if $\mathbf{\Sigma}^0_1$ is the class of open sets then $\langle \mathrm{Bo}_\nu(\mathbf{\Sigma}^0_1) : \nu \in \Omega\rangle$ is the classical Borel hierarchy $\langle$open, F_σ, $G_{\delta\sigma}$, $F_{\sigma\delta\sigma}, \ldots\rangle$; in the standard terminology of the theory of definability

$$\mathrm{Bo}_\nu(\mathbf{\Sigma}^0_1) = \mathbf{\Sigma}^0_{\nu+1} \quad \text{and} \quad \mathrm{Bo}^-_\nu(\mathbf{\Sigma}^0_1) = \mathbf{\Pi}^0_{\nu+1} \qquad \text{for finite } \nu \text{ and}$$
$$\mathrm{Bo}_\nu(\mathbf{\Sigma}^0_1) = \mathbf{\Sigma}^0_{\nu} \quad \text{and} \quad \mathrm{Bo}^-_\nu(\mathbf{\Sigma}^0_1) = \mathbf{\Pi}^0_{\nu} \qquad \text{for infinite } \nu.$$

This hierarchy has the typical bilateral or two-sided feature which is associated with the role of denial in the defining language (or with the operation of complementation). Note that we write "$\mathrm{Bo}_\nu(\mathscr{C})$" for "$\mathrm{Bo}(\mathscr{C})_\nu$" and "$\mathrm{Bo}^-_\nu(\mathscr{C})$" for "$\mathrm{Bo}_\nu(\mathscr{C})^-$." We call Bo and Bo^-, or $\mathrm{Bo}(\mathscr{C})$ and $\mathrm{Bo}^-(\mathscr{C})$, *duals* of each other. It is often interesting in the context of such hierarchies to consider, for a class $\mathscr{C}$ of subsets of ${}^{\omega}\omega$, the classes $\mathscr{C} \cap \mathscr{C}^-$ and $\mathscr{C} \cup \mathscr{C}^-$; if we think of a class $\mathscr{C}$ also as "$\mathscr{C}^+$" then it is natural to write $\mathscr{C}^*$ ("$\mathscr{C}$ plus *and* minus" or "$\mathscr{C}$ star") for $\mathscr{C} \cap \mathscr{C}^-$ and $\mathscr{C}^\pm$ ("$\mathscr{C}$ plus *or* minus") for $\mathscr{C} \cup \mathscr{C}^-$. In this terminology $\mathrm{Bo}^*_\nu(\mathbf{\Sigma}^0_1)$ is what is called in the theory of definability $\mathbf{\Delta}^0_{\nu+1}$ or $\mathbf{\Delta}^0_\nu$ according as ν is finite or infinite.

It was Baire who first set out to show that the Borel "hierarchy" is actually a hierarchy. In 1906 he made important contributions to the problem of characterizing the classes $\mathrm{Bo}_\nu(\mathbf{\Sigma}^0_1)$ and $\mathrm{Bo}^*_\nu(\mathbf{\Sigma}^0_1)$ for $\nu = 1$ and 2. In particular he worked on the test problem of classifying the set $\{\alpha : \forall m \exists n \forall p > n\alpha(p) > m\}$ of eventually arbitrarily large functions and developed machinery to prove that it is in $\mathbf{\Pi}^0_3 \sim \mathbf{\Sigma}^0_3$. Baire's work was reformulated and extended to higher classes by Lebesgue, Hausdorff, Lavrent'ev, and Luzin. Out of this work came the first example of a *construction principle* in which the sets in a given class are characterized as those constructed in a certain way out of "simpler" sets.

Let $\bigoplus$ be the operation of alternating sum—i.e., for any ordinal ν and any ν-family A of sets let $A_\nu = \varnothing$ and let

$$\bigoplus A = \bigcup\{A_\rho \sim A_{\rho+1} : \rho \in \nu \;\&\; \rho \text{ is even}\}.$$

Call a ν-family A of sets $\subseteq$-*reversing* or *antitone* if and only if for any $\rho, \sigma \in \nu$ if $\rho \leqq \sigma$ then $A_\rho \supseteq A_\sigma$.

Then for any class $\mathscr{C}$ of subsets of ${}^{\omega}\omega$ we can define the Hausdorff hierarchy generated by $\mathscr{C}$ as follows. For any ordinal ν,

$$\mathrm{Ha}_\nu(\mathscr{C}) = \{\bigoplus A : A \in {}^{\nu}\mathscr{C} \;\&\; A \text{ is antitone}\}.$$

In terms of this we have the following construction principle for $\mathbf{\Delta}^0_2$:

$$\mathbf{\Delta}^0_2 = \mathrm{Ha}_{(\Omega)}(\mathbf{\Pi}^0_1),$$

or equivalently,

$$\mathrm{Bo}^*_1(\mathbf{\Sigma}^0_1) = \mathrm{Ha}_{(\Omega)}\mathrm{Bo}^-_0(\mathbf{\Sigma}^0_1).$$

This has turned out to be a very typical phenomenon. Later the principle was extended to all $\nu \in \Omega$:

$$\mathrm{Bo}^*_{\nu+1}(\mathbf{\Sigma}^0_1) = \mathrm{Ha}_{(\Omega)}\mathrm{Bo}^-_\nu(\mathbf{\Sigma}^0_1).$$

This gives an elegant classification of all the Bo^*_ν classes of the classical Borel hierarchy when ν is a successor ordinal. The case of the limit-ordinal classes Bo^*_λ has long remained open and is singled out for attention by Luzin in his classic memoir of 1930. In the past several years there has been renewed attention to this problem, and several nonexhausting hierarchies in $\mathrm{Bo}^*_\lambda(\mathbf{\Sigma}^0_1)$ have been discovered and studied.

The remaining case, when $\nu = 0$, has been given several successful independent treatments, however. The first full solution was by Luzin in 1930, although Brouwer in 1927 had suggested something similar. Another solution was found by Kalmàr in 1953, and in the effective case by Kleene also in 1953. Each of the four was apparently unaware of the work of the other three. These various solutions, both in the classical and in the effective case, have all been studied and compared by Robert F. Barnes, Jr. in his doctoral dissertation (Berkeley, 1965), who showed that two fundamental approaches—genetic and relational—lead essentially to the same classification.

For $k \in \omega$ and $\alpha \in {}^\omega\omega$ let $k\alpha$ be the sequence obtained by "sticking k on the front of the sequence α"—i.e., for any $n \in \omega$, $(k\alpha)(n) = k$ if $n = 0$ and $\alpha(n-1)$ otherwise. For any $X \subseteq {}^\omega\omega$ we let $X_{(k)}$ be the kth *detail* of X, i.e., $X_{(k)} = \{\alpha : k\alpha \in X\}$. Then for any class $\mathscr{C}$ of subsets of ${}^\omega\omega$ we define the Kalmar hierarchy $\mathrm{Ka}(\mathscr{C})$ generated by $\mathscr{C}$ as follows:

$$\begin{aligned}\mathrm{Ka}_\nu(\mathscr{C}) &= \mathscr{C} && \text{if } \nu = 0,\\ &= \{X : \forall k\ X_{(k)} \in \mathrm{Ka}^-_{(\nu)}(\mathscr{C})\} && \text{otherwise.}\end{aligned}$$

In terms of this there is the following construction principle for $\mathbf{\Delta}^0_1$:

$$\mathbf{\Delta}^0_1 = \mathrm{Ka}_{(\Omega)}(\{\varnothing\}^{\pm}),$$

or equivalently,

$$\mathrm{Ha}^*_1(\mathbf{\Pi}^0_1) = \mathrm{Ka}_{(\Omega)}\mathrm{Ha}^{\pm}_0(\mathbf{\Pi}^0_1).$$

In the course of trying to discover if there was any uniform way of considering Ka and Ha, Barnes and I found, to our surprise, the following generalization of the Kalmar construction principle: for any ordinal $\nu \in \Omega$,

$$\mathrm{Ha}^*_{\nu+1}(\mathbf{\Pi}^0_1) = \mathrm{Ka}_{(\Omega)}\mathrm{Ha}^{\pm}_\nu(\mathbf{\Pi}^0_1).$$

Since $\mathrm{Ha}_0(\mathscr{C}) = \{\varnothing\}$ for any $\mathscr{C}$, this includes the Kalmar principle as the case when $\nu = 0$. Whether there is a suitable generalization of this from $\mathbf{\Pi}^0_1$ to $\mathbf{\Pi}^0_\nu$ seems likely and is under investigation. In fact still greater generalizations appear to be a possibility.

Out of this emerges the fundamental question, not yet completely understood: *Why should there be these construction principles for $\mathscr{C}^*$ classes?* Obviously their existence is more than just a coincidence.

The above two families of construction principles for Bo* and Ha* classes are not the only ones, of course. One of the most productive periods in the history of descriptive set theory was in the famous seminar of Nikolaĭ Luzin, founder of the Moscow mathematical school, held at the University of Moscow exactly fifty years ago this academic year. (Among the students were Aleksandrov, Suslin, and Urysohn.) It was in this seminar that Aleksandrov discovered *fusion*—or as it was called in the quaint terminology of yesteryear "the operation (A)" ("A" was for "analytic").

Let Sq be the set $\bigcup\{{}^n\omega : n \in \omega\}$ of finite sequences of natural numbers. Fusion is an operation on families of sets indexed by Sq. For such a family A let

$$*A = *\langle A_s : s \in \mathrm{Sq}\rangle = \bigcup\{\bigcap\{A_{\alpha|n} : n \in \omega\} : \alpha \in {}^\omega\omega\}.$$

Fusion and complementation, like countable union and complementation, can be used to generate hierarchies of sets. For any class $\mathscr{C}$ of subsets of ${}^\omega\omega$ we can define the Suslin hierarchy generated by $\mathscr{C}$ as follows:

$$\begin{aligned} \mathrm{Su}_\nu(\mathscr{C}) &= \mathscr{C} && \text{if } \nu = 0,\\ &= \{*A : A \in {}^{\mathrm{Sq}}\mathrm{Su}^{\neg}_{(\nu)}(\mathscr{C})\} && \text{otherwise.}\end{aligned}$$

Suslin showed that $\mathrm{Su}(\mathbf{\Sigma}^0_1)$ is a genuine hierarchy at least at the first level—i.e., he showed that there is a $\mathrm{Su}_1(\mathbf{\Sigma}^0_1)$, or analytic or $\mathbf{\Sigma}^1_1$, set whose complement is not analytic. He also established one of the most famous of the construction principles, namely

$$\mathbf{\Sigma}^{1*}_1 = \mathrm{Bo}_{(\Omega)}(\mathbf{\Sigma}^0_1),$$

or equivalently,

$$\mathrm{Su}^*_1(\mathbf{\Sigma}^0_1) = \mathrm{Bo}_{(\Omega)}\mathrm{Su}_0(\mathbf{\Sigma}^0_1).$$

We might naturally wonder if this does not generalize to $\nu > 1$, i.e., if $\mathrm{Su}^*_{\nu+1}(\mathbf{\Sigma}^0_1) = \mathrm{Bo}_{(\Omega)}\mathrm{Su}_\nu(\mathbf{\Sigma}^0_1)$. Actually Suslin never really investigated the $\mathrm{Su}(\mathbf{\Sigma}^0_1)$ hierarchy beyond the first level. These investigations were carried out in 1927 by Selivanovskiĭ. Effective analogs of this hierarchy have been studied by Guy M. Benson in his doctoral dissertation (Berkeley, 1965). Selivanovskiĭ noted that the easy $\supseteq$-half (the *inclusion principle*) holds for all ν but was unable to settle the $\subseteq$-half (the *exhaustion principle*). This problem was solved negatively by Kunugui in 1937—the exhaustion principle fails for all $\nu > 0$. It is possible to pinpoint quite closely the reason for this failure: at these levels we are beyond the point where the concept of a well-ordering of ω is definable, i.e., the set of elements of ${}^\omega\omega$ representing (in any "simple" way) well-orderings of ω is $\mathrm{Su}^{\neg}_1(\mathbf{\Sigma}^0_1)$ (or a complement of an analytic set), but not $\mathrm{Su}_1(\mathbf{\Sigma}^0_1)$.

Soon after fusion was introduced and it was shown that the analytic sets could be characterized either as fusions of closed sets or as *projections* of closed sets,

Luzin wondered if projection and complementation, like countable union and complementation, can be used to generate hierarchies of sets.

Unlike the other operations considered here, projection is awkward because it makes it necessary to consider powers of ${}^{\omega}\omega$ as well as ${}^{\omega}\omega$ itself. But because of the existence of simple maps between ${}^{\omega}\omega \times {}^{\omega}\omega$ and ${}^{\omega}\omega$ this presents no real problem. Let δ be the unpairing map from ${}^{\omega}\omega$ onto ${}^{\omega}\omega \times {}^{\omega}\omega$ taking α in ${}^{\omega}\omega$ to $\langle\langle\alpha(2n):n \in \omega\rangle, \langle\alpha(2n+1):n \in \omega\rangle\rangle$ and let π be the projection map from ${}^{\omega}\omega \times {}^{\omega}\omega$ onto ${}^{\omega}\omega$ taking $\langle\alpha, \beta\rangle$ in ${}^{\omega}\omega \times {}^{\omega}\omega$ to α. For any function f and set X let $f[X] = \{f(x):x \in X\}$. Then for any $\mathscr{C} \subseteq P^{\omega}\omega$ we can define the Luzin or projective hierarchy generated by $\mathscr{C}$ as follows:

$$\begin{aligned} \mathrm{Lu}_\nu(\mathscr{C}) &= \mathscr{C} && \text{if } \nu = 0, \\ &= \{\pi[\delta[X]] : \forall k\ X_{(k)} \in \mathrm{Lu}^{-}_{(\nu)}(\mathscr{C})\} && \text{otherwise.} \end{aligned}$$

The hierarchy $\langle \mathrm{Lu}_\nu(\mathbf{\Sigma}^0_1) : 0 < \nu < \omega\rangle$ is the classical projective hierarchy; in the standard terminology of the theory of definability

$$\mathrm{Lu}_\nu(\mathbf{\Sigma}^0_1) = \mathbf{\Sigma}^1_\nu, \quad \mathrm{Lu}^{-}_\nu(\mathbf{\Sigma}^0_1) = \mathbf{\Pi}^1_\nu \quad \text{and} \quad \mathrm{Lu}^*_\nu(\mathbf{\Sigma}^0_1) = \mathbf{\Delta}^1_\nu$$

for positive finite ν.

Now $\mathrm{Lu}_1(\mathbf{\Sigma}^0_1) = \mathrm{Su}_1(\mathbf{\Sigma}^0_1)$ so the Suslin construction principle can also be written

$$\mathrm{Lu}^*_1(\mathbf{\Sigma}^0_1) = \mathrm{Bo}_{(\Omega)}\mathrm{Lu}_0(\mathbf{\Sigma}^0_1).$$

But since projection is stronger than fusion this principle has no chance at all of being extended from 0 to higher ν. In fact one natural possibility would be that

$$\mathrm{Lu}^*_2(\mathbf{\Sigma}^0_1) = \mathrm{Su}_{(\Omega)}\mathrm{Lu}_1(\mathbf{\Sigma}^0_1).$$

Here the $\supseteq$ or inclusion half holds, as was first shown by Kantorovič and Livenson in 1934, but the $\subseteq$ or exhaustion half fails—again for the same reason as before that this is beyond the level of the concept of well-ordering of ω.

Indeed no construction principle has ever been found for $\mathrm{Su}^*_\nu(\mathbf{\Sigma}^0_1)$ or $\mathrm{Lu}^*_\nu(\mathbf{\Sigma}^0_1)$ for $\nu > 1$, or indeed for any *-class arising in descriptive set theory beyond the point where the concept of a well-ordering of ω is definable. In fact all proofs of the failure of a proposed exhaustion principle for a class $\mathscr{D}$ by a hierarchy $\mathfrak{H}(\mathscr{C})$ have used the idea of Kunugui of using the $\mathbf{\Pi}^1_1$ definition of a well-ordering of ω to construct a universal set for $\mathfrak{H}_{(\Omega)}(\mathscr{C})$ which is in $\mathscr{D}$. Although the vague idea of what constitutes a "reasonable" or "nice" construction principle is a prerequisite to a proof that no reasonable or nice construction principle for these classes can be given, the Kunugui technique is clearly sufficiently broad to rule out a wide class of principles based on hierarchies of type Ω. Some interesting work along the lines of proposing a satisfactory definition of a "nice" hierarchy has recently been done by Moschovakis. Hierarchies of type Ω_2 are not ruled out by the Kunugui technique at least until the level where the concept of a well-ordering of Ω is passed, but such hierarchies would have to be generated by Ω-ary set operations, and thus far no very natural Ω-ary operations have been proposed for study.

Many operations intermediate in power between fusion and projection have been studied. In fact in 1928 Kolmogorov undertook a general study of such operations. He considered the proportion

$$\frac{\text{countable union}}{\text{fusion}} = \frac{\text{fusion}}{?}$$

and found a solution. Analyzing the process of going from fusion to the solution of this proportion he discovered the operator R which maps set operations into set operations and which, roughly speaking, satisfies the condition

$$R\,(\text{countable union}) = \text{fusion}.$$

Letting Γ be countable union and beginning with $R^0(\Gamma) = \Gamma$, we can iterate R to obtain for each $\rho \in \Omega$ an operation $R^\rho(\Gamma)$. Each operation $R^\rho(\Gamma)$ together with complementation induces the hierarchy $\mathrm{Ko}^\rho(\mathscr{C})$ generated by a class $\mathscr{C}$ of subsets of ${}^\omega\omega$. These hierarchies satisfy the inclusion principles

$$\mathrm{Ko}^{\rho+1*}_{\nu+1}(\mathbf{\Sigma}^0_1) \supseteq \mathrm{Ko}^\rho_{(\Omega)}\mathrm{Ko}^{\rho+1}_\nu(\mathbf{\Sigma}^0_1)$$

but again the corresponding exhaustion principles fail (except for the case of Suslin's theorem). Moreover the class $\mathrm{Lu}^*_2(\mathbf{\Sigma}^0_1)$ of $\mathbf{\Delta}^1_2$ sets properly includes the class $\bigcup\{\mathrm{Ko}^\rho_{(\Omega)}(\mathbf{\Sigma}^0_1) : \rho \in \Omega\}$ of the so-called R-sets. A modern study of these hierarchies has recently been carried out by Peter G. Hinman in his doctoral dissertation (Berkeley, 1966), who has also investigated their effective analogs and shown illuminating interconnections with the theory of recursive functions of higher type.

Construction principles in descriptive set theory have long been associated with separation principles. Recall that for any class $\mathscr{C}$ of subsets of ${}^\omega\omega$: (i) $\mathscr{C}$ has the *first separation property* (relative to ${}^\omega\omega$) iff for any disjoint $X, Y \in \mathscr{C}$ there exists a Z in $\mathscr{C}^*$ such that $X \subseteq Z$ and $Z \cap Y = \varnothing$, and (ii) $\mathscr{C}$ has the *reduction property* if and only if for any $X, Y \in \mathscr{C}$ there exist X', Y' in $\mathscr{C}$ such that $X' \subseteq X$, $Y' \subseteq Y$, $X' \cup Y' = X \cup Y$, and $X' \cap Y' = \varnothing$. If $\mathscr{C}$ has the *coreduction property* (i.e., $\mathscr{C}^-$ has the reduction property) then $\mathscr{C}$ has the first separation property. If there is a class $\mathscr{D}$ of subsets of ${}^\omega\omega$ such that for any $X, Y \in \mathscr{C}$ there exists a Z in $\mathscr{D}$ such that $X \subseteq Z$ and $Z \cap Y = \varnothing$ then $\mathscr{D}$ is a separation class for $\mathscr{C}$. $\mathscr{C}^*$ is included in any separation class for $\mathscr{C}$. Thus if there is a so-called *strong separation principle* stating that a subclass $\mathscr{D}$ of $\mathscr{C}^*$ is a separation class for $\mathscr{C}$ then we know both that $\mathscr{D}$ exhausts $\mathscr{C}^*$ and that $\mathscr{C}$ has the first separation property. These connections can be summarized in the following chart:

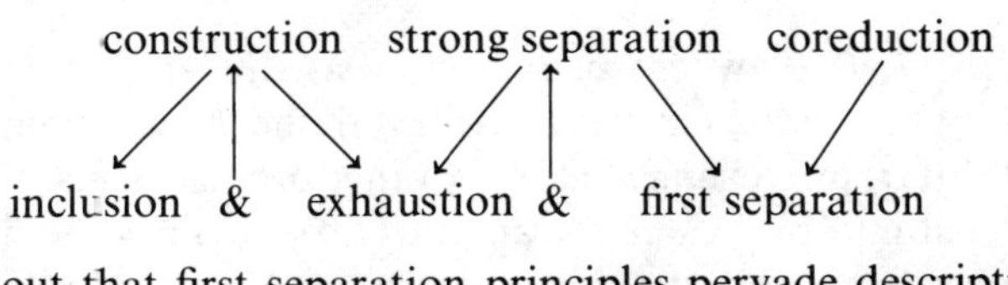

Now it turns out that first separation principles pervade descriptive set theory. In fact for every class $\mathscr{C}$ in any of the hierarchies generated from closed or open sets and discussed above either $\mathscr{C}$ or $\mathscr{C}^-$ has the first separation property or it is at least consistent with Zermelo-Fraenkel set theory that $\mathscr{C}$ or $\mathscr{C}^-$ has the first separation property. Of course for selfdual classes such as those appearing in the Kalmar hierarchies this is trivial, but in the Hausdorff, Borel, Suslin, Kolmogorov, and Luzin hierarchies it is not. The reason for this phenomenon is not well understood: It constitutes one of the fundamental problems of descriptive set theory.

Another interesting phenomenon is that in nonselfdual classes in these hierarchies it never happens that both $\mathscr{C}$ and $\mathscr{C}^-$ have the first separation property. To a certain extent this phenomenon is understood, at least in the sense that a uniform style of proof can be given for all cases. But even here there is evidence—in terms of the behavior of hierarchies outside of descriptive set theory—that the reason lies deeper.

The roots of some of these one-but-not-both-sides phenomena may someday be traced in part to the Separation Urprinciple:

$$\{\varnothing\}^- \text{ has the first separation property but } \{\varnothing\} \text{ does not.}$$

Of course it follows by definition that when there is a first separation principle for $\mathscr{C}$ together with an exhaustion principle for $\mathscr{C}^*$ then there is automatically a strong separation principle for $\mathscr{C}$. What is interesting is that in all cases discussed above more is true—namely, that a proof for the strong separation principle can be obtained by a slight generalization of the proof of the exhaustion principle. The reason for this is not clearly understood.

On which side of the two-sided hierarchies does the first separation principle hold? For any $\nu \in \Omega$, $\mathrm{Bo}_\nu^-(\mathbf{\Sigma}_1^0)$ has it. For any $\nu \in \Omega$, $\mathrm{Su}_\nu(\mathbf{\Pi}_1^0)$ has it. For any $\nu \in \Omega$, $\mathrm{Ha}_{2\nu}(\mathbf{\Pi}_1^0)$ and $\mathrm{Ha}_{2\nu+1}^-(\mathbf{\Pi}_1^0)$ have it. What is the explanation for this shifting around?

If we compare the definitions of the hierarchies discussed thus far, we note that they are all given by recursion except one—namely Ha. Some step in the direction of a unified understanding of these hierarchies and of explaining the behavior of the separation principle has recently been made by the following recasting of the definition of the Hausdorff hierarchies.

For any class $\mathscr{D}$ of subsets of ${}^\omega\omega$ let an ω-family A of sets be *$\mathscr{D}$-separated* if and only if there exists a B in ${}^\omega\mathscr{D}$ such that, for any distinct m, n in ω, $A_n \subseteq B_n$ and $B_m \cap B_n = \varnothing$. Then, for any $\mathscr{C} \subseteq P^\omega\omega$ and any ordinal ν, let

$$\begin{aligned} \mathfrak{H}_\nu(\mathscr{C}) &= \mathscr{C} && \text{if } \nu = 0, \\ &= \{\textstyle\bigcup \mathrm{Rn}\, A : A \in {}^\omega\mathfrak{H}_{(\nu)}^-(\mathscr{C})\ \&\ A \text{ is } \mathscr{C}^*\text{-separated}\} && \text{otherwise.} \end{aligned}$$

Now the hierarchy $\mathfrak{H}(\mathbf{\Sigma}_1^0)$, whose definition is so similar to that of the Borel hierarchy, turns out to be the same as $\mathrm{Ha}(\mathbf{\Pi}_1^0)$ with the 0th class omitted and with $\mathrm{Ha}_\nu(\mathbf{\Pi}_1^0)$ and $\mathrm{Ha}_\nu^-(\mathbf{\Pi}_1^0)$ interchanged at every other ordinal. Thus the pieces of the puzzle begin to fall into place and, for any $\nu \in \Omega$, $\mathfrak{H}_\nu^-(\mathbf{\Sigma}_1^0)$ has the first separation property. Whether a similar situation prevails for $\mathbf{\Sigma}_\nu^0$ in place of $\mathbf{\Sigma}_1^0$ is under investigation.

The definition of the Kalmar hierarchies can also be reformulated to make it look more like the definition of the Borel hierarchy. For any subset X of ${}^{\omega}\omega$ and any $k \in \omega$ let $kX = \{k\alpha : \alpha \in X\}$. Then the Kalmar hierarchies could have been defined as follows: For any ordinal ν and any class $\mathscr{C}$ of subsets of ${}^{\omega}\omega$,

$$\begin{aligned} \mathrm{Ka}_{\nu}(\mathscr{C}) &= \mathscr{C} && \text{if } \nu = 0, \\ &= \{\textstyle\bigcup\{kA_k : k \in \omega\} : A \in {}^{\omega}\mathrm{Ka}^{-}_{(\nu)}(\mathscr{C})\} && \text{otherwise.} \end{aligned}$$

In a similar spirit it is possible to normalize the separated unions mentioned in the definition of the $\mathfrak{H}(\mathbf{\Sigma}^0_1)$ hierarchy above. Doing this leads to the following equivalent definition of $\mathfrak{H}(\mathbf{\Sigma}^0_1)$. Let ζ be the constant zero function on ω, i.e., $\zeta = \langle 0 : n \in \omega\rangle$. For any $X \subseteq {}^{\omega}\omega$ let $X^+ = \{(n+1)\alpha : n\alpha \in X\}$. For any finite sequence s in ${}^{k}\omega$ and any $X \subseteq {}^{\omega}\omega$ let $sX = \{s\alpha : \alpha \in X\}$, where $(s\alpha)(n) = s(n)$ if $n \in k$ and $\alpha(n-k)$ otherwise. Then, for any ordinal ν,

$$\begin{aligned} \mathfrak{H}_{\nu}(\mathbf{\Sigma}^0_1) &= \mathbf{\Sigma}^0_1 && \text{if } \nu = 0, \\ &= \{f^{-1}(\textstyle\bigcup\{(\zeta|k)A^+_k : k \in \omega\}) : \\ &\qquad A \in {}^{\omega}\mathfrak{H}^{-}_{(\nu)}(\mathbf{\Sigma}^0_1), f : {}^{\omega}\omega \to {}^{\omega}\omega, f \text{ continuous}\} && \text{otherwise.} \end{aligned}$$

(Draw a picture and almost all will be clear.)

Actually, although the first separation property does not logically imply the coreduction property, in the case of all first separation principles known to hold in the hierarchies above, the coreduction property also holds. This fact is important because it is needed in the uniform proofs that either $\mathscr{C}$ or $\mathscr{C}^-$ fails to have the first separation property.

As mentioned above the separation problem has been solved, either outright or at least up to consistency in Zermelo-Fraenkel set theory (ZF), for all of the hierarchies generated by the closed and open sets and discussed above. Actually all of the solutions are outright solutions except in the projective hierarchy $\mathrm{Lu}(\mathbf{\Sigma}^0_1)$. Here the pattern is as follows:

$$\begin{matrix} \mathrm{Lu}_0(\mathbf{\Sigma}^0_1)^{\times} & \mathrm{Lu}_1(\mathbf{\Sigma}^0_1)^{\surd} & \mathrm{Lu}_2(\mathbf{\Sigma}^0_1)^{\times} \\ \mathrm{Lu}^-_0(\mathbf{\Sigma}^0_1)^{\surd} & \mathrm{Lu}^-_1(\mathbf{\Sigma}^0_1)^{\times} & \mathrm{Lu}^-_2(\mathbf{\Sigma}^0_1)^{\surd} \end{matrix}$$

Beyond this point it is known to be consistent with ZF that the $\times, \surd$ pattern at level 2 continues. Whether or not such an $\times, \surd$ pattern at level 3 and higher levels is independent of ZF is an important open problem. This pattern at level 3 follows from the existence of a $\mathbf{\Delta}^1_3$ well-ordering of ${}^{\omega}\omega$, which follows in turn from the axiom of constructibility. One possibility is that higher "axioms of infinity" such as the axiom of measurable cardinals might imply this pattern at the third level. From results of Silver it is known that this pattern at the third level is at least consistent with the axiom of measurable cardinals. On the other hand the axiom of measurable cardinals wipes out some nice well-orderings of ${}^{\omega}\omega$ and it is thought by some that still higher axioms of infinity may be found which wipe out all projective well-orderings of ${}^{\omega}\omega$. Although nice well-orderings can be viewed as pushing in the direction of a $\times, \surd$ pattern, weaker principles not ruled out by higher axioms of infinity might still be enough to force it.

On the other hand if there is indeed some pressure, not yet understood, pushing for the separation principle to hold on one side or the other then it might be sufficient, if higher axioms of infinity wipe out enough nice well-orderings of ${}^{\omega}\omega$ and other regular phenomena of this kind, to push through a $\surd, \times$ pattern at level 3. At first this might look surprising, but at least one respected logician has suggested it. It has the advantage of prolonging the alternation of the $\times, \surd$ and $\surd, \times$ patterns. Such an alternation may itself first look unnatural. But recall the case of $\mathrm{Ha}(\mathbf{\Pi}_1^0)$ in which there was the same alternating pattern. Recall too that its definition could be recast in terms of the $\mathfrak{H}(\mathbf{\Sigma}_1^0)$ hierarchy, where the alternating separation pattern was straightened out and the $\times, \surd$ pattern persists throughout. Perhaps the definition of the $\mathrm{Lu}(\mathbf{\Sigma}_1^0)$ hierarchy can be similarly recast. Note furthermore that if an "inner quantifier" rather than the "outer quantifier" $\mathbf{\Sigma}_n^1$ notation is used for the projective hierarchy then this has exactly the effect of reversing every other pair of classes. To the 3 levels—0, 1, and 2—of alternating pattern in the $\mathrm{Lu}(\mathbf{\Sigma}_1^0)$ hierarchy one might argue that the first is too arbitrary to count. For example, $\mathrm{Lu}(\mathbf{\Pi}_6^0)$ agrees with $\mathrm{Lu}(\mathbf{\Sigma}_1^0)$ except at 0 and the separation pattern at 0 for it is $\surd, \times$ rather than $\times, \surd$. On the other hand $\mathbf{\Sigma}_1^0$ is the simplest class of its kind which can be used in this way—for example, $\mathbf{\Pi}_1^0$ would not work. Some slight linguistic evidence can also be used to argue for its naturalness.[2]

Of course still another possibility is that new axioms of length, which seem at least in part to be removing the tools needed to solve the problem, may be insufficient to resolve it and new axioms of width will be required. Still other possibilities are perhaps too distasteful for open discussion at this Institute.

Whatever one's predictions about the problem we can perhaps agree—at least if the notion of "analyst" is sufficiently broadly interpreted (or alternatively if we remember Cohen's background)—with Luzin's 1930 statement: "C'est un problème qui mérite d'attirer l'attention des analystes malgré sa difficulté."

UNIVERSITY OF CALIFORNIA, BERKELEY

[2] (ADDED 31 JULY 1970.) This paragraph turned out to be prophetic. Only a month or so after the talk was given it was shown that "the axiom of definable determinateness" does indeed give the alternation of the $\times, \surd$ and $\surd, \times$ patterns discussed here. (See Addison and Moschovakis, Proc. Nat. Acad. Sci. **59** (1968), 708–712, and Martin, Bull. Amer. Math. Soc. **74** (1968), 687–689.) Moreover the reasons mentioned above for the plausibility of this hypothesis actually lie behind the proof of the alternating pattern from the axiom of definable determinateness. Furthermore the idea of considering $\mathbf{\Sigma}_1^0$ and $\mathbf{\Pi}_1^0$ as the first level of the projective hierarchy is not only completely vindicated by the outcome but indeed actually suggested the structure of the proof of the alternating pattern. Incidentally the "respected logician" (who made his suggestion in 1956) was Kurt Gödel.

Proceedings of Symposia in Pure Mathematics
Volume 13, Part II, 1974

PREDICATIVELY REDUCIBLE SYSTEMS OF SET THEORY

SOLOMON FEFERMAN*

1. Introduction. *Predicatively reducible*[1] subsystems of Z_2 (the usual formalization of classical analysis) were studied in [**1**]. In particular, a system IR (contained in Z_2) was found, every finite subsystem of which is predicatively reducible, but not the whole. Thus each theorem proved in IR has a predicative interpretation. It turns out that IR provides a reasonably convenient means for formulating large parts of classical mathematics without any *ad hoc* devices, and thus a means for showing their predicative reducibility.

The generality of arguments in mathematics has been made apparent by rephrasing them in set-theoretical terms. Moreover, familiar results in modern algebra and in topology are *prima-facie* essentially set-theoretical. It is therefore natural to look for a system of set theory which, like IR, is *locally* predicatively reducible. A solution to this problem was announced in [**2**]; the system found there was denoted by PS. The main purpose of this paper is to present the corresponding work for a closely related system PS_1. (The differences between PS and PS_1 are discussed at the end of § 2.4 below.)

The local predicative reducibility of PS_1 is established by taking advantage of the result already known for IR and by the following:

MAIN THEOREM. PS_1 *is a conservative extension of* IR, *modulo the usual translation of the language of* Z_2 *into set theory.*

This is proved in Theorem 4.6 below. The proof makes use of the hereditary countability of predicative sets which allows sets to be treated as well-founded

---

AMS (MOS) subject classifications (1970). Primary 02K10.
* Research supported by grant DA-ARO(D)-31-124-G655.

[1] A precise analysis of this notion has been given in terms of the autonomous progression of systems of ramified analysis; cf. [**4**] for a brief review of work on predicativity. The present paper can be read independently of this.

trees in the natural numbers. We cannot deal adequately with such trees in general in IR. However, we can conveniently deal with the trees of rank bounded by any given predicative ordinal in IR. The argument automatically covers the addition of an axiom E, which expresses that every set is enumerable and which thus implies various principles of choice.

A possible alternative approach to the problem solved here would have been to make a fresh start, to consider autonomous progressions of ramified (constructible) set theory (cf. footnote 1), and to follow out the research which led to IR. There might still be some independent interest in carrying this out.

The plan of the paper is as follows. § 2.1 describes PS_1 and an interesting subsystem PS_0; the system IR is recalled in § 2.2, along with some technical facts. Some special definitional extensions of these systems, which are even more convenient to use, are dealt with in the remainder of § 2. § 3.1 gives model-theoretic preliminaries on hereditarily countable sets in various ranks; it contains a direct model-theoretic proof that PS_0 has a minimum model which consists of all sets constructible before $\kappa^{(\omega)}(0)$. § 3.2 introduces *representation structures*, which are ordinally coded well-founded trees in ω. The proof of the Main Theorem in § 4 proceeds by formalization of the work in § 3. As a corollary (to be expected from [**1**]), it is seen that PS_1 has a minimum model consisting of all sets constructible before Γ_0.[2,3] In § 5 we compare PS_1 with systems of set theory which play a role in current work on generalized recursion theory.

It will be evident to the reader how to formalize a good deal of modern algebra in PS_0 + AC, which is contained in PS_0 + E.

2. The formal systems.

2.1. PS_i *and* E. These systems of axioms and rules are given in the basic language $\mathscr{L}$ of set theory, i.e., with just the relation symbols $=$ and $\in$; the variables are $a, b, c, \ldots, x, y, z$. Unfamiliar abbreviations and syntactic notions are explained below. PS_0 is given by the following axioms and axiom and rule schemata I–VI.

I. (*Extensionality*). $\bigwedge x(x \in a \leftrightarrow x \in b) \rightarrow a = b$.

II. (Δ_0-*separation*). *For each* Δ_0-*formula* $\psi(x)$,

$$\bigwedge a \bigvee b \bigwedge x[x \in b \leftrightarrow x \in a \wedge \psi(x)].$$

III. (Σ-*reflection rule*). *For each* Σ-*formula* ψ,

$$\psi / \bigvee a \psi^{(a)}.$$

IV. (*Infinity*). $\bigvee a \operatorname{Lim}(a)$.

V. (*Principle of proof by* $\in$-*induction*).

$$\bigwedge x[\bigwedge y(y \in x \rightarrow \varphi(y)) \rightarrow \varphi(x)] \rightarrow \bigwedge x \varphi(x).$$

---

[2] The definitions of the ordinal functions $\kappa^{(\alpha)}(\beta)$ and the ordinal Γ_0 are recalled in § 2.4.

[3] We leave open whether (and, hence, in what sense) the formal system PS_1 or PS is distinguished among subsystems of set theory with the same minimum model.

VI. (*Rule for definition by* $\in$-*recursion*). *For each* Σ-*formula* $\psi(x, y, z)$,

$$\frac{\bigwedge x, y \bigvee ! z \psi(x, y, z)}{\bigwedge a\{\bigcup a \subseteq a \to \bigvee f[\mathrm{Fn}(f) \wedge \mathscr{D}(f) = a \wedge \bigwedge x \in a \psi(x, f \upharpoonright x, f(x))]\}}.$$

PS_1 is PS_0 together with the following:

VII. (*Ordinal sufficiency rule*). *For any provably* $\Sigma \cap \Pi$ *formula* $(x \prec y)$ *and* Σ-*formula* $\theta(x)$, *without parameters*,

$$\frac{\mathrm{Wf}_{\prec}, \mathrm{Bd}^{\theta}_{\prec}}{\bigvee f[\mathrm{Fn}(f) \wedge \mathrm{Ord}(\breve{\mathscr{D}}(f)) \wedge \bigwedge x, y(x \prec y \to f(x) < f(y))]}.$$

(The hypothesis of VI expresses that $\prec$ is well-founded and that its field is θ-bounded, in a sense explained below; the conclusion expresses that there is an order-preserving map of $\prec$ into a von Neumann ordinal.)

PS_iE denotes the system PS_i together with the following axiom:

VIII. (*Enumerability*). $\bigwedge a \bigvee f[\mathrm{Fn}(f) \wedge \text{1-1}(f) \wedge \mathscr{D}(f) \subseteq \omega \wedge \breve{\mathscr{D}}(f) = a]$.

Unless otherwise indicated, all of the formulas $\varphi, \psi, \theta, \ldots$ above and in the following may contain free variables (*parameters*) besides those explicitly shown. Any initial universal quantifiers in the axioms are superfluous by the rule of generalization; however, some of these are included for special attention.

It is convenient to assume that in addition to the usual logical connectives and quantifiers, the syntax has *restricted* quantifiers $\bigvee x \in a(\cdots)$ and $\bigwedge x \in a(\cdots)$ (x, a distinct) with the logical laws:

$$\bigvee x \in a \varphi(x) \leftrightarrow \bigvee x(x \in a \wedge \varphi(x)), \qquad \bigwedge x \in a \varphi(x) \leftrightarrow \bigwedge x(x \in a \to \varphi(x)).$$

The usual quantifiers $\bigvee x(\cdots)$, $\bigwedge x(\cdots)$ are said to be *unrestricted*. A Δ_0-*formula* is one built up using only the propositional connectives and restricted quantifiers. A Σ-*formula* (Π-*formula*) is one which has no unrestricted universal (existential) quantifiers when put in prenex form; in this case it is said to be in Σ-*form* (Π-*form*). A formula φ is said to be *provably* $\Sigma \cap \Pi$ if there are ψ_1 in Σ-form, and ψ_2 in Π-form such that $(\varphi \leftrightarrow \psi_1) \wedge (\varphi \leftrightarrow \psi_2)$ is provable (in the given system). The significance of these classes of formulas for predicativity is discussed in § 2.3 below. The *relativization* $\varphi^{(a)}$ is obtained by replacing each unrestricted quantifier $\bigvee x(\cdots)$, $\bigwedge x(\cdots)$ in φ by $\bigvee x \in a(\cdots)$, $\bigwedge x \in a(\cdots)$, respectively.

LEMMA 2.1. *The following are logically provable for any* Σ-*formula* ψ:

(i) $a \subseteq b \wedge \psi^{(a)} \to \psi^{(b)}$,

(ii) $\psi^{(a)} \to \psi$.

Further useful properties of Σ-formulas will be developed in § 2.3.

LEMMA 2.2. *The following are derived rules of inference in* PS(I–III).

(i) (Σ-*replacement rule*). *For any* Σ-*formula* $\psi(x, y)$,

$$\frac{\bigwedge x \bigvee ! y \psi(x, y)}{\bigwedge a \bigvee b \bigwedge y[y \in b \leftrightarrow \bigvee x \in a \psi(x, y)]}.$$

(ii) ($\Sigma \cap \Pi$-*separation rule*). *For any provably* $\Sigma \cap \Pi$-*formula* $\psi(x)$,

$$\bigwedge a \bigvee b \bigwedge x[x \in b \leftrightarrow x \in a \wedge \psi(x)].$$

PROOF. (i) By reflection applied to $\bigwedge x \in a \bigvee y \psi(x, y)$ we can infer that for any a there exists c with

$$(1) \qquad \bigwedge x \in a \bigvee y \in c \psi^{(c)}(x, y).$$

Using Δ_0-separation we get b such that

$$\bigwedge y[y \in b \leftrightarrow y \in c \wedge \bigvee x \in a \psi^{(c)}(x, y)].$$

It is sufficient then to show

$$\bigwedge y[y \in c \wedge \bigvee x \in a \psi^{(c)}(x, y) \leftrightarrow \bigvee x \in a \psi(x, y)].$$

The forward implication is by Lemma 2.1. Suppose $x \in a$ and $\psi(x, y)$. By (1) there is $z \in c$ with $\psi^{(c)}(x, z)$, hence $\psi(x, z)$; but then by hypothesis $y = z$, so $y \in c \wedge \bigvee x \in a \psi^{(c)}(x, y)$.

(ii) is an easy consequence of (i).

On the basis of I–III and Lemma 2.2 we can introduce as usual 0, $\{a, b\}$, $\{a\}$, $\langle a, b \rangle$, $\bigcup a$, $a \cup b$, $a \cap b$, $a - b$, $a \times b$, and $\breve{a}$. We take $a' = a \cup \{a\}$. $\bigcup a \subseteq a$ is equivalent to the Δ_0-formula which expresses that a is $\in$-transitive. The notation $\{x : \varphi(x)\}$ is used whenever $\bigvee b \bigwedge x(x \in b \leftrightarrow \varphi(x))$ has been proved. $\mathscr{D}(a) = \{x : \bigvee y(\langle x, y \rangle \in a)\} = \{x : \bigvee y \in \bigcup\bigcup a(\langle x, y \rangle \in a)\}$, $\breve{\mathscr{D}}(a) = \mathscr{D}(\breve{a})$ and $\mathfrak{F}(a) = \mathscr{D}(a) \cup \breve{\mathscr{D}}(a)$. Fn$(f)$ expresses that f is a function; $f(x)$ is the value of f at x if Fn$(f) \wedge x \in \mathscr{D}(f)$, otherwise 0. $f \restriction a = f \cap (a \times \breve{\mathscr{D}}(f))$. 1-1$(f)$ expresses that f is one–one. Throughout the following, "f" ranges over functions.

Let $(x \prec y)$ be any formula $\chi(x, y)$. The following explains notations used in PS_1 and below. $\mathrm{Trans}_\prec$ expresses that $\prec$ is transitive, $\mathrm{So}_\prec$ that it is a simple ordering. $\mathrm{Trans}_\prec(a)$ is

$$\bigwedge x, y(x \prec y \wedge y \in a \rightarrow x \in a).$$

$\mathrm{Wf}_\prec$ expresses that $\prec$ is *well-founded* in the sense that

$$\bigwedge a[a \neq 0 \rightarrow \bigvee x \in a \sim \bigvee y \in a(y \prec x)].$$

$\mathrm{TI}(\prec, \varphi)$ is the formula

$$\bigwedge x[\bigwedge y(y \prec x \rightarrow \varphi(y)) \rightarrow \varphi(x)] \rightarrow \bigwedge x \varphi(x)$$

expressing *transfinite induction* for φ with respect to $\prec$. $\mathrm{TI}(\prec)$ is the following rule:

(2.3) *For any formula* φ,

$$\mathrm{Wf}(\prec)/\mathrm{TI}(\prec, \varphi).$$

Suppose $\prec$ is provably $\Sigma \cap \Pi$. $\mathrm{Bd}_\prec$ is $\bigwedge x \bigvee a \bigwedge y(y \prec x \rightarrow y \in a)$. If $\mathrm{Bd}_\prec$ has been proved we can introduce $\mathrm{Pd}_\prec(x) = \{y : y \prec x\}$. $\mathrm{Bd}^\theta_\prec$ is the formula

$$\bigvee a \theta(a) \wedge \bigwedge a, x, y[\theta(a) \wedge y \prec x \rightarrow x \in a \wedge y \in a].$$

If θ is a Σ-formula and $\mathrm{Bd}^{\theta}_{\prec}$ has been proved we can introduce $\{\prec\} = \{\langle x, y\rangle : x \prec y\}$. TR($\prec$) is the following *rule*.

(2.4) *For any Σ-formula ψ,*

$$\frac{\mathrm{Wf}_{\prec}, \mathrm{Bd}_{\prec}, \bigwedge x, y \bigvee !z\psi(x, y, z)}{\bigwedge a\{\mathrm{Trans}_{\prec}(a) \to \bigvee f[\mathscr{D}(f) = a \wedge \bigwedge x \in a\psi(x, f \restriction \mathrm{Pd}_{\prec}(x), f(x))]\}}.$$

The conclusion here is denoted by TR($\prec, \psi$).

Ord(a) expresses that a is an ordinal, i.e., $\mathrm{So}_{\in \cap (a \times a)}$. Lim($a$) is the formula

$$\mathrm{Ord}(a) \wedge 0 \in a \wedge \bigwedge x \in a(x' \in a).$$

(Note that both Ord(a) and Lim(a) are Δ_0-formulas.) ($x < y$) is $\mathrm{Ord}(x) \wedge \mathrm{Ord}(y) \wedge (x \in y)$. The axiom schema V is equivalent to $\mathrm{Wf}_{\in}$ plus the rule TI($\in$); this schema is also denoted by TI($\in$). It follows from TI($\in$) and the axiom of infinity that there is a least limit ordinal ω. Rule VI is equivalent to TR($\in$). Rule VII is a consequence of TR($\prec$).

Let $\mathrm{Nds}_{\prec}$ be the formula

$$\sim \bigvee f[\mathscr{D}(f) = \omega \wedge \bigwedge x \in \omega(f(x') \prec f(x))].$$

This follows from an instance TI($\prec, \varphi$). However, the inference of even $\mathrm{Wf}_{\prec}$ from $\mathrm{Nds}_{\prec}$ seems to require some form of the axiom of choice AC.

LEMMA 2.5. (i) $\mathrm{PS}_0\mathrm{E} \vdash \mathrm{AC}$.

(ii) *For $\prec$ provably $\Sigma \cap \Pi$*, $\mathrm{PS}_0\mathrm{E} \vdash (\mathrm{Nds}_{\prec} \to \mathrm{Wf}_{\prec})$.

(iii) $\mathrm{PS}_i\mathrm{E}$ *is closed under the following rule* (Σ-*choice*), *for any Σ-formula* $\psi(x_1, \ldots, x_n, y_1, \ldots, y_n)$:

$$\frac{\bigwedge x_1 \bigvee y_1 \ldots \bigwedge x_n \bigvee y_n \psi(x_1, \ldots, x_n, y_1, \ldots, y_n)}{\bigwedge a \bigvee f_1, \ldots, f_n \bigwedge x_1, \ldots, x_n \in a\psi(x_1, \ldots, x_n, f_1(x_1), \ldots, f_n(x_1, \ldots, x_n))}.$$

To prove (iii) one combines Σ-reflection with AC.

Let ZF (ZF^-) be the system of Zermelo-Fraenkel axioms (resp., ZF without the power set axiom). By familiar work, ZF is an extension of PS_1. It is not evident that this holds with ZF^- in place of ZF (since the usual arguments for reflection make use of either the ranks R_α or a well-ordering of the universe, even in the simplest cases). However, it follows from Theorem 4.7 below that $\mathrm{PS}_1\mathrm{E}$ has a model in ZF^-. Note that TR($\in$) is derivable from TI($\in$) in ZF^-, but this makes use of an instance of the replacement *axiom* which is not available in PS_1 (even for ψ in Σ-form).

There are some immediate consequences of AC in ZF^- whose status in the systems $\mathrm{PS}_i\mathrm{E}$ is not evident; for example, I do not know whether Zorn's Lemma is provable in $\mathrm{PS}_1\mathrm{E}$.

2.2. IR. This system is formulated in the language $\mathscr{L}_2$ of second-order number theory, with *numerical* variables $n, m, p, \ldots, x, y, z$, and *set* variables $A, B, C, \ldots, X, Y, Z$. The basic relations are $s = t$, $X = Y$, and $t \in X$, where s, t

are terms built up from numerical variables and 0 by primitive recursive operations $', +, \cdot, \ldots$. Primitive recursive relations $<, \ldots$ are defined as usual from these. $\mathfrak{F}, \mathfrak{G}, \ldots$ range over formulas of $\mathcal{L}_2$; unless specified, each formula $\mathfrak{F}(x), \mathfrak{G}(x, X), \ldots$ may have numerical and set parameters. Other syntactic notions and abbreviations are explained below.

(1) (*Axioms of elementary number theory*). Z.

(2) (*Extensionality*). $\bigwedge x(x \in X \leftrightarrow x \in Y) \to X = Y$.

(3) (*Foundation*). $\mathrm{Wf}_<$.

(4) (*Hyperarithmetic comprehension rule*). *For each provably* $\Sigma \cap \Pi$*-formula* $Q(x)$, $\bigvee X \bigwedge x[x \in X \leftrightarrow Q(x)]$.

(5) (*Transfinite induction rule*). *For each arithmetic formula* $(x \prec y)$ *without parameters, and each formula* $\mathfrak{F}(x)$,

$$\frac{\mathrm{Wf}_\prec, \mathrm{So}_\prec}{\mathrm{TI}(\prec, \mathfrak{F})}$$

(6) (*Transfinite recursion rule*). *For each formula* $(x \prec y)$ *satisfying the hypotheses of* (5) *and any arithmetic* $\mathfrak{A}(x, n, X)$,

$$\bigvee X \bigwedge n, x[x \in (X)_n \leftrightarrow \mathfrak{A}(x, n, X \restriction_\prec n)].$$

Except for the instances needed to derive simple ordering of $<$, the induction schema of Z is superfluous by (3), (5). Alternatively, (3) can be dropped and (1) extended to include all instances:

$$\mathfrak{F}(0) \wedge \bigwedge x(\mathfrak{F}(x) \to \mathfrak{F}(x')) \to \bigwedge x \mathfrak{F}(x).$$

A formula is *arithmetic* if it contains no set quantifiers; $\mathfrak{A}$, $\mathfrak{B}$ range over these in the following. A Σ*-formula* (Π*-formula*) here is one which contains no universal (existential) set quantifiers when put in prenex form. $(X)_n$ denotes the set $\{x:(n, x) \in X\}$ where $(x, y) = 2^x \cdot 3^y$; this notation is extended below to $(X)_{n_1,\ldots,n_k}$ where $(X)_{n_1,n_2} = ((X)_{n_1})_{n_2}$, etc. $X \restriction_\prec n = \{(m, y): m \prec n \wedge y \in (X)_m\}$.

With these explanations, the above axioms and rules (1)–(6) are evidently equivalent to those of the system IR as described in [**1**, p. 26]. We shall need to make use of the following facts about IR. Proofs of these can be gleaned from [**1**, § 6]; full proofs will appear in a subsequent publication. The definition of the ordinal Γ_0 is recalled in the next section; $\prec_{\Gamma_0}$ is the natural primitive recursive ordering of order-type Γ_0, and, for each $\alpha < \Gamma_0$, $\prec_\alpha$ is its initial segment of type α.

(2.6) FACTS ABOUT IR. (i) *For each provably* $\Sigma \cap \Pi$*-formula* $(x \prec y)$ *without parameters such that* $\mathrm{IR} \vdash \mathrm{Wf}_\prec$ *there is an* $\alpha < \Gamma_0$ *such that*

$$\mathrm{IR} \vdash \bigvee X\{\bigwedge x \bigvee ! n(x, n) \in X \wedge \bigwedge x, y, n, m[(x, n) \in X \wedge (y, m) \in X \wedge x \prec y \to n \prec_\alpha m]\}.$$

(ii) *For each* $\alpha < \Gamma_0$, $\mathrm{IR} \vdash \mathrm{Wf}_{\prec_\alpha}$.

(iii) $\mathrm{IR} \nvdash \mathrm{Wf}_{\prec_{\Gamma_0}}$.

(iv) *The collection* HYP_{Γ_0} *of sets recursive in* H_α *for some* $\alpha < \Gamma_0$ *is the minimum* ω*-model of* IR, *and the collection* HYP *of all hyperarithmetic sets is an* ω*-model of* IR.

(iii) is, of course, an immediate consequence of (i). It follows from (i), (ii) and IR(5) that also TI($\prec$, $\mathfrak{F}$) can be proved in IR for any $\prec$ satisfying (i).

FACTS ABOUT IR (CONTINUED). (v) (Σ-TR *rule*). *For each formula* $(x \prec y)$ *satisfying the hypotheses of* IR(5) *and each* Σ-*formula* $Q(n, X, Y)$ *for which* $\mathrm{IR} \vdash \bigwedge n, X \bigvee ! Y Q(n, X, Y)$ *we have*

$$\mathrm{IR} \vdash \bigvee X \bigwedge n Q(n, X \upharpoonright_{\prec} n, (X)_n).$$

(vi) (Σ-*uniformization rule*). *For each* Σ-*formula* $Q(X, Y)$ *such that* $\mathrm{IR} \vdash \bigwedge X \bigvee Y Q(X, Y)$ *we can find a* Σ-*formula* $Q_1(X, Y)$ *such that*

$$\mathrm{IR} \vdash \bigwedge X \bigvee ! Y Q_1(X, Y) \wedge \bigwedge X, Y[Q_1(X, Y) \rightarrow Q(X, Y)].$$

(vii) (Σ-*choice rule*). *If* $Q(x_1, \ldots, x_k, X_1, \ldots, X_k)$ *is a* Σ-*formula and*

$$\mathrm{IR} \vdash \bigwedge n_1 \bigvee X_1 \cdots \bigwedge n_k \bigvee X_k Q(n_1, \ldots, n_k, X_1, \ldots, X_k)$$

then

$$\mathrm{IR} \vdash \bigvee X^{(1)} \cdots X^{(k)} \bigwedge n_1 \cdots n_k Q(n_1, \ldots, n_k, (X^{(1)})_{n_1}, \ldots, (X^{(k)})_{n_1, \ldots, n_k}).$$

2.3. *Σ-formulas and rules of definition.* Let $\mathscr{L}^+$ be $\mathscr{L}$ or any extension of $\mathscr{L}$ by additional relation symbols $R, \ldots$ and function symbols $F, G, \ldots$ of various numbers of arguments ($\geqq 0$). Restricted quantifiers may now take the form $\bigvee x \in t(\cdots)$ and $\bigwedge x \in t(\cdots)$ with t any term not containing x. The notions of Δ_0-, Σ- and Π-formulas are extended to any such $\mathscr{L}^+$. Let S be any system of axioms and rules in $\mathscr{L}^+$. φ is said to be in $\Sigma \cap \Pi$-*form relative to S* if there are ψ_1 in Σ-form, ψ_2 in Π-form with $S \vdash (\varphi \leftrightarrow \psi_1) \wedge (\varphi \leftrightarrow \psi_2)$. In particular, this holds whenever $\varphi = \varphi(x, y)$ is in Σ-form and $S \vdash \bigwedge x \bigvee ! y \varphi(x, y)$, by $S \vdash \varphi(x, y) \leftrightarrow \bigwedge z(\varphi(x, z) \rightarrow y = z)$.

The significance of these classes of formulas for predicativity comes from some results established in [**6**], [**5**] for the following notion for $\mathscr{L}^+$-structures $\mathscr{M} = (M, E_{\mathscr{M}}, R_{\mathscr{M}}, \ldots, F_{\mathscr{M}}, \ldots)$: $\mathscr{M}' = (M', E_{\mathscr{M}'}, \ldots)$ is said to be an *outer* or $\in$-*extension* of $\mathscr{M} = (M, E_{\mathscr{M}}, \ldots)$ if $\mathscr{M} \subseteq \mathscr{M}'$ and $\forall x \in M \forall y \in M'[y E_{\mathscr{M}'} x \Rightarrow y \in M]$. $\varphi(v_1, \ldots, v_n)$ is said to be *persistent for* (or preserved under) $\in$-*extensions relative to S* if whenever $\mathscr{M}, \mathscr{M}'$ are models of S, $\mathscr{M}'$ is an $\in$-extension of $\mathscr{M}$, and $x_1, \ldots, x_n \in M$,

$$\mathscr{M} \models \varphi(x_1, \ldots, x_n) \Rightarrow \mathscr{M}' \models \varphi(x_1, \ldots, x_n).$$

φ is said to be *invariant for* $\in$-*extensions relative to S* if both it and $\sim\varphi$ are persistent for these. The results are the following:

1. *φ is persistent for $\in$-extensions relative to S iff $S \vdash (\varphi \leftrightarrow \psi)$ for some Σ-formula ψ.*

2. *φ is invariant for $\in$-extensions relative to S' iff φ is in $\Sigma \cap \Pi$-form relative to S.*

In certain special cases, we can replace the conclusion in 2 by

φ is equivalent to a Δ_0-formula in S.

Using the completeness theorem for many-sorted logic, these results can be expressed in a syntactic form. First enlarge the language by a new sort of variable

$a', b', c', \ldots, x', y', z'$. Let φ' be obtained from φ by replacing all quantifiers $\bigvee x(\cdots), \bigwedge x(\cdots)$ in φ by $\bigvee x'(\cdots), \bigwedge x'(\cdots)$; define S' similarly. Then, for example, $\varphi(v_1, \ldots, v_n)$ is invariant for $\in$-extensions relative to S iff

$$S, S', \bigwedge x \bigwedge y' \in x \bigvee y(y' = y) \vdash [\varphi(v_1, \ldots, v_n) \leftrightarrow \varphi'(v_1, \ldots, v_n)].$$

In this case we say φ is *provably definite relative to* S; by 2 above these are exactly the same as the $\Sigma \cap \Pi$-formulas relative to S. Thinking of the variables of $\mathscr{L}^+$ as ranging over sets, the notion of definiteness corresponds to the informal idea that the meaning of φ is unambiguous when individual sets are regarded as completed totalities, though the universe of sets (satisfying S) is only regarded as a potential totality.

The notion of Σ-formula depends on $\mathscr{L}^+$ and, in addition, that of $\Sigma \cap \Pi$-formula depends on S. We shall now show that these notions are essentially unchanged when one passes to a definitional extension $S + D(S)$ where the definitions in $D(S)$ are given by provably definite formulas in S. More precisely, we extend the language $\mathscr{L}^+$ and the system S in the following way:

(i) *For each Σ-formula ψ with just the free variables $x_1, \ldots, x_n$, y ($n \geqq 0$) such that*

$$S \vdash \bigwedge x_1, \ldots, x_n \bigvee ! y \psi(x_1, \ldots, x_n, y),$$

adjoin a new n-ary function symbol (or constant symbol when $n = 0$) F_ψ and the axiom

$$\bigwedge x_1, \ldots, x_n \psi(x_1, \ldots, x_n, F_\psi(x_1, \ldots, x_n)).$$

(ii) *For each Σ-formula ψ_1 and Π-formula ψ_2 with just the free variables $x_1, \ldots, x_n$ ($n > 0$) such that*

$$S \vdash \bigwedge x_1, \ldots, x_n[\psi_1(x_1, \ldots, x_n) \leftrightarrow \psi_2(x_1, \ldots, x_n)],$$

adjoin a new n-ary relation symbol R_{ψ_1,ψ_2} and the axiom

$$\bigwedge x_1, \ldots, x_n[R_{\psi_1,\psi_2}(x_1, \ldots, x_n) \leftrightarrow \psi_1(x_1, \ldots, x_n)].$$

Let $S\nabla$ be the closure under this procedure, i.e., $S\nabla = S + D(S) + D(S + D(S)) + \cdots$, and let $\mathscr{L}^+_\nabla$ be the language of $S\nabla$. (No new instances of the nonlogical rules of S are added in this extension.) The usual elimination procedure for definitional extensions in this case yields the following:

LEMMA 2.7. *With each formula φ of $\mathscr{L}^+_\nabla$ is associated a formula φ^- of $\mathscr{L}^+$ with the same free variables such that the following hold:*

(i) $S\nabla \vdash (\varphi \leftrightarrow \varphi^-)$.

(ii) *If $S\nabla \vdash \varphi$ then $S \vdash \varphi^-$.*

(iii) *If φ is in Σ-form then so also is φ^-.*

Developments in PS_1, for example, are simplified by use of Lemma 2.7; this is considered in the next section.

All of the notions of this section and the preceding result can be taken to apply equally well to systems S in second-order languages $\mathscr{L}_2^+$, with numerical quantifiers taking the place of restricted quantifiers. We shall take advantage of this in § 4 when proving the reduction of PS_1 to IR by working with IR∇ instead.

2.4. $PS_i\nabla$. Apply § 2.3 to the case $\mathscr{L}^+ = \mathscr{L}$ and $S = PS_0$. The notations for sets determined by the Σ-replacement or $\Sigma \cap \Pi$-separation rules, previously introduced as abbreviations in PS_0, can now be introduced explicitly by function symbols in $PS_0\nabla$. For, $b = \{y : \bigvee x \in a\psi(x, y)\}$ is the unique solution of $\bigwedge y \in b \bigvee x \in a\psi(x, y) \wedge \bigwedge x \in a \bigvee y \in b\psi(x, y)$, when $\bigwedge x \bigvee ! y\psi(x, y)$ has been proved, and $b = \{x : x \in a \wedge \psi(x)\}$ is the unique solution of $\bigwedge x \in b\psi(x) \wedge \bigwedge x \in a(\psi(x) \to x \in b)$. Thus, for suitable function symbols $F_0, F_1, \ldots$ of $\mathscr{L}_\nabla$ we have $F_0 = 0, F_1(x_1, x_2) = \{x_1, x_2\}$, $F_2(x) = \bigcup x$, etc. The formula $\mathrm{Fn}(f)$ is now equivalent to a Δ_0-formula of $\mathscr{L}_\nabla$, namely,

$$\bigwedge z \in f \bigvee x, y \in \bigcup z (z = \langle x, y \rangle)$$
$$\wedge \bigwedge x, y_1, y_2 \in \bigcup\bigcup f (\langle x, y_1 \rangle \in f \wedge \langle x, y_2 \rangle \in f \to y_1 = y_2).$$

Hence $PS\nabla \vdash [\mathrm{Fn}(f) \leftrightarrow R(f)]$ for a relation symbol R of $\mathscr{L}_\nabla$. Similarly, $\mathscr{D}(f)$, $\breve{\mathscr{D}}(f)$, $\mathrm{Val}(f, x) (= f(x))$ and $\mathrm{Res}(f, x) (= f \upharpoonright x)$ are given by function symbols in $\mathscr{L}_\nabla$. For any unary F of $\mathscr{L}_\nabla$ we can prove $\bigwedge a \bigvee ! f[\mathrm{Fn}(f) \wedge \bigwedge x \in a(f(x) = F(x))]$; we also write $F \upharpoonright a$ for this unique f.

By Lemma 2.7, we get no new theorems if we replace "formula" by "formula of $\mathscr{L}_\nabla$" throughout in PS_0(I–VI). Then some of the axioms and rules can be simplified. For example, the rule TR($\in$) can be recast in the following form:

LEMMA 2.8. *With each symbol G of* $PS_0\nabla$ *is associated a symbol F such that*

$$PS_0\nabla \vdash \bigwedge x[F(x) = G(x, F \upharpoonright x)].$$

(G has $n + 2$ arguments and F $n + 1$ arguments, where $n \geqq 0$; the parameters are suppressed.)

PROOF. By TR($\in$), any function can be iterated up to any ordinal. In particular, starting with any set a and iterating $\bigcup$ up to ω we get the transitive closure TC(a). TC can be regarded as a function symbol of $\mathscr{L}_\nabla$. Now let $\psi(a, f)$ be $\{\mathscr{D}(f) = \mathrm{TC}(a) \wedge \bigwedge x \in \mathscr{D}(f)[f(x) = G(x, f \upharpoonright x)]\}$. By TI($\in$) and TR($\in$) in $PS_0\nabla$ we prove $\bigwedge a \bigvee ! f \; \psi(a, f)$ and hence

$$\bigwedge x \bigvee ! y \bigvee f[\psi(\mathrm{TC}(\{x\}), f) \wedge f(x) = y].$$

F gives y as a function of x.

The schema of Lemma 2.8 is also denoted by TR($\in$). A simple special case is the definition of the rank function $\rho(x)$. We can get schemes of slightly more general form from the given one, for example: $F(x) = G(x, F \upharpoonright \mathrm{TC}(x))$. On the other hand, the usual form of definition by recursion on the ordinals TR($<$) is obtained from TR($\in$) simply by restricting attention to ordinal arguments. In particular, one can introduce the primitive recursive functions of ordinals $+, \cdot$, exp, etc. in $PS_0\nabla$. We shall return to the associated critical functions below.

Of course, all of the preceding applies equally well to $\mathrm{PS}_1\nabla$ in place of $\mathrm{PS}_0\nabla$ (as well as to the adjunction of any particular axioms, such as E). Consider any formula $(x \prec y)$ satisfying the hypotheses of the ordinal sufficiency rule VII. Since it is provably $\Sigma \cap \Pi$, we can regard $\prec$ as given by a relation symbol of $\mathrm{PS}_1\nabla$. The set $d = \mathfrak{F}(\{\prec\})$ is the unique solution of the Σ-formula

$$\bigwedge x \in d \bigvee y(x \prec y \vee y \prec x) \wedge \bigvee a[\theta(a) \wedge \bigwedge x, y \in a(x \prec y \rightarrow x, y \in d)],$$

so that d is a constant of $\mathrm{PS}_1\nabla$. The same holds for $r = \{\prec\} = \{\langle x, y\rangle : x, y \in d \wedge (x \prec y)\}$; further $\mathrm{Pd}_{\prec}(x)$ is a function of the system.

LEMMA 2.9. *Suppose the hypotheses of* VII *and* $\mathrm{So}_{\prec}$ *have been proved in* $\mathrm{PS}_1\nabla$.

(i) *For any formula* $\varphi(x)$, $\mathrm{PS}_1\nabla \vdash \mathrm{TI}(\prec, \varphi)$.

(ii) *With any symbol G of* $\mathrm{PS}_1\nabla$ *we can associate a symbol F such that*

$$\mathrm{PS}_1\nabla \vdash \bigwedge x \in \mathfrak{F}(\{\prec\})[F(x) = G(x, F \restriction \mathrm{Pd}_{\prec}(x))].$$

PROOF. By the conclusion of VII there is an f such that

$$\mathscr{D}(f) = \mathfrak{F}(\{\prec\}) \wedge \mathrm{Ord}(\breve{\mathscr{D}}(f)) \wedge \bigwedge x, y \in \mathfrak{F}(\{\prec\})(x \prec y \rightarrow f(x) < f(y)).$$

Since $\prec$ is a simple ordering, f is 1-1. It is seen by $\mathrm{TI}(<)$ that f is the unique solution of this formula. Hence f can be regarded as a constant of $\mathrm{PS}_1\nabla$. $\mathrm{TI}(\prec, \varphi)$ follows from $\mathrm{TI}(<, \varphi_1)$ where $\varphi_1(z) \leftrightarrow \varphi(\breve{f}(z))$. The required F is given by $F(x) = F_1(f(x))$ where $\bigwedge z[F_1(z) = G(\breve{f}(z), F_1 \restriction z)]$.

The hypothesis of simple ordering can be eliminated, but this would require a slightly more complicated formulation and argument.

Using the explicit definitions of the primitive recursive functions on ω in $\mathrm{PS}_1\nabla$, we associate with each sentence $\mathfrak{F}$ of $\mathscr{L}_2$ a translation $\mathfrak{F}'$ in $\mathscr{L}_\nabla$ obtained by replacing all numerical quantifiers $\bigvee x(\cdots), \bigwedge x(\cdots)$ in $\mathfrak{F}$ by $\bigvee x \in \omega(\cdots)$, $\bigwedge x \in \omega(\cdots)$, resp., and all set quantifiers $\bigvee A(\cdots), \bigwedge A(\cdots)$ in $\mathfrak{F}$ by $\bigvee a(a \subseteq \omega \wedge \cdots)$, $\bigwedge a(a \subseteq \omega \rightarrow \cdots)$, resp. This also determines a translation $\mathfrak{F}^* = (\mathfrak{F}')^-$ of $\mathscr{L}_2$ into $\mathscr{L}$. If $\mathfrak{F}$ is a Σ-formula then so is $\mathfrak{F}^*$. The following part of our Main Theorem is now obvious by Lemma 2.9.

LEMMA 2.10. PS_1 *is an extension of* IR *under the translation of* $\mathscr{L}_2$ *into* $\mathscr{L}$; *i.e., if* $\mathrm{IR} \vdash \mathfrak{F}$ *then* $\mathrm{PS}_1 \vdash \mathfrak{F}^*$.

We now give some preliminary indications on the provable ordinals in the PS_i. Suppose F is a function symbol of $\mathscr{L}_\nabla$ which has been proved to be *normal* on Ord in $\mathrm{PS}_i\nabla$. Then the *derived function* F' can be proved to exist and be normal in the same system by $\mathrm{TR}(<)$. Hence each of the critical functions $\kappa^{(n)}$ can be proved to be normal in $\mathrm{PS}_0\nabla$ where $\kappa^{(0)}(\alpha) = \omega^\alpha$, $\kappa^{(n+1)} = (\kappa^{(n)})'$.

Speaking informally, the extension of these functions into the transfinite is given by

(2.11)
(i) $\kappa^{(\alpha+1)} = (\kappa^{(\alpha)})'$,
(ii) $\kappa^{(\alpha)}$ *enumerates* $\bigcap_{\beta<\alpha}\breve{\mathscr{D}}(\kappa^{(\beta)})$ *for* α *a limit number.*[4]

[4] Cf. [3] for more information on these functions.

The ordinals generated in a finite number of steps using $0, +, \cdot, \kappa^{(0)}, \ldots, \kappa^{(n)}, \ldots$ ($n < \omega$) are exactly the ordinals $< \kappa^{(\omega)}(0)$. The ordinals generated in a finite number of steps using $0, +, \cdot, \lambda\alpha\beta\kappa^{(\alpha)}(\beta)$ are exactly the ordinals $< \Gamma_0$, where Γ_0 is the least solution of $\kappa^{(\alpha)}(0) = \alpha$. The function $\lambda\alpha\beta\kappa^{(\alpha)}(\beta)$ is defined by a form of transfinite recursion which is not available in $\mathrm{PS}_1\nabla$, namely as a double recursion:

$$F(\alpha, \beta) = G(\alpha, \beta, \lambda\beta_1 < \beta F(\alpha, \beta_1), \lambda\alpha_1 < \alpha\lambda\beta_1 F(\alpha_1, \beta_1)).$$

LEMMA 2.12.
(i) *Each ordinal* $< \kappa^{(\omega)}(0)$ *can be proved to exist in* $\mathrm{PS}_0\nabla$.
(ii) *Each ordinal* $< \Gamma_0$ *can be proved to exist in* $\mathrm{PS}_1\nabla$.

PROOF. (i) uses the definitions of each $\kappa^{(n)}$ in $\mathrm{PS}_0\nabla$. (ii) uses the ordinal sufficiency rule for the $\prec_\alpha$ where $\alpha < \Gamma_0$, and 2.6(ii), 2.10.

We conclude this section with a comparison of PS_1 and the system PS of [**2**]. The introduction of Σ-definable functions was incorporated in PS, so we can write $\mathrm{PS} = \mathrm{PS}\nabla$. The basic axioms I-VII of PS are slightly weaker than I-IV of PS_1, since the Σ-replacement rule is used instead of the Σ-reflection rule as here. PS(VIII) is $\mathrm{Nds}_\in$, PS(IX, X) are $\mathrm{TI}(\prec)$ and $\mathrm{TR}(\prec)$, resp., for any provably $\Sigma \cap \Pi$-formula $\prec$ without parameters, except that $\mathrm{Nds}_\prec$ is used instead of $\mathrm{Wf}_\prec$ as hypothesis. By 2.5(ii), $\mathrm{PS}\mathrm{E}$ certainly includes $\mathrm{TI}(\in)$, $\mathrm{TR}(\in)$ and the ordinal representatives rule VII. However, it also covers relations which do not immediately fall under those which can be treated in PS_1.

3. Model-theoretic considerations.

3.1. *Countable models for theorems of* $\mathrm{PS}_1\mathrm{E}$. Let HC be the collection of hereditarily countable sets. The usual method of reducing a subsystem S_1 of ZF^- to a second order system S_2 (cf., e.g., [**11**, p. 375]) is to find a transitive model $M \subseteq \mathrm{HC}$ of S_1 for which the verification that M is a model of S_1 can be formalized in S_2. For this, discussion of sets x in HC is translated into second order terms by dealing with suitable well-founded relations X in the natural numbers. The formula $\mathrm{Rep}(X)$ of $\mathscr{L}_2$ which expresses that X represents a set x in HC is in Π-form. However, with this method every Σ-formula of set theory is translated into a $\Sigma\Pi$-formula ("essentially Σ_2^1-formula") of analysis. Because of this, the method cannot be applied to reduce PS_1 to IR; it first breaks down at the Σ-reflection rule, but also at the rules TI and TR.

Given α, let $M_\alpha = \mathrm{HC} \cap \{x : \rho(x) < \alpha\}$. The notion $\mathrm{Rep}_\alpha(X)$: X represents a set in M_α, can be defined by transfinite recursion up to α; hence, for each $\alpha < \Gamma_0$, $\mathrm{Rep}_\alpha(X)$ is expressed by a relation $R(X)$ of $\mathrm{IR}\nabla$. Given any statement φ we can associate with it a translation φ^α in $\mathscr{L}_\nabla$, which is true in $(\omega, \mathscr{P}\omega)$ just in case $M_\alpha \models \varphi$. This translation will be given explicitly in § 4; it will be shown there that for each theorem φ of $\mathrm{PS}_1\mathrm{E}$ there is an $\alpha < \Gamma_0$ such that $\mathrm{IR}\nabla \vdash \varphi^\alpha$. This will be done by formalizing a procedure, to be given in this section, which associates with each derivation of a theorem φ of $\mathrm{PS}_1\mathrm{E}$ an $\alpha < \Omega$ (= first uncountable ordinal) such that $M_\alpha \models \varphi$ (i.e., φ is satisfied by all assignments in M_α).

Actually, more must be done to handle the rules of Σ-reflection and transfinite recursion. To illustrate the problem with the former, suppose ψ is a Δ_0-formula with free variables x, y and that $M_\alpha \models \bigwedge x \bigvee y \psi(x, y)$. We wish to find a β such that $M_\beta \models \bigwedge a \bigvee b \bigwedge x \in a \bigvee y \in b \psi(x, y)$. The best that can be done with the given information is that for each $a \in M_\alpha$ there is $b \in M_{\alpha+1}$ such that $\bigwedge x \in a \bigvee y \in b \psi(x, y)$. This is overcome by associating with each (derivation of a) theorem φ of $\mathrm{PS_1E}$ a *normal sequence* $\langle \alpha_\xi \rangle_{\xi < \Omega}$ of countable ordinals such that, *for each* ξ, $M_{\alpha_\xi} \models \varphi$. When this holds, we shall say that φ is *satisfied along* $\langle \alpha_\xi \rangle$. The procedure is given by the following lemma, which is presented in several parts.

LEMMA 3.1. (i) *Each of the axioms of* $\mathrm{PS_0E}$ *is satisfied along the sequence* $\langle \omega \cdot (2 + \xi) \rangle$.

PROOF. The axioms of extensionality, Δ_0-separation and $\in$-induction are satisfied in any M_α with $\alpha > 0$, and of infinity in any M_α with $\alpha > \omega$. The axiom of enumerability is satisfied in any M_α with α a limit number $> \omega$; for if $\rho(a) = \beta$ and $f : a \to \omega$, $\rho(f) \leqq \max(\beta + 2, \omega)$.

We assume throughout the following that $\langle \alpha_\xi \rangle$ *is any normal sequence of limit numbers* $> \omega$.

LEMMA 3.1. (ii) *Suppose* ψ *is a* Σ*-formula which is satisfied along* $\langle \alpha_\xi \rangle$. *Let* $\beta_\xi = \alpha_{\omega \cdot \xi}$; *then* $\bigvee a \psi^{(a)}$ *is satisfied along* $\langle \beta_\xi \rangle$.

PROOF. A typical case is for $\psi = \psi(x)$ of the form

$$\bigwedge y \in x \bigvee z_1 \bigwedge w \in z_1 \bigvee z_2 \theta(x, y, z_1, w, z_2)$$

where θ is Δ_0. Consider any ξ and any $x \in M_{\beta_\xi} = M_{\alpha_{\omega \cdot \xi}} = \bigcup_{\nu < \omega \cdot \xi} M_{\alpha_\nu}$. For some $\nu < \omega \cdot \xi$, $x \in M_{\alpha_\nu}$ and by hypothesis,

$$\bigwedge y \in x \bigvee z_1 \in M_{\alpha_\nu} \bigwedge w \in z_1 \bigvee z_2 \in M_{\alpha_\nu} \theta(x, y, z_1, w, z_2).$$

Using the axiom of choice, we can find f_1, f_2 such that

$$\bigwedge y \in x \{ f_1(y) \in M_{\alpha_\nu} \wedge \bigwedge w \in f_1(y) [f_2(y, w) \in M_{\alpha_\nu} \wedge \theta(x, y, f_1(y), w, f_2(y, w))]\}.$$

Let $a_1 = \{ f_1(y) : y \in x\}$, $a_2 = \{ f_2(y, w) : y \in x \wedge w \in f_1(y)\}$, $a = a_1 \cup a_2$. Then a is countable and $a \subseteq M_{\alpha_\nu}$ so $a \in M_{\alpha_\nu + 1} \subseteq M_{\alpha_{\nu+1}} \subseteq M_{\beta_\xi}$ and

$$\bigwedge y \in x \bigvee z_1 \in a \bigwedge w \in z_1 \bigvee z_2 \in a \theta(x, y, z_1, w, z_2),$$

i.e., $M_{\beta_\xi} \models \bigvee a \psi^{(a)}$.

LEMMA 3.1. (iii) *Suppose* $\psi(x, y, z)$ *is a* Σ*-formula such that* $\bigwedge x, y \bigvee ! z \psi(x, y, z)$ *is satisfied along* $\langle \alpha_\xi \rangle$. *Suppose further that each* α_ξ *is closed under* $+$. *Let* $\langle \beta_\xi \rangle$ *enumerate the fixed points* $\alpha_\eta = \eta$. *Then* $\mathrm{TR}(\in, \psi)$ *is satisfied along* $\langle \beta_\xi \rangle$.

PROOF. Consider any choice of the parameters of ψ in any $M_{\alpha_{\xi_0}}$; these are fixed in the following: For each $\xi \geqq \xi_0$, ψ determines a function $G_\xi : M^2_{\alpha_\xi} \to M_{\alpha_\xi}$ with $M_{\alpha_\xi} \models \psi(x, y, G_\xi(x, y))$. Moreover, for $\eta \geqq \xi$, $G_\xi = G_\eta \restriction M_{\alpha_\xi}$, since M_{α_η} is an $\in$-extension of M_{α_ξ}. Let $G = \bigcup_\xi G_\xi$, so $G : \mathrm{HC}^2 \to \mathrm{HC}$. Let F be the unique function such that, for each $x \in \mathrm{HC}$, $F(x) = G(x, F \restriction x)$, and let $F_a = F \restriction (a \cup \{a\})$. We claim that

(1) *for each* $\xi \geqq \xi_0$ *and* $a \in M_{\alpha_\xi}$ *we have* $F_a \in M_{\alpha_{\xi + \rho(a)+1}}$.
This is proved by $\in$-induction on a. Suppose it holds for each $x \in a$; then also $x \in M_{\alpha_\xi}$ and $F(x) \in M_{\alpha_{\xi + \rho(x)+1}}$ for each such x. Hence

$$F \restriction a = \{\langle x, F(x)\rangle : x \in a\} \subseteq \bigcup_{\zeta < \rho(a)} M_{\alpha_{\xi + \zeta + 1}} \subseteq M_{\alpha_{\xi + \rho(a)}} \quad \text{and} \quad F \restriction a \in M_{\alpha_{\xi + \rho(a)+1}}.$$

Since also $a \in M_{\alpha_{\xi + \rho(a)-1}}$, we can apply closure of the M_{α_ν} under G to conclude that $F(a) \in M_{\alpha_{\xi+\rho(a)+1}}$; this gives the desired conclusion for F_a.

It follows from (1) that

(2) *for each* ξ *with the parameters of* ψ *and* $a \in M_{\beta_\xi}$ *we have* $F_a \in M_{\beta_\xi}$.
For, β_ξ is of the form α_η where $\alpha_\eta = \eta$. Since η is a limit number, we can choose $\xi_0 < \eta$ such that the parameters of ψ and a both lie in $M_{\alpha_{\xi_0}}$. By (1), $F_a \in M_{\alpha_{\xi_0 + \rho(a)+1}}$. From $\rho(a) < \alpha_{\xi_0} < \alpha_\eta$ and the closure of α_η under $+$ we have $(\xi_0 + \rho(a) + 1) < \alpha_\eta = \eta$ and $F_a \in M_{\alpha_\eta} = M_{\beta_\xi}$. Applying (2) to transitive a gives the desired conclusion $M_{\beta_\xi} \models \mathrm{TR}(\in, \psi)$.

One is necessarily led to sequences $\langle \alpha_\xi \rangle$ where the α_ξ are closed under $+$, if one is to satisfy the instance of $\mathrm{TR}(\in)$ giving the existence of $+$.

Suppose the hypotheses of the ordinal sufficiency rule for $\prec$ are satisfied along a sequence $\langle \alpha_\xi \rangle$. For each ξ, let $\prec^{(\xi)}$ be the relation $\{\langle x, y\rangle : x, y \in M_\alpha$ and $M_\alpha \models (x \prec y)\}$. By hypothesis, there are Σ-formulas ψ_1, ψ_2, θ such that

$$[x \prec y \leftrightarrow \psi_1(x, y)] \wedge [\psi_1(x, y) \leftrightarrow \sim\psi_2(x, y)]$$

and

$$\bigvee a \theta(a) \wedge \bigwedge a, x, y[\theta(a) \wedge x \prec y \rightarrow x, y \in a]$$

are satisfied along $\langle \alpha_\xi \rangle$. By the first of these $(\prec^{(\xi)}) = (\prec^{(\eta)}) \restriction M_{\alpha_\xi}$ for each $\eta \geqq \xi$. Choose $a \in M_{\alpha_0}$ satisfying θ; then also $M_\alpha \models \theta(a)$ for each ξ, so the field d_ξ of $\prec^{(\xi)}$ is $\subseteq a$. Hence all the relations $\prec^{(\xi)}$ are identical with $\prec^{(0)}$. This is actually a well-founded relation, for if $b \subseteq d_0$ then $b \in M_{\alpha_1}$ and $M_{\alpha_1} \models \mathrm{Wf}_\prec$. Hence there is a function f mapping d_0 onto an ordinal δ, by

$$f(x) = \sup\{f(y) + 1 : y \prec^{(0)} x\}.$$

We write $|\prec| = \delta$. This proves the following:

LEMMA 3.1. (iv) *Suppose the hypotheses of* PS_1(VII) *are satisfied along a sequence* $\langle \alpha_\xi \rangle$. *Then the conclusion*

$$\bigvee f[\mathrm{Ord}(\breve{\mathscr{D}}(f)) \wedge \bigwedge x, y(x \prec y \rightarrow f(x) < f(y))]$$

is satisfied along the sequence $\langle \alpha_\xi \rangle_{\xi \geqq \xi_0}$ *where* α_{ξ_0} *is the least term with* $|\prec| < \alpha_{\xi_0}$.

The problem with trying to apply this part of Lemma 3.1 to find models of PS_1 is that we have no *a priori* means of bounding $|\prec|$. Such bounds will be found in § 4 by formalizing the present proof in IR.

On the other hand, 3.1(i)–(iii) can be applied directly to give the following:

COROLLARY 3.2. $\mathrm{PS}_0\mathrm{E}$ *has a model in* $M_{\kappa^{(\omega)}(0)}$.

PROOF. Suppose $\mathscr{D}$ is a derivation of a theorem φ of $\mathrm{PS}_0\mathrm{E}$. We assign to each θ in $\mathscr{D}$ a sequence $\langle\alpha_\xi\rangle$ along which it is satisfied, and for which the α_ξ are closed under $+$ and multiplication by ω. We can take $\langle\alpha_\xi\rangle = \langle\omega^{\omega+\xi}\rangle$ for each axiom θ. By 3.1(ii), (iii), if the hypotheses of a rule are satisfied along $\langle\alpha_\xi\rangle$ and $\langle\beta_\xi\rangle$ enumerates the fixed points of $\langle\alpha_\xi\rangle$ then the conclusion can be satisfied along $\langle\beta_\xi\rangle$. (Reflection is covered by 3.1(ii), since this $\langle\beta_\xi\rangle$ is a subsequence of $\langle\alpha_{\omega\cdot\xi}\rangle$.) Thus for some $n \geqq 1$, φ is satisfied along the sequence $\langle\kappa^{(n)}(\xi)\rangle$. $\kappa^{(\omega)}(0) = \kappa^{(n)}(\kappa^{(\omega)}(0))$ is in every such sequence.

By 2.12(i) this shows $\kappa^{(\omega)}(0)$ is the least ordinal which cannot be proved to exist in $\mathrm{PS}_0\mathrm{E}$. However, the result is not best possible in terms of models. Although we have been working only with the $M_\alpha = \mathrm{HC} \cap \{x : \rho(x) < \alpha\}$, examination of the proof of 3.1(i)–(iii) shows that only weaker closure conditions need be satisfied to produce models M_α. Indeed, we can take the collection L_α of sets constructible before α instead of M_α throughout. Moreover, by $\mathrm{TR}(<)$ we can prove the existence of L_β in PS_0 whenever we can prove the existence of β. Hence

COROLLARY 3.3. $L_{\kappa^{(\omega)}(0)}$ *is the minimum model of* PS_0 $(+\mathrm{E})$.

3.2. *Representation structures.* This subsection takes up informally the details concerning representations of sets $x \in \mathrm{HC}$ by subsets X of ω which will be needed (formally) in the next section. As in § 2.2 we use capital letters A, B, E, R, X, Y, Z to range over subsets of ω. Relations and functions are identified with sets by the pairing function (,). We write $p\,E\,q$ for $(p, q) \in E$, and $\mathscr{D}(E)$, $\breve{\mathscr{D}}(E)$, $\mathfrak{F}(E)$ for the domain, range and field of E, resp. E^* is the transitive closure of E and the identity, i.e.,

$$q\,E^*\,p \quad \text{iff} \quad \begin{array}{l}\text{there is a finite sequence } r_0, \ldots, r_n\,(n \geqq 0)\\ \text{with } r_0 = q, r_i\,E\,r_{i+1} \text{ for each } i < n \text{ and } r_n = p.\end{array}$$

Partial functions are identified with their graphs. Pairing is extended to sets by $X = (A, B) = \{(p, q) : p \in A \;\&\; q \in B\}$, similarly for n-tupling of sets and natural numbers; unpairing is denoted by $A = (X)_0$, $B = (X)_1$.

Let $\prec$ be a well-ordering in the natural numbers of limit order type α; $\beta, \gamma, \ldots, \xi, \eta$ range over the field of $\prec$. We also write $\beta < \alpha$ for $\beta \in \mathfrak{F}(\prec)$, and $\beta < \gamma$ for $\beta \prec \gamma$. We write $R : A \to \alpha$ to mean R is a partial function with $\mathscr{D}(R) = A$ and $R(p) < \alpha$ for each $p \in A$. $\beta + 1$ denotes the successor of β and $\beta = \sup B$ expresses that the supremum of $B \subseteq \mathfrak{F}(\prec)$ exists and equals β.

By an α- (or $\prec$-) *representation structure* is meant a set $X = (A, E, m, R)$ satisfying the following conditions:

(i) $A = \mathfrak{F}(E)$ and $m \in A$,

(ii) $\bigwedge p \in A[p\,E^*\,m]$,

(iii) $R : A \to \alpha$,

(iv) $\bigwedge p \in A[R(p) = \sup\{R(q) + 1 : q\,E\,p\}]$.

By the *rank* $R(X)$ of X is meant $R(m)$. X is said to be *extensional* if, in addition,
(v) $\bigwedge p, q \in A[\bigwedge r(r\, E\, p \leftrightarrow r\, E\, q) \rightarrow p = q]$.
Note that if $n \in A$ then $X \upharpoonright n = (A \upharpoonright n, E \upharpoonright n, n, R \upharpoonright n)$ is an α-representation structure, where $A \upharpoonright n = \{p : p\, E^*\, n\}$, $p(E \upharpoonright n)q$ iff $p, q \in (A \upharpoonright n)$ and $p\, E\, q$, and $(R \upharpoonright n)(p) = R(p)$ for $p \in (A \upharpoonright n)$.

Since in any representation structure both E and E^* are well-founded by (iv), we can apply transfinite induction and recursion to these. For any representation structure $X = (A, E, m, R)$ we define a function F^X from A into sets by the following recursion:

$$F^X(p) = \{F^X(q) : q\, E\, p\} \quad \text{for each } p \in A.$$

Then evidently

LEMMA 3.4. (i) $F^X : A \rightarrow \mathrm{HC}$,
(ii) $|R(p)| = \rho(F^X(p))$ *for each* $p \in A$,
(iii) $\breve{\mathscr{D}}(F^X) = \mathrm{TC}(\{F^X(m)\})$,
(iv) *if* $p \in (A \upharpoonright n)$ *then* $F^{(X \upharpoonright n)}(p) = F^X(p)$, *and*
(v) *if* X *is extensional* F^X *is one-one.*

Let $x = F^X(m)$; we say X *represents* x. For $n \in A$ and $y = F^X(n)$ we say that n *represents* y, which is equivalent to $X \upharpoonright n$ representing y. Conversely to Lemma 3.4 every $x \in \mathrm{HC}$ is representable by some X, given any function F with $\mathscr{D}(F) \subseteq \omega$ and $\breve{\mathscr{D}}(F) = \mathrm{TC}(\{x\})$. Namely take $A = \mathscr{D}(F)$, $q\, E\, p$ if $F(q) \in F(p)$, $m = F^{-1}(x)$ and $|R(p)| = \rho(F(p))$ for each p; then $X = (A, E, m, R)$ represents x and $F^X = F$. If F is one-one then X is extensional.

Suppose $X_i = (A_i, E_i, m_i, R_i)$ are representation structures ($i = 1,2$). We shall need an intrinsic n.a.s.c. for $p_1 \in A_1$ to represent the same set as $p_2 \in A_2$. This is given in terms of the following relation:

$$\begin{aligned} p_1 \equiv p_2 \quad \text{iff} \quad & R_1(p_1) = R_2(p_2) \quad \text{and} \\ & \bigwedge q_1[q_1\, E_1\, p_1 \rightarrow \bigvee q_2(q_2\, E_2\, p_2 \wedge q_1 \equiv q_2)] \quad \text{and} \\ & \bigwedge q_2[q_2\, E_2\, p_2 \rightarrow \bigvee q_1(q_1\, E_1\, p_1 \wedge q_1 \equiv q_2)]. \end{aligned}$$

This can be considered to be defined by transfinite recursion up to α. We define $X_1 \equiv X_2$ if $m_1 \equiv m_2$ and $X_1 \mathscr{E} X_2$ if there is some $p_2\, E_2\, m_2$ with $m_1 \equiv p_2$; or what comes to the same $X_1 \equiv X_2 \upharpoonright p_2$. Transfinite induction gives

LEMMA 3.5. (i) $\equiv$ *is an equivalence relation with* $\mathscr{D}(\equiv) \subseteq A_1, \breve{\mathscr{D}}(\equiv) \subseteq A_2$,
(ii) $p_1 \equiv p_2$ *iff* $F^{X_1}(p_1) = F^{X_2}(p_2)$,
(iii) X_1 *and* X_2 *represent the same set iff* $X_1 \equiv X_2$, *and*
(iv) X_1 *represents a member of the set represented by* X_2 *iff* $X_1 \mathscr{E} X_2$.

When X_1 and X_2 are the same structure X, $\equiv$ is an equivalence relation with $\mathscr{D}(\equiv) = A$. Let $\bar{m}$ be the least p with $m \equiv p$, and $\bar{A} = \{\bar{m} : m \in A\}$. For $m, m_1, m_2 \in \bar{A}$ define $m_1\, \bar{E}\, m_2$ iff $\bigvee p_1, p_2\{m_1 \equiv p_1 \wedge m_2 \equiv p_2 \wedge p_1\, E\, p_2\}$, $\bar{R}(m) = R(m)$, and $\bar{X} = (\bar{A}, \bar{E}, \bar{n}, \bar{R})$.

LEMMA 3.6. *For each representation structure X, the structure $\bar{X}$ is extensional and $X \equiv \bar{X}$.*

The following *collection* or *join* operation $J(\langle X_i \rangle_{i \in I})$ is defined for any sequence $X_i = (A_i, E_i, m_i, R_i)$ of α-representation structures for which $\sup \{R_i(m_i) + 1 : i \in I\} < \alpha$. It is sufficient to define this for the case that the A_i are disjoint and $0 \notin \bigcup_{i \in I} A_i$. Take $J(\langle X_i \rangle_{i \in I})$ to be $(A, E, 0, R)$ where

(i) $A = \{0\} \cup \bigcup_{i \in I} A_i$,

(ii) $p\, E\, q$ iff, for some i, $p \in A_i$ and ($q \in A_i$ and $p\, E_i\, q$ or $q = 0$),

(iii) $R(p) = R_i(p)$ for $p \in A_i$ and $R(0) = \sup \{R_i(m_i) + 1 : i \in I\}$.

In general, define $J(\langle X_i' \rangle) = J(\langle X_i \rangle)$ where X_i are obtained in some canonical way satisfying the disjointness condition and $X_i' \cong X_i$.

LEMMA 3.7. *$Y \mathscr{E} J(\langle X_i \rangle_{i \in I})$ iff for some $i \in I$, $Y \equiv X_i$.*

4. Proof of the main theorem. In the case that $\alpha < \Gamma_0$ we can fix $\prec$ to be the initial segment $\prec_\alpha$ of the primitive recursive ordering $\prec_{\Gamma_0}$ (cf. § 2.2). Let $\mathrm{Rep}_\alpha(X)$ be the arithmetical formula of $\mathscr{L}_2$ expressing that the set X of natural numbers is an α-representation system, $X = ((X)_0, (X)_1, (X)_2, (X)_3)$. The relation $X_1 \equiv X_2$ is defined by TR with respect to $\prec_\alpha$ and so $\equiv$ can be considered to be a relation symbol of IR∇. The relation $X_1 \mathscr{E} X_2$ is arithmetically defined from $\equiv$, hence can also be taken to be a relation symbol of IR∇. The symbols $\equiv$, $\mathscr{E}$ both depend on α; this should be shown explicitly by $\equiv_\alpha, \mathscr{E}_\alpha$. However, we can prove in IR∇ that

$$\mathrm{Rep}_\alpha(X_1) \wedge \mathrm{Rep}_\alpha(X_2) \wedge \alpha < \beta \rightarrow [X_1 \equiv_\alpha X_2 \leftrightarrow X_1 \equiv_\beta X_2] \wedge [X_1 \mathscr{E}_\alpha X_2 \leftrightarrow X_1 \mathscr{E}_\beta X_2].$$

There is an operation symbol of IR∇ whose value at X, n with $n \in (X)_0$ is $X \restriction n$. Then with each formula $\varphi(x_1, \cdots, x_k)$ of $\mathscr{L}$ is associated a formula $\varphi^{(\alpha)}(X_1, \cdots, X_k)$ of $\mathscr{L}_2\nabla$, as follows:

(i) $(x_1 = x_2)^\alpha$ *is* $(X_1 \equiv X_2)$ *and* $(x_1 \in x_2)^\alpha$ *is* $(X_1 \mathscr{E} X_2)$,

(ii) $(\ )^\alpha$ *preserves the propositional connectives,*

(iii) $(\bigvee x \varphi)^\alpha$ *is* $\bigvee X[\mathrm{Rep}_\alpha(X) \wedge \varphi^\alpha]$,
$(\bigwedge x \varphi)^\alpha$ *is* $\bigwedge X[\mathrm{Rep}_\alpha(X) \rightarrow \varphi^\alpha]$,

(iv) $(\bigvee y \in x \varphi(x, y, \ldots))^\alpha$ *is* $\bigvee n[n \in (X)_0 \wedge \varphi^\alpha(X, X \restriction n, \ldots)]$ *and*
$(\bigwedge y \in x \varphi(x, y, \ldots))^\alpha$ *is* $\bigwedge n[n \in (X)_0 \rightarrow \varphi^\alpha(X, X \restriction n, \ldots)]$.

In the following, $\bigvee X^\alpha[\cdots]$ abbreviates $\bigvee X[\mathrm{Rep}_\alpha(X) \wedge \cdots]$ and $\bigwedge X^\alpha[\cdots]$ abbreviates $\bigwedge X[\mathrm{Rep}_\alpha(X) \rightarrow \cdots]$. Clearly if φ is a Δ_0-formula then φ^α is arithmetical (in $\equiv_\alpha$) and if φ is a Σ-, resp. Π-formula, the same holds for φ^α.

Informally, when φ is a sentence, φ^α holds just in case $M_\alpha \models \varphi$. This can be expressed formally in PS$_0\nabla$ as follows. Let

$$M_\alpha(a) \leftrightarrow \rho(a) < \alpha \wedge \bigvee f[\mathscr{D}(f) \subseteq \omega \wedge \breve{\mathscr{D}}(f) = \mathrm{TC}(\{a\})].$$

(In PS$_0$E∇ this is simply equivalent to $\rho(a) < \alpha$.) Let $\mathrm{Rep}_\alpha(x, a)$ express that $x \subseteq \omega$ and x represents a, i.e., $F^X(m) = a$ where $m = (x)_2$. In PS$_0\nabla$ we can prove

$$\bigwedge a[M_\alpha(a) \leftrightarrow \bigvee x(\mathrm{Rep}_\alpha(x, a))].$$

Let $\varphi^{(M_\alpha)}$ be the result of relativizing all unbounded quantifiers in φ to $M_\alpha(\ )$. Then

LEMMA 4.1. *For any formula $\varphi(a_1, \ldots, a_k)$ of $\mathscr{L}$,*

$$\mathrm{PS}_0\nabla \vdash \mathrm{Rep}_\alpha(x_1, a_1) \wedge \cdots \wedge \mathrm{Rep}_\alpha(x_k, a_k) \rightarrow \{\varphi^{(M_\alpha)}(a_1, \ldots, a_k) \leftrightarrow \varphi^\alpha(x_1, \ldots, x_k)\}.$$

This is proved by induction on φ using 3.5 and the preceding remark.

For subsets X and A of ω, the formula $\mathrm{Rep}_\alpha(X, A)$ can be considered to be directly in IR∇. Moreover with each $A \subseteq \omega$ is associated in a natural way an X such that $\mathrm{Rep}_\alpha(X, A)$ for any $\alpha > \omega$; namely $X = J(\langle \boldsymbol{n} \rangle_{n \in A})$ where for each $n \in \omega$, $\boldsymbol{n}$ represents the ordinal n. We easily adapt the proof of 4.1 to obtain

LEMMA 4.2. *Suppose $\omega < \alpha$, and $\mathfrak{F}(A_1, \ldots, A_k)$ is any formula of $\mathscr{L}_2$. Let $\varphi = \mathfrak{F}^*$ be the translation of $\mathfrak{F}$ in $\mathscr{L}$. Then*

$$\mathrm{IR}\nabla \vdash \mathrm{Rep}_\alpha(X_1, A_1) \wedge \cdots \wedge \mathrm{Rep}_\alpha(X_k, A_k) \rightarrow [\mathfrak{F}(A_1, \ldots, A_k) \leftrightarrow \varphi^\alpha(X_1, \ldots, X_k)].$$

In particular, if $\mathfrak{F}$ is a sentence then IR$\nabla \vdash \mathfrak{F} \leftrightarrow (\mathfrak{F}^*)^\alpha$.

The persistence of Σ-formulas under $\in$-extensions in the M_α's is translated into the following:

LEMMA 4.3. *Suppose $\alpha < \beta$ and $\varphi(x_1, \ldots, x_k)$ is a Σ-formula of $\mathscr{L}$. Then*

$$\mathrm{IR}\nabla \vdash \mathrm{Rep}_\alpha(X_1) \wedge \cdots \wedge \mathrm{Rep}_\alpha(X_k) \wedge \varphi^\alpha(X_1, \ldots, X_k) \rightarrow \varphi^\beta(X_1, \ldots, X_k).$$

The main work now is to formalize in IR∇ as far as possible the argument for Lemma 3.1. By that result (cf. also the argument for Corollary 3.2) we shall only need to deal with certain *end* sequences $\langle \kappa^{(n)}(\xi) \rangle_{\xi_0 \leqq \xi < \Omega}$ of the sequences of critical numbers $\kappa^{(n)}(\xi)$. However, these cannot be dealt with in full, but only for segments $\xi_0 \leqq \xi < \gamma$ for each $\gamma < \Gamma_0$. We make use of the following, which is established in [3, § 3]:

4.4. *The functions $\lambda\xi, \eta(\xi + \eta)$, $\lambda\xi, \eta(\xi \cdot \eta)$ and $\lambda\xi, \eta\kappa^{(\xi)}(\eta)$ are recursively represented on $\prec_{\Gamma_0}$, and their elementary order relations are derivable in Z.*

These recursive representations and elementary properties depend only on how the natural ordering $\prec_{\Gamma_0}$ is built up, and do not require any verification of well-ordering properties. The following examples illustrate the content of 4.4: We have a recursive function $\oplus$ such that if $\alpha, \beta \in \mathfrak{F}(\prec_{\Gamma_0})$ then $\alpha \oplus \beta \in \mathfrak{F}(\prec_{\Gamma_0})$ and $|\alpha \oplus \beta| = |\alpha| + |\beta|$. We can prove in Z that (for the function $\lambda\alpha\omega^\alpha$ representing $\kappa^{(0)}$) $\alpha \prec_{\Gamma_0} w^\gamma \wedge \beta \prec_{\Gamma_0} w^\gamma \rightarrow \alpha \oplus \beta \prec_{\Gamma_0} w^\gamma$. We can also prove there (for the function $\lambda\alpha, \beta k^{(\alpha)}(\beta)$ representing $\lambda\xi, \eta\kappa^{(\xi)}(\eta)$) $\beta \prec_{\Gamma_0} \alpha \rightarrow k^{(\beta)}(k^{(\alpha)}(\gamma)) = k^{(\alpha)}(\gamma)$. The reader is referred to [3] for more details on this. To simplify notation, we shall use the same symbols for these number-theoretic functions on $\prec_{\Gamma_0}$ as for the corresponding ordinal functions, and $<$ for $\prec_{\Gamma_0}$. The restriction to segments $\prec_\alpha$ for $\alpha < \Gamma_0$ is necessary only when it comes to dealing with notions introduced by TR in IR∇ such as Rep_α.

LEMMA 4.5. *With each derivation $\mathscr{D}$ of a sentence φ in* PS$_1$E *can be associated an $i < \omega$, a $\nu < \Gamma_0$, and an end sequence $\langle \alpha_\xi \rangle = \langle \kappa^{(i)}(\nu + \xi) \rangle$ of the sequence $\langle \kappa^{(i)}(\xi) \rangle$*

such that, for each $\gamma < \Gamma_0$,

$$\mathrm{IR}\nabla \vdash \bigwedge\xi[\xi < \gamma \to \varphi^{\alpha_\xi}].$$

PROOF. First replace each formula in $\mathscr{D}$ by its universal closure, so that we are dealing throughout with sentences. As in the proof of 3.2, start out with the sequence $\langle\alpha_\xi\rangle = \langle\omega^{\omega+\xi}\rangle$ assigned to the axioms in $\mathscr{D}$. The arguments are by straightforward formalization of 3.1(i), using formal versions of 3.6, 3.7 in IR. We illustrate only with Δ_0-separation; given a Δ_0-formula $\psi(x, y, z_1, \cdots, z_k)$ this translates for any $\alpha = \alpha_\xi$ with $\alpha < \gamma$ to

$$(1) \qquad \bigwedge Z_1^\alpha \ldots Z_k^\alpha \bigwedge X^\alpha \bigvee W^\alpha \bigwedge Y^\alpha [Y \in W \leftrightarrow Y \in X \wedge \psi^\alpha(X, Y, Z_1, \ldots, Z_k)].$$

ψ^α is arithmetical. Let

$$X = (A, E, m, R) \quad \text{and} \quad I = \{n : n\, E\, m \wedge \psi^\alpha(X, X \restriction n, Z_1, \ldots, Z_k)\}.$$

Then we can take $W = J(\langle X \restriction n\rangle_{n \in I})$.

For the axiom of enumerability it is more convenient to pass from any α-representation structure X to an extensional $\bar{X} \equiv X$.

Next proceed to the rules in $\mathscr{D}$. Suppose $\langle\alpha_\xi\rangle = \langle\kappa^{(i)}(\nu + \xi)\rangle$ has been assigned to the conjunction φ of the hypotheses of a rule so that, for each $\gamma < \Gamma_0$, IR∇ $\vdash \bigwedge\xi[\xi < \gamma \to \varphi^{\alpha_\xi}]$.

The case of the Σ-reflection rule is illustrated as in 3.1(ii) by

$$(2) \qquad \varphi = \bigwedge x \bigwedge y \in x \bigvee z_1 \bigwedge w \in z_1 \bigvee z_2 \theta(x, y, z_1, w, z_2),$$

where θ is Δ_0. Let $\beta_\xi = \alpha_{\omega\cdot\xi}$, and consider any $\beta = \beta_\xi$ with $\xi < \gamma$. Let X be any β-representation structure; there is $\nu < \omega \cdot \xi$ with X an $\alpha = \alpha_\nu$ representation structure. The hypothesis for φ is applied to $\gamma_1 = \omega \cdot \gamma$, so that

$$(3) \quad \bigwedge n \bigvee Z_1^\alpha \bigwedge p \bigvee Z_2^\alpha \{(X \restriction n)\, \mathscr{E}\, X \wedge (Z_1 \restriction p)\, \mathscr{E}\, Z_1 \to \theta^\alpha(X, X \restriction n, Z_1, Z_1 \restriction p, Z_2)\}.$$

The argument for this case can be completed by using the Σ-choice rule 2.6(vii) in IR∇ and the J operation. It is then a simple step to shift up to the sequence $\langle\alpha'_\xi\rangle = \langle\kappa^{(i+1)}(\nu + \xi)\rangle$.

Now consider an instance of the $\in$-recursion rule, with hypothesis $\varphi = \bigwedge x, y \bigvee !z\psi(x, y, z)$ where ψ is a Σ-formula (for simplicity, without parameters). Given any $\gamma < \Gamma_0$ we have, for $\xi < \gamma$ and $\alpha = \alpha_\xi$,

$$(4) \quad \begin{array}{l} \text{(i)} \ \bigwedge X^\alpha, Y^\alpha \bigvee Z^\alpha \psi^\alpha(X, Y, Z), \text{ and} \\ \text{(ii)} \ \bigwedge X^\alpha, Y^\alpha, Z_1^\alpha, Z_2^\alpha\{\psi^\alpha(X, Y, Z_1) \wedge \psi^\alpha(X, Y, Z_2) \to Z_1 \equiv Z_2\} \end{array}$$

established in IR∇. By the Σ-uniformization rule 2.6(vi) and the Σ-rule of definition in IR∇ we can introduce a function symbol $G(X, Y)$ (depending on the choice of γ) such that

$$(5) \qquad \bigwedge X^\alpha, Y^\alpha[\mathrm{Rep}_\alpha(G(X, Y)) \wedge \psi^\alpha(X, Y, G(X, Y))].$$

Let $X = (A, E, m, R)$ be any α-representation structure which represents a transitive set; taking X to be extensional this simply means that E is transitive. Define P

from J to represent the ordered pair operation. By TR(E) (2.6(v)) we obtain a function F_1 from numbers to sets such that, for each $n \in A$,

$$F_1(n) = G(X \upharpoonright n, J(\langle P(X \upharpoonright q, F_1(q)) \rangle_{qEn})). \tag{6}$$

Let $F(X \upharpoonright n) = F_1(n)$. As in 3.1(iii) it is proved that, for each $\xi < \gamma$ and each $n \in A$, $F(X \upharpoonright n)$ is an $\alpha_{\xi+R(n)+1}$-representation structure. This allows us to show that TR($\in, \psi$) is provably satisfied along $\langle \kappa^{(i+1)}(\nu + \xi) \rangle$ for $\xi < \gamma$.

Consider finally the ordinal sufficiency rule. Here

$$\varphi = \bigwedge x, y[(x \prec y \leftrightarrow \psi_1(x, y)) \wedge (\psi_1(x, y) \leftrightarrow \sim\psi_2(x, y))] \wedge \mathrm{WF}_{\prec} \wedge \mathrm{Bd}^{\theta}_{\prec}$$

where $\psi_1(x, y)$, $\psi_2(x, y)$ and $\theta(z)$ are Σ-formulas without parameters. Let $\mathrm{Rep}_{\alpha_0}(Z)$ with $\theta^{\alpha_0}(Z)$. The field of $\prec$ is represented by a subset of Z and $\prec$ is isomorphic to the relation $\prec^*$:

$$n \prec^* q \quad \text{iff} \quad \psi_1^{\alpha_0}(Z \upharpoonright n, Z \upharpoonright q). \tag{7}$$

$\prec^*$ is provably $\Pi \cap \Sigma$ and we have established $\mathrm{Wf}_{\prec^*}$. By 2.6(i) we can find $\mu < \Gamma_0$ and an order-preserving map of $\prec^*$ into $\prec_\mu$; take $\nu = \mu$. By definiteness, it follows that the conclusion of PS_1(VII) is provably satisfied along $\langle \kappa^{(i)}(\mu + \xi) \rangle$ for $\xi < \gamma$.

This completes the proof.

THEOREM 4.6. $\mathrm{PS}_1\mathrm{E}$ *is a conservative extension of* IR *under the translation of* $\mathscr{L}_2$ *into* $\mathscr{L}$.

PROOF. By 2.10, 4.2, 4.5.

We can now complete the model-theoretic work of § 3.1.

THEOREM 4.7. (i) M_{Γ_0} *is a model of* $\mathrm{PS}_1\mathrm{E}$.

(ii) L_{Γ_0} *is the minimal model of* PS_1 (+E).

PROOF. By extension of the arguments for 3.2, 3.3 using the last part of the proof of 4.5. Minimality is by 2.12(ii).

We can also extract models of $\mathrm{PS}_1\mathrm{E}$ from ω-models $\mathscr{M}$ of IR. Let $\mathrm{Rep}_\alpha[\mathscr{M}]$ be the collection of all $X \in \mathscr{M}$ with $\mathrm{Rep}_\alpha(X)$, and let $\mathrm{Rep}[\mathscr{M}] = \mathrm{Rep}_\Omega[\mathscr{M}]$.

THEOREM 4.8. (i) *If* $\mathscr{M}$ *is an* ω*-model of* IR *and* $\alpha \geqq \Gamma_0$ *and* $\kappa^{(i)}(\alpha) = \alpha$ *for all* $i < \omega$ *then* $\mathrm{Rep}_\alpha[\mathscr{M}]$ *is a model of* $\mathrm{PS}_1\mathrm{E}$.

(ii) Rep[HYP] *is a model of* $\mathrm{PS}_1\mathrm{E}$.

(iii) $\mathrm{Rep}_{\Gamma_0}[\mathrm{HYP}_{\Gamma_0}]$ *is the minimal model of* PS_1 (+E).

PROOF. (i) By 4.5. (ii), (iii) are corollaries by 2.6. (Note that the equality $\mathrm{Rep}_\alpha[\mathrm{HYP}_\alpha] = L_\alpha$ can also be proved directly for suitable limit α including $\alpha = \Gamma_0$.)

Various independence results for $\mathrm{PS}_1\mathrm{E}$ follow simply from 4.8(ii) and known results for HYP. For example, let CWO_1 be the statement expressing that *any two well-ordering relations* r_0, r_1 *are comparable*, and let CWO_2 be the statement expressing that *any well-ordering relation r is isomorphic to an ordinal.*

COROLLARY 4.9. (i) $\mathrm{PS}_1\mathrm{E} \nvdash \mathrm{CWO}_1$, *and hence*
(ii) $\mathrm{PS}_1\mathrm{E} \nvdash \mathrm{CWO}_2$.

PROOF. This is by an example (cf. [**11**, p. 373]) of two recursive simple ordering relations $\prec_0, \prec_1$ such that, for $j = 0, 1$,

(1) $\mathrm{HYP} \models \mathrm{Wf}_{\prec_j}$ and

(2) there is no HYP order-preserving map of $\prec_j$ onto an initial segment of $\prec_{1-j}$.

Note that for (ii) it would be sufficient to use the simpler example of [**8**] of a recursive simple-ordering relation $\prec$ for which $\mathrm{HYP} \models \mathrm{Wf}_{\prec}$, but which is not a well-ordering.

The main results of §§ 4.6–4.8 also hold for PSE instead of $\mathrm{PS}_1\mathrm{E}$, where PS is the system of [**2**] described in § 2.4 above. However, the proof for PSE is somewhat more delicate and a presentation of it would obscure the main points. There does not seem to be any advantage of PSE compared to $\mathrm{PS}_1\mathrm{E}$ for the formulation of predicative portions of mathematics, since in practice all of the work can already be formalized in a weak subsystem of $\mathrm{PS}_0 + \mathrm{AC}$.

5. Comparison with some formally related systems of set theory. In connection with their work on generalizations of recursion theory to various collections of sets, Kripke [**9**] and Platek [**12**] have considered models of certain weak systems of set theory which are formally related to PS_0. These are denoted here by S_i ($i = 1, 2, 3$). Each of these contains the axioms (either directly, or as a theorem) of extensionality, Δ_0-separation, pairing, union and regularity (i.e., $\mathrm{Wf}_{\in}$) and, in addition, one of the following axiom schemes for Σ-formulas ψ:

S_1 (*Σ-replacement axiom*).

$$\bigwedge x \in a \bigvee ! y \psi(x, y) \rightarrow \bigvee b \bigwedge y[y \in b \leftrightarrow \bigvee x \in a \psi(x, y)].$$

S_2 (*Σ-reflection axiom*).

$$\psi \rightarrow \bigvee a \psi^{(a)}.$$

S_3 (*Σ-choice axiom*).

$$\bigwedge x \in a \bigvee y \psi(x, y) \rightarrow \bigvee f \bigwedge x \in a \psi(x, f(x)).$$

These are written in order of increasing strength. We write $S_i(\infty)$ for S_i + (axiom of infinity). For any of the S-theories considered, a set M is said to be *S-admissible* if it is transitive and $(M, \in | M)$ is a model of S. For $M = L_\alpha$, S_1-admissibility already implies S_3-admissibility, and we simply say M is admissible in this case. The least admissible set is L_ω, i.e., the *hereditarily finite sets*. The least admissible set $\neq L_\omega$ (or, the least $S_1(\infty)$-admissible set) is L_{ω_1}, where ω_1 is the least non-recursive ordinal. In the notation of the preceding section, we can also write $L_{\omega_1} = \mathrm{Rep}_{\omega_1}[\mathrm{HYP}] = \mathrm{Rep}[\mathrm{HYP}]$, and refer to it as the collection of *hereditarily hyperarithmetic sets*. The following can be obtained from 4.5; the interest here is in a direct model-theoretic proof.

THEOREM 5.1. *Every $S_2(\infty)$-admissible set is a model of* PS_1.

PROOF. We show by induction (on derivations of) theorems φ of PS_1 that

$$\text{for every } S_2(\infty)\text{-admissible } M,\ M \models \varphi. \tag{1}$$

The only cases to consider are the rules VI, VII; these are both special cases of the following: Suppose $\prec$ is a formula such that (1) holds for the sentence

$$\bigwedge x, y[(x \prec y \leftrightarrow \psi_1(x, y)) \wedge (x \prec y \leftrightarrow \sim\psi_2(x, y))] \tag{2}$$

where ψ_1, ψ_2 are Σ-formulas without parameters. Suppose (1) also holds for the sentence

$$\bigwedge x \bigvee a[\mathrm{Trans}_{\prec}(a) \wedge x \in a]. \tag{3}$$

The existence of a set $\mathrm{Pd}_{\prec}(x) = \{y : y \prec x\}$ follows from (2), (3) in S_1. For each M let $(\prec_M) = \{\langle x, y\rangle : x, y \in M \text{ and } M \models (x \prec y)\}$. Thus for $M \subseteq M'$ with M, M' S_2-admissible, $(\prec_M) = (\prec_{M'}) \restriction M$. Suppose finally that (1) holds for the sentence

$$\mathrm{Wf}_{\prec}. \tag{4}$$

Let M be any $S_2(\infty)$-admissible set. We claim $\prec_M$ *must actually be well-founded.* For, suppose A is any nonempty subset of the field of $\prec_M$. A need not belong to M; however there exists an $S_2(\infty)$-admissible $M' \supseteq M$ such that $A \in M'$ (for example, M' consisting of all the sets hereditarily of cardinality $\leqq \mathrm{card}\, M$). Since by the assumptions $M' \models \mathrm{Wf}_{\prec}$, there is $x \in A$ which is minimal in A with respect to the relation $\prec_{M'}$; the same x is then minimal in A with respect to $\prec_M$.[5] Hence, for any formula $\varphi(x)$,

$$M \models \mathrm{TI}(\prec, \varphi). \tag{5}$$

Now suppose $\psi(x, y, z)$ is a Σ-formula (possibly with parameters). Then

$$M \models \bigwedge x, y \bigvee !z\psi(x, y, z) \rightarrow \mathrm{TR}(\prec, \psi). \tag{6}$$

This is carried out by the usual argument by $\mathrm{TI}(\prec)$ using the Σ-replacement axiom.[6]

4.8(ii) is a corollary of 5.1. It would be interesting to see if there are also direct model-theoretic arguments for any one of 4.7(i), (ii), 4.8(iii).

As noted, the minimal model of $S_1(\infty)$ is Rep[HYP] and hence it has *no standard predicative interpretations.*[7] Nevertheless it has been shown[8] by Friedman that $S_3(\infty)$ is proof-theoretically reducible to a predicative system, using an adaptation

---

[5] Note that by the example [**8**] of a recursive linear non-well-founded $\prec$ for which $L_{\omega_1} \models \mathrm{Wf}_{\prec}$, this argument makes essential use of the quantifier "for every M" in (1).

[6] The proof essentially establishes that every $S_1(\infty)$-admissible set is a model of the system PS of [**2**].

[7] This is to be expected because of the use of the Σ-replacement *axiom* rather than *rule*. The *prima facie* impredicativity of such axioms and predicative reducibility of corresponding rules is discussed in Kreisel [**10**].

[8] Private communication.

of his argument in [7] to show that IR proves the consistency of the subsystem (Σ_1^1-AC) of analysis.[9] More precisely, $S_3(\infty)$ as well as (Σ_1^1-AC) is consistent relative to a system $(\Pi_\infty^0\text{-CA})_{<\epsilon_0}$ which is of the same strength as ramified analysis in all levels $<\epsilon_0$. The argument gives a *nonstandard predicative interpretation* of $S_3(\infty)$. It makes essential use of an instance of $\mathrm{TI}(\prec_{\epsilon_0})$, and hence of an instance of $\mathrm{TI}(\in)$, which is *not derivable* in $S_3(\infty)$ (though it holds in all $S_3(\infty)$-admissible sets).

It is not known whether also $S_3(\infty) + \mathrm{TI}(\in)$ is weaker than PS_1. However, the theory S which results from this by adding the statement CWO_2 (cf. 4.9) is definitely stronger than PS_1. For, $S \vdash \mathrm{Wf}_{<\Gamma_0}$ by [**3**, § 4] (cf. also [**11**, p. 346]), so L_{Γ_0} can be proved to exist in S and hence $S \vdash \mathrm{Con}_{\mathrm{PS}_1}$.

BIBLIOGRAPHY

1. S. Feferman, *Systems of predicative analysis*, J. Symbolic Logic **29** (1964), 1–30. MR **33** #1228.

2. ———, *Predicative provability in set theory*, Bull. Amer. Math. Soc. **72** (1966), 486–489. MR **33** #1230.

3. ———, *Systems of predicative analysis*. II. *Representations of ordinals*, J. Symbolic Logic **33** (1968), 193–220. MR **41** #5213.

4. ———, *Autonomous transfinite progressions and the extent of predicative mathematics*, Logic, Methodology and Philosophy of Science. III (Proc. Third Internat. Congr., Amsterdam, 1967), North-Holland, Amsterdam, 1968, pp. 121–135. MR **40** #5417.

5. ———, *Persistent and invariant formulas for outer extensions*, Compositio Math. **20** (1968), 29–52. MR **37** #6170.

6. S. Feferman and G. Kreisel, *Persistent and invariant formulas relative to theories of higher order*, Bull. Amer. Math. Soc. **72** (1966), 480–485. MR **33** #1229.

7. H. Friedman, *Iterated inductive definitions and* Σ_2^1-AC, Intuitionism and Proof Theory, North-Holland, Amsterdam, 1970, pp. 435–442. MR **44** #1555.

8. R. O. Gandy, *Proof of Mostowski's conjecture*, Bull. Acad. Pol. Sci. Sér. Sci. Math. Astronom. Phys. **8** (1960), 571–575. MR **23** #A3679.

9. S. Kripke, *Transfinite recursions on admissible ordinals*. I, II, J. Symbolic Logic **29** (1964), 161–162. Abstracts.

10. G. Kreisel, *The axiom of choice and the class of hyperarithmetic functions*, Nederl. Acad. Wetensch. Proc. Ser. A **65** = Indag. Math. **24** (1962), 307–319. MR **35** #3838.

11. ———, *A survey of proof theory*, J. Symbolic Logic **33** (1968), 321–388. MR **43** #7295.

12. R. Platek, *Foundations of recursion theory*, Dissertation, Stanford University, 1966.

STANFORD UNIVERSITY

[9] For the record: This result disproved my conjecture in [**1**, footnote 11].

Proceedings of Symposia in Pure Mathematics
Volume 13, Part II, 1974

ELEMENTARY EMBEDDINGS OF MODELS OF SET-THEORY AND CERTAIN SUBTHEORIES

HAIM GAIFMAN

INTRODUCTION

The two main themes which are the subject of the present work are the one indicated in the title and the iterations of the mentioned embeddings, what, in previous terminology, I called "the dynamo process." Only the second theme is related to the paper which I presented, six years ago, at the AMS summer institute on axiomatic set-theory which was held at UCLA. But even there, most of this paper presents work which has been done since then, some of it quite recently, in the summer of 1973. The two subjects are, of course, connected and this connection is brought out in Part II. But the first half of Part II and the whole of Part IV are independent of the rest of the work and do not involve the iteration method. They are best read separately.

I find it more convenient to review this work not in its given order but first to describe some of the ideas and the results which do not involve the iteration method and then go back to the other parts.

Let $Z =$ Zermelo's set theory and let Z^+ be obtained from Z by adding the axiom $\forall v \exists \alpha\, [R\alpha \text{ exists} \wedge v \in R\alpha]$, where $R\alpha$ is, as usual, the set of all sets of rank $< \alpha$. '$R\alpha$ exists' says formally what it appears to say intuitively (the details are given in Part II). Thus, the natural models of Z^+ are the V_α's with α a limit ordinal, where $V =$ Cantor's universe and V_α is its substructure whose universe is $R\alpha$. In general, the models to be discussed are not assumed to be well-founded.

Theorem II.1 provides a useful, necessary and sufficient condition for a cofinal embedding $j\colon \mathfrak{A} \to \mathfrak{B}$ to be elementary, where $\mathfrak{A} \models Z^+$. The condition is:

j is Σ_0-elementary (where $\Sigma_0 =$ set of all formulas in which all quantifiers are bounded—$\exists x \in y, \forall x \in y$) and, whenever $x \in \mathfrak{A}$, and, in $\mathfrak{A}$, x is an ordered set which some mapping, definable in $\mathfrak{A}$ (from parameters), maps cofinally into the

---

AMS (MOS) subject classifications (1970). Primary 02H13; Secondary 02K35.

ordinals of $\mathfrak{A}$, then j maps x cofinally into jx. (If $\mathfrak{A} \models$ ZF then the condition is vacuously satisfied and the theorem is known, especially for transitive models, and is easily derivable from the reflection principle.)

This theorem is used later in Part II in connection with the iteration method. Further use is made of it in Part IV. In this part we treat models of ZC^+, where $ZC^+ = Z^+ \cup \{\text{Axiom of Choice}\}$. We introduce the concept of a $<\gamma$-cofinal embedding:

$j:\mathfrak{A} \to \mathfrak{B}$ is $<\gamma$-cofinal if γ is an infinite cardinal in $\mathfrak{A}$ and every $x \in \mathfrak{B}$ is a member, in $\mathfrak{B}$, of some jy where y is, in $\mathfrak{A}$, of cardinality $<\gamma$.

(*Note.* Since $\mathfrak{A}$ need not be a model of ZF, not all its cardinals can be identified in it with ordinals. The theorems are formulated as if the cardinals were initial ordinals, but they hold in general. See Postscript A for details.)

Theorem IV.1 asserts the following: Let γ be an infinite cardinal of $\mathfrak{A}$ (where $\mathfrak{A} \models ZC^+$) and let $j:\mathfrak{A} \to \mathfrak{B}$ be cofinal and elementary. If $\mathfrak{B}^*$ is the submodel of $\mathfrak{B}$ consisting of those x which satisfy

$$\exists y[\mathfrak{A} \models \text{`cardinality } (y) < \gamma\text{'} \wedge \mathfrak{B} \models \text{`}x \in jy\text{'}],$$

then $\mathfrak{B}^*$ is an elementary submodel.

If $\mathfrak{A}$ and $\mathfrak{B}$ are transitive and $\tilde{\mathfrak{B}}$ is the transitive isomorph of $\mathfrak{B}^*$ we get a decomposition

$$\mathfrak{A} \xrightarrow{j} \mathfrak{B} = \mathfrak{A} \xrightarrow{\tilde{j}} \tilde{\mathfrak{B}} \xrightarrow{k} \mathfrak{B} \quad (\tilde{j} \text{ and } k\text{–elementary}).$$

This decomposition (i.e., $\tilde{\mathfrak{B}}, \tilde{j}$ and k) is uniquely determined by: $x \in \tilde{\mathfrak{B}} \Leftrightarrow$ for some y of cardinality $<y$ in $\mathfrak{A}$, $kx \in jy$. It is also uniquely determined by another simple property:

$\tilde{j}$ is $<\gamma$ cofinal and for every $\alpha < \gamma$ if $\lambda \leqq \tilde{j}\alpha$ then $k\lambda = \lambda$ (Theorem IV.3).

The embedding $\tilde{j}: \mathfrak{A} \to \tilde{\mathfrak{B}}$ is called the $<y$-approximation of j. The point is that by choosing γ large enough we can recapture in the $<\gamma$-approximation many of the properties of the original embedding. This is useful in view of the following connection between $<\gamma$-cofinal embeddings and limit ultrapowers (Theorem IV.4):

Let $j: V \to \mathfrak{B}$ be an elementary embedding satisfying the condition:

$\forall x\{y:\mathfrak{B} \models y \in x\}$ is a set (i.e., not a proper class).

Then j is $<\gamma$-cofinal for some γ iff it is a limit ultrapower extension, that is to say, $j: V \to \mathfrak{B} = V \to V^I_{P,\mathfrak{F}} \simeq \mathfrak{B}$, for some set I, some ultrafilter $\mathfrak{F}$ over I and some filter P of partitions of I, where $V \to V^I_{P,\mathfrak{F}}$ is the usual canonical embedding. (See Part IV for the terminology. Theorem IV.4 can be applied within any model of ZC^+.)

It now easily follows that, in many cases, quantification of the form: "$\exists$ elementary embedding $V \to M$ such that ..." is equivalent to first-order quantification. For the property denoted by "..." can be, in many cases, recaptured by some $<\gamma$-approximation of the embedding, and then one can quantify over I, $\mathfrak{F}$ and P. For example, let M range over transitive classes and consider the property:

$\exists$ elementary embedding $j: V \to M$ such that $\mu =$ first ordinal moved, $R(j^3\alpha + \beta) \subset M$ and $M^{j^2\mu+\gamma} \subset M$ (where $j^1\alpha = \alpha, j^{n+1}\alpha = j(j^n\alpha)$ and $M^\theta = \{\langle x_\xi \rangle_{\xi<\theta} : x_\xi \in M\}$).

This is a first-order condition on α, β, γ and μ, for, as it is shown, if j has the property so will its $<\gamma$-approximation for some appropriate γ.

The existence of certain large cardinals is well known to be equivalent to the existence of certain elementary embeddings $j: V \to M$. Vice versa, the existence of certain elementary embeddings can be taken as the property which defines certain large cardinals. Reinhardt [**16**] employed this idea to define extendible cardinals. He used embeddings of the form $V_\alpha \to V_\beta$. It is also well known that the α-supercompactness of μ is equivalent to: $\exists$ elementary $j: V \to M$ with $\mu =$ first ordinal moved and $M^\alpha \subset M$ (Reinhardt-Solovay [**17**]). We can now use a general schema of the form:

$\exists$ elementary embedding $j: V \to M$ such that $\mu =$ first ordinal moved and such that $\Phi(j, M)$.

Here Φ is some property of $j: V \to M$, possibly involving additional parameters. If Φ has a "local" character then some $<\gamma$-approximation will share it; hence, the whole schema will be equivalent to a first-order schema and can be used to define a property of μ. One can then quantify over the parameters in Φ, use it in combination with other first-order properties, etc. If C is the smallest set which includes all formulas of the following form and is closed under disjunctions, conjunctions and existential quantification, then every formula in C is for our purposes "local":

$$M^\alpha \subset M, \quad \varphi(j, M, \ldots), \quad \psi(\mu, \ldots),$$

where φ is a Σ_1-formula (in Lévy's hierarchy) in which names for the mapping j and the class M can occur (and any other first-order parameters) and ψ is any formula, not involving $j: V \to M$.

Taking $\Phi =$ '$R(j(\mu + \alpha)) \subset M$' we get a concept which is the natural analogue of α-extendibility, it implies it and is, as we later show, implied by $(\alpha + 1)$-extendibility. Other variants of extendibility are discussed (e.g., take $\Phi =$ '$R(j^n\mu + \alpha) \subset M$' or $\Phi =$ '$R(j^n(\mu + \alpha)) \subset M$,' or require that the resulting property holds for all α).

The last section of Part IV treats the subject of lifting a given $j: M \to N$, where N is a transitive set, i.e., extending it to an elementary embedding $j^*: V \to N^*$ with N^* transitive and $N \subset N^*$. (If "$N \subset N^*$" is dropped we speak of a partial lifting.) Once an embedding has been lifted it can be iterated.

There is some overlap between the ideas and some of the results (in particular, parts of Theorem IV.8) of this section, and work done independently by W. C. Powell which has reached me later in a preprint [**15**]. (Most of this section reflects work done by me in the summer of 1971 and Powell's analogous results have been worked out off and on during 1970 and 1971.) Whereas I employ limit ultrapowers and apply them to transitive models, Powell employs a somewhat different system of canonical direct limits of ultrapowers and works with arbitrary models. The actual coinciding results are indicated in the text. I'd like to thank him for pointing out certain errors which I made in an earlier hasty formulation and for his illuminating observations. I should also like to thank Solovay for his useful comments (clause (III) of Theorem IV.9 has been pointed out to me by him) and Reinhardt for many interesting discussions concerning these problems.

The lifting (or partial lifting) of $j: M \to N$ is done by means of a limit ultrapower extension. (By the previous results, if j can be lifted at all it can be lifted by a limit ultrapower.) The main point is to verify that the lifting is well-founded.

Theorem IV.8 gives an almost complete solution to the problem: Which cofinal elementary embedding $j: V_\alpha \to V_\beta$ can be lifted? The answer is:

If $\alpha = \alpha' + 1$, j can be partially lifted to $V \to N$ with $V_{j\alpha'} \subset N$, and for limit ordinals of cofinality $>\omega$ the embedding can be lifted, except for certain pathological cases where $V_\alpha \models$ ZF and j does not map α^* cofinally into $j\alpha^*$ where $\alpha^* = \text{cof}(\alpha) < \alpha$; for these cases the question is still open. Theorem IV.9 shows that the results cannot be improved by having in general a full lifting of $V_\alpha \to V_\beta$, where $\alpha = \alpha' + 1$, or a partial lifting if $\text{cof}(\alpha) = \omega$.

We go on from there to investigate the possibility of lifting of a noncofinal elementary embedding $V_\alpha \to V_\beta$ and we end by comparing the requirement that μ is α-extendible with the analogous global requirement that there is an elementary embedding $V \to M$ with μ = first ordinal moved and $V_{j(\mu+\alpha)} \subset M$.

Self-extensions and the method of iterations. The second main subject of the paper, self-extensions and their iterations, is treated in Parts I, III and the second half of Part II. A self-extension of a structure $\mathfrak{A}$ is an embedding $j: \mathfrak{A} \to \mathfrak{B}$, where $\mathfrak{B}$ and j are defined (possibly from parameters) inside $\mathfrak{A}$. (As indicated in Part I, "defined" can be taken here in a very general sense.) Sometimes the same defining formulas can be used inside $\mathfrak{B}$ to get another embedding $\mathfrak{B} \to \mathfrak{C}$, and so on. If one can continue the process any finite number of times one can take the direct limit of the structures thus obtained and so on.

The iteration method has been found and applied at the beginning of 1963. The operator to be iterated was Scott's ultrapower extension of V in which a measurable cardinal is used. The well-foundedness of all the universes $O^\alpha V$ (or $O^{\text{OR}} V$) which one gets after α (or ∞) iterations was proved at that time, using essentially the same argument which is now used to prove the much more general Theorem II.5. Applying the whole process to a countable model in which there is a measurable cardinal μ, I tried, at that time, to get results concerning the constructible universe L which are by now well known, such as a truth definition for L, $L_\alpha \prec L$ for some countable α and others. But for countable $\mathfrak{A}$'s the well-foundedness of all $O^\alpha\mathfrak{A}$ cannot, in general, be established. Only a year later, in 1964, I realized that by choosing a countable elementary submodel of V_α, where $\text{cof}(\alpha) > \mu$ (μ = a measurable cardinal), and utilizing the fact that the iterated operator is a functor, the well-foundedness can be obtained. The results have been announced in [**2**]. (Meanwhile, Rowbottom [**18**] derived weaker claims in the same direction, such as $\omega_1^L < \omega_1$, from the weaker assumption that Ramsey cardinals exist. Later, in 1965, Silver [**21**] using, like Rowbottom, partition techniques derived all the results concerning L from Ramsey cardinals, as well as the existence of a set of formulas which completely describe the structure of L, what is now known as $O^\#$. This can be also easily obtained, from measurable cardinals, by the iteration method. $O^\#$ was also defined later and independently by Solovay. The upward-type results such as getting well-founded elementary extensions of V which push μ to arbitrary high ordinals, or even to OR (class of all ordinals) and higher [**4**] can be obtained, so far, only by iterations. The same applies to those major results which Kunen [**11**] got in his dissertation utilizing this method.)

Part I outlines the basic concepts and results concerning iterated self-extensions. Proofs are mostly omitted. (The subject is treated in detail in [**6**].) This part belongs to general model theory. It should also be noted that self-extensions are special cases of the still more general concept of a natural extension operator, which does not rely on any definability conditions. Only a few basic requirements are postulated for this concept, such as functorialness and the commutativity of certain diagrams. There are numerous examples of such operators in algebra and model theory, and one can develop a general theory of their iterations, in which the main tool is the use of direct limits ([**3**] and [**6**]). Particular cases of iterations, such as ultralimits, are well known. The general approach has useful additional applications.

The main new result of Part II is Theorem II.4 which enables one to iterate any elementary self-extension $j: V \to OV$, provided that some limit ordinal λ is mapped by j cofinally onto its image $j\lambda$. In particular, the iteration is definable whenever OV is well-founded (not necessarily isomorphic to a transitive class) for one can put $\lambda = \omega$. If instead of V we have a model which is a set, the iteration is easily definable by induction (Definitions I.1, I.2). For proper classes this is no longer a possibility. In trying to define the iteration of $V \to OV$ our main goal is to show that the iteration can be defined inside the model itself; the same definition can be applied inside any model of ZF. The proof of the general version of the well-foundedness theorem (II.5) rests on this possibility. Thus, operating with proper classes is here very essential and not a question of generalizing only. Since nothing is presupposed concerning the formulas which define the self-extension $V \to OV$, showing that the iteration is definable is not easy and takes some work. Then, we go on to prove (again without assuming anything on the nature of the self-extension) the well-foundedness theorem.

Part III describes some general properties of $O^I\mathfrak{A}$ where $\mathfrak{A} \models Z^+$ and I is any ordered class ($\mathfrak{A}$ is not assumed to be well-founded).

Each $I' \subset I$ determines an elementary submodel $\mathfrak{A}^{I'}$, of $O^I\mathfrak{A}$, which is naturally isomorphic to $O^{I'}\mathfrak{A}$. Its members are said to be contributed by I'. With every order preserving embedding $I' \to I''$ an elementary embedding $\mathfrak{A}^{I'} \to \mathfrak{A}^{I''}$ is correlated and this correlation is a functor.

The question which members and, in particular, which ordinals of $O^I\mathfrak{A}$ are contributed by which subsets of I (every member is contributed by some finite subset) is of great importance. It is closely connected with well-foundedness questions, the phenomenon of the "pushing up" of ordinals as a result of the iteration, the special status of the first ordinal which is moved by the embedding, and other problems. Some basic properties concerning these questions are proved in Part III.

A special section of Part III outlines in brief the "blowing up" via iterations of small models and the way in which the results concerning L can be derived. Some more recent observations are made concerning this method.

The last section of Part III shows how properties of OV are passed on to the structures $O^\alpha V$. One theorem shows what can be inferred concerning $O^\alpha V$ if $R\beta \subset OV$, where $\mu \leqq \beta \leqq \mu_\omega$. ($\mu_\omega = \sup\{j^n\mu : n < \omega\}$.) The second theorem shows what can be inferred if we assume

$$\{\langle x, jx\rangle : x \in R\beta\} \subset OV, \quad \text{where } \mu \leqq \beta \subseteq j\mu.$$

Postscript B belongs together with Part II. It is quite a recent work, motivated by Kunen's generalization of iterated ultrapowers in which $\mathfrak{A}$-ultrafilters are used, i.e., "ultrafilters" not necessarily in the structure $\mathfrak{A}$ [11]. Such an "ultrapower," it is pointed out, can be construed as a self-extension operator, provided that we operate not on $\mathfrak{A}$ but on $(\mathfrak{A}, \mathfrak{F})$ where $\mathfrak{F}$, the $\mathfrak{A}$-ultrafilter, is an additional 1-ary relation. However, by adding $\mathfrak{F}$ we, in general, no longer have a model of ZF or even of Z. Furthermore, the self-extension $(\mathfrak{A}, \mathfrak{F}) \to O(\mathfrak{A}, \mathfrak{F})$ need no longer be elementary. The possibility of iterating such a self-extension is, in general, far from obvious because one has to show that any finite number of applications of the operator will yield models and embeddings which are still in its domain. Yet, in the special case of "ultrapowers" the iteration can be defined. In the postscript we outline a general approach which does not rely on the concept of an ultrapower and enables one to iterate a wide class of self-extensions (not necessarily elementary) of models (not necessarily of Z). Kunen's "ultrapowers" are a special case within the framework.

The problem of formulating and proving an analogue of the well-foundedness theorem for such self-extensions is open, as it is, indeed, open for "ultrapowers" if $\mathfrak{F} \notin \mathfrak{A}$.

General Terminology and Notation

"Structure" and "model" are used synonymously. "$\mathfrak{A}$", "$\mathfrak{B}$", "$\mathfrak{C}$", "$\mathfrak{A}_1$",... denote structures.

If $\mathfrak{A} = (A, R_i)_{i \leqq n}$ then $|\mathfrak{A}| =_{\mathrm{Df}} A$. $|\mathfrak{A}|$ is referred to as the universe of $\mathfrak{A}$, or its underlying domain. Sometimes, when confusion is unlikely, the same letter is used to denote both the structure and its domain; "V", "M", "N" are employed for this purpose.

$\mathfrak{A}$ is said to be a set or a proper class according to $|\mathfrak{A}|$ being a set or a proper class. When dealing with models for the language of set theory we say that $\mathfrak{A}$ is transitive if $|\mathfrak{A}|$ is transitive and the $\in$ of $\mathfrak{A}$ is the real $\in$ restricted to $|\mathfrak{A}|$.

An embedding is a triple $(j, \mathfrak{A}, \mathfrak{B})$, where $\mathfrak{A}$ and $\mathfrak{B}$ have the same language, j is a mapping defined on $|\mathfrak{A}|$ and having values in $|\mathfrak{B}|$ and, for all $x_1, \ldots, x_n \in |\mathfrak{A}|$ and all atomic formulas φ,

$$\mathfrak{A} \models \varphi(x_1, \ldots, x_n) \leftrightarrow \mathfrak{B} \models \varphi(jx_1, \ldots, jx_n).$$

(When dealing with satisfaction we automatically use the members of the structures as names for themselves.)

We employ "$j: \mathfrak{A} \to \mathfrak{B}$", "$\mathfrak{A} \xrightarrow{j} \mathfrak{B}$" or simply "$\mathfrak{A} \to \mathfrak{B}$" to denote embeddings. Sometimes, when no confusion is likely we let "j" itself stand for the embedding as well as for the mapping, thus putting $j = j: \mathfrak{A} \to \mathfrak{B}$.

If h is a mapping then $h''A = \{h(x): x \in A\}$. If $j = j: \mathfrak{A} \to \mathfrak{B}$ then $j''\mathfrak{A}$ is the submodel $\mathfrak{B}^*$ of $\mathfrak{B}$ such that $|\mathfrak{B}^*| = j''(|\mathfrak{A}|)$.

Composition of embeddings is defined as usual. "$\mathfrak{A} \xrightarrow{j} \mathfrak{B} \xrightarrow{h} \mathfrak{C}$" denotes the composition of $\mathfrak{A} \xrightarrow{j} \mathfrak{B}$ and $\mathfrak{B} \xrightarrow{h} \mathfrak{C}$; "$h \circ j$" is also used for the same purpose.

An isomorphism is an embedding, $j: \mathfrak{A} \to \mathfrak{B}$, in which $\mathfrak{B} = j''\mathfrak{A}$. Sometimes "$\mathfrak{A} \simeq \mathfrak{B}$" is employed to denote an isomorphism of $\mathfrak{A}$ onto $\mathfrak{B}$ as well as the claim that $\mathfrak{A}$ and $\mathfrak{B}$ are isomorphic.

$1_{\mathfrak{A}}$ = identity embedding of $\mathfrak{A}$ onto itself. "$\mathfrak{A} \subset \mathfrak{B}$" means that $\mathfrak{A}$ is a submodel of $\mathfrak{B}$, but sometimes it is also used to denote the inclusion embedding (i.e., the embedding $\mathfrak{A} \to \mathfrak{B}$ which sends every $x \in |\mathfrak{A}|$ to itself).

"$\mathfrak{A} \prec \mathfrak{B}$" means that $\mathfrak{A}$ is an elementary submodel of $\mathfrak{B}$. "$\mathfrak{A} \prec_{\Phi} \mathfrak{B}$" means that $\mathfrak{A}$ is a Φ-elementary submodel of $\mathfrak{B}$, i.e., $\mathfrak{A} \models \varphi(x_1, \ldots, x_n) \Leftrightarrow \mathfrak{B} \models \varphi(x_1, \ldots, x_n)$ for all $\varphi \in \Phi$.

Elementary (or Φ-elementary) embeddings are defined in the obvious way.

"$\equiv$" denotes elementary equivalence.

If $\mathfrak{A}$ is obtained from $\mathfrak{B}$ by dropping some of $\mathfrak{B}$'s relations then it is a reduct of $\mathfrak{B}$, and $\mathfrak{B}$ is an expansion of $\mathfrak{A}$.

$(\mathfrak{A}, R)$ is the structure obtained from $\mathfrak{A}$ by adding the relation R.

Other conventions are explained in the text.

The numbering of theorems, claims and definitions in each part does not depend on the other parts. For cross-reference the number of the part is included, e.g., Theorem II.4.

PART I

Self-extensions. All embeddings of the form $j: V \to M$ are assumed here to be definable in V. This is a natural assumption; since V is Cantor's universe, everything should be there already.

In general, the definability of an embedding $f: \mathfrak{A} \to \mathfrak{B}$ in the structure $\mathfrak{A}$ means (i) that the universe of $\mathfrak{B}$ and all its relations and individual constants are defined in $\mathfrak{A}$ and (ii) that the mapping f is defined in $\mathfrak{A}$.

An extension of this kind is called a *self-extension*.

Members of $\mathfrak{A}$ can occur in the definitions of $\mathfrak{B}$ and f as parameters, but in that case we add their names to the language; they become distinguished members, and their images under f (which are, of course, definable because f is definable) are the corresponding members in the second structure. We pass from $\mathfrak{A} \to \mathfrak{B}$ to $\mathfrak{A}^* \to \mathfrak{B}^*$, where the second embedding is definable in the language of $\mathfrak{A}^*$ (without parameters) and has all the worthwhile properties of the first (e.g., if $\mathfrak{A} \to \mathfrak{B}$ is Φ-elementary so is $\mathfrak{A}^* \to \mathfrak{B}^*$). Consequently, with no loss of generality, we assume that the definitions are in the language of $\mathfrak{A}$ to start with. This means that for some formulas $\sigma, \varphi_1, \ldots, \varphi_l$, if R_i = ith relation of $\mathfrak{B}$, we have

$$|\mathfrak{B}| = \{x: \mathfrak{A} \models \sigma(x)\}, \quad R_i = \{\boldsymbol{x}: \mathfrak{A} \models \varphi_i(\boldsymbol{x})\}, \qquad i = 1, \ldots, k, \text{ and}$$

if $\mathfrak{B}$ has distinguished members then the jth distinguished member is the unique x such that $\mathfrak{A} \models \varphi_{k+j}(x)$.

Put $t = \langle k, \sigma, \varphi_1, \ldots, \varphi_k, \varphi_{k+1}, \ldots, \varphi_l \rangle$ and call t a *defining schema*.

(All structures to be considered have finitely many relations and distinguished members.)

Furthermore, the mapping f is defined by some formula, ψ say, that is, $f(x) = y \Leftrightarrow \mathfrak{A} \models \psi(x, y)$.

Put $e = \langle t, \psi \rangle$ and call e a *self-extending schema.*

Given $\mathfrak{A}$ we let $O_e\mathfrak{A}$ be the model which is defined in $\mathfrak{A}$ by the defining schema t, and let $f_{e,\mathfrak{A}}$ be the mapping defined in $\mathfrak{A}$ by ψ.

In order that $O_e\mathfrak{A}$ be well-defined as a structure for the given language, $\mathfrak{A}$ should satisfy certain sentences; namely, $\exists v\sigma(v)$ (guaranteeing the nonemptiness of $|O_e\mathfrak{A}|$) and, if we have distinguished members in $O_e\mathfrak{A}$ (i.e., $l > k$), $\exists! v\varphi_{k+i}(u)$, $1 \leqq i \leqq l - k$. Furthermore, for ψ to define a mapping, $\forall u \exists! v\psi(u, v)$ should be satisfied in $\mathfrak{A}$, and if this mapping is to be an embedding of $\mathfrak{A}$ into $O_e\mathfrak{A}$, other finitely many sentences should be satisfied, such as $\forall u, v[\psi(u, v) \to \sigma(u)]$, $\forall u, v[\psi(u, v) \wedge P_1(u) \to \varphi_1(u)]$ (assuming that P_1 is a one-place predicate), and similar sentences for each n-place predicate, as well as sentences dealing with the individual constants. *Let the set of all these sentences be* S_e. Note that S_e is a finite set. From now on O_e is to be applied only to models of S_e.

Let Γ_e = set of all subformulas of $\sigma, \varphi_1, \ldots, \varphi_l, \exists u\psi(u, v)$.

It is obvious that if $\mathfrak{A} \prec_{\Gamma_e} \mathfrak{B}$ and both are models of S_e then $O_e\mathfrak{A} \subset O_e\mathfrak{B}$. For example, $x \in |O_e\mathfrak{A}| \Rightarrow \mathfrak{A} \models \sigma(x) \Rightarrow \mathfrak{B} \models \sigma(x) \Rightarrow x \in |O_e\mathfrak{B}|$, and a similar argument shows that the relations of $O_e\mathfrak{A}$ are the restrictions of the corresponding relations of $O_e\mathfrak{B}$ and the distinguished members are the same. (Here, the fact that $\psi, \exists u\psi \in \Gamma_e$ is not used. The reasons for including these and for closing Γ_e under subformulas will appear later.)

In general, if $\mathfrak{A}$ and $\mathfrak{B}$ satisfy S_e then every Γ_e-elementary embedding $h: \mathfrak{A} \to \mathfrak{B}$ induces an embedding $O_e\mathfrak{A} \to O_e\mathfrak{B}$. This last embedding is obtained simply by restricting h to $|O_e\mathfrak{A}|$. *We will denote it by "O_eh".*

By a *category of structures* we mean a class $\boldsymbol{C}$, consisting of structures $\mathfrak{A}$ and embeddings $\mathfrak{A} \to \mathfrak{B}$ (where $\mathfrak{A}, \mathfrak{B} \in \boldsymbol{C}$) such that $\boldsymbol{C}$ is closed under compositions ($\mathfrak{A} \to \mathfrak{B}, \mathfrak{B} \to \mathfrak{C} \in \boldsymbol{C} \Rightarrow \mathfrak{A} \to \mathfrak{B} \to \mathfrak{C} \in \boldsymbol{C}$) and whenever $\mathfrak{A} \in \boldsymbol{C}$, every isomorphic $\mathfrak{A}'$ and every isomorphism $\mathfrak{A} \simeq \mathfrak{A}'$ are in $\boldsymbol{C}$. (All structures in a category are assumed to be sets.)

Let $\boldsymbol{C}_e$ = category of all $\mathfrak{A}$ such that $\mathfrak{A} \models S_e$ and all embeddings $h: \mathfrak{A} \to \mathfrak{B}$ which are Γ_e-elementary (with $\mathfrak{A}, \mathfrak{B} \in \boldsymbol{C}_e$).

We put $\boldsymbol{O}_e = (O_e, f_e)$; we refer to it as a self-extension operator and regard it as an operator acting on the structures and embeddings of $\boldsymbol{C}_e$. Altogether, we get the following properties of $\boldsymbol{O}_e$:

(1) *For every* $\mathfrak{A} \in \boldsymbol{C}_e$, $O_e\mathfrak{A}$ *is a structure.*

(2) *For every* $h: \mathfrak{A} \to \mathfrak{B} \in \boldsymbol{C}$, O_eh *is an embedding of* $O_e\mathfrak{A}$ *into* $O_e\mathfrak{B}$.

(3) $O_e 1_{\mathfrak{A}} = 1_{O_e\mathfrak{A}}$ *and* $O_e(h_2 \circ h_1) = O_eh_2 \circ O_eh_1$.

(4) *For every* $\mathfrak{A} \in \boldsymbol{C}_e, f_{e,\mathfrak{A}}$ *is an embedding of* $\mathfrak{A}$ *into* $O_e\mathfrak{A}$ *and for every* $h:\mathfrak{A} \to \mathfrak{B}$ *in* $\boldsymbol{C}_e$ *the following diagram is commutative*:

$$\begin{array}{ccc} O_e\mathfrak{A} & \xrightarrow{O_eh} & O_e\mathfrak{B} \\ {\scriptstyle f_{e,\mathfrak{A}}}\uparrow & & \uparrow{\scriptstyle f_{e,\mathfrak{B}}} \\ \mathfrak{A} & \xrightarrow{h} & \mathfrak{B} \end{array}$$

(1)–(3) mean that O_ϵ is a functor. The property of self-extension enters in (4). The commutativity of the diagram follows from the fact that h is Γ_e-elementary and $\psi \in \Gamma_e$ ($y = f_{e,\mathfrak{A}}(x) \Leftrightarrow \mathfrak{A} \models \psi(x, y) \Leftrightarrow \mathfrak{B} \models \psi(hx, hy) \Leftrightarrow hy = f_{e,\mathfrak{B}}(hx)$, but $hy = (O_eh)(y)$, $hx = (O_eh)(x)$).

Furthermore, because we include $\exists u\psi(u, x)$ in Γ_e, we have

(5) *In the diagram of* (4), *for every* $x \in |O_e\mathfrak{A}|$, *if* $x \notin f''_{e,\mathfrak{A}}(|\mathfrak{A}|)$ *then* $hx \notin f''_{e,\mathfrak{B}}(|\mathfrak{B}|)$.

(This is so because $x \notin f''_{e,\mathfrak{A}}(|\mathfrak{A}|) \Leftrightarrow \mathfrak{A} \models \neg\exists u\psi(u, x)$.)

In [**6**] general operators, not necessarily based on a self-extension schema, are treated. Those satisfying (1)–(4) are called natural extension operators. Those satisfying (5) are called uniform.

Self-extension operators have also the following continuity property, which can be easily verified:

(6) *If* $\mathfrak{A} = \bigcup_{x\in X} \mathfrak{A}_x$, *where* $\{\mathfrak{A}_x\}_{x\in X}$ *is directed under inclusion*, $\mathfrak{A} \in \boldsymbol{C}_e$, $\mathfrak{A}_x \in \boldsymbol{C}_e$, *for all* $x \in X$, *and all the inclusions* $\mathfrak{A}_x \subset \mathfrak{A}$ *are in* $\boldsymbol{C}_e$, *then* $O_e\mathfrak{A} = \bigcup_{x\in X} O_e\mathfrak{A}_x$.

Concerning the category $\boldsymbol{C}_e$ we have:

(7) *If* $X = (X, \leqq)$ *is a directed set, for all* $x \in X$, $\mathfrak{A}_x \in \boldsymbol{C}_e$ *and whenever* $x \leqq y$, $\mathfrak{A}_x \subset \mathfrak{A}_y$ *and the inclusion is in* $\boldsymbol{C}_e$, *then* $\bigcup_{x\in X} \mathfrak{A}_x$ *and all the inclusions* $\mathfrak{A}_x \subset \bigcup_{x\in X} \mathfrak{A}_x$ *are in* $\boldsymbol{C}_e$. *If, moreover, Y is a directed subset of X then* $\bigcup_{y\in Y} \mathfrak{A}_y \subset \bigcup_{x\in X} \mathfrak{A}_x$ *is in* $\boldsymbol{C}_e$.

(7) follows from the fact that Γ_e is closed under subformulas; this implies that $\mathfrak{A}_x \prec_{\Gamma_e} \bigcup_{x\in X} \mathfrak{A}_x$. Every sentence of S_e is easily seen to be equivalent to some universal generalization of a sentential combination of formulas of Γ_e, from which it follows that $\bigcup_{x\in X} \mathfrak{A}_x \models S_e$. (If this were not so we could have added all subformulas of the sentences of S_e to Γ_e.) (7) can be rephrased in general terms of direct limits (instead of direct unions). We shall say that a category which satisfies (7) is *closed under direct limits*.

Γ_e was chosen to be closed under subformulas in order to make $\boldsymbol{C}_e$ closed under direct limits. This is convenient, although not essential to the later development.

Before proceeding further, note that the concept of a defining schema which we employed is a special case of a more general concept of definability which could have been used equally well. In the general case the members of the structure, $\mathfrak{B}$ say, which is defined in $\mathfrak{A}$, are n-tuples of members of $\mathfrak{A}$, for some fixed n. Moreover, one can have m kinds of n-tuples, for some fixed m. That is to say, there are m formulas $\sigma_1(\boldsymbol{v}), \ldots, \sigma_m(\boldsymbol{v})$ ($\boldsymbol{v} = (v_1, \ldots, v_n)$) and m functions $I_1, \ldots, I_m$ such that $|\mathfrak{B}| = \{I_i(x_1, \ldots, x_n): \mathfrak{A} \models \sigma_i(x_1, \ldots, x_n),\ i = 1, \ldots, m\}$. Furthermore, if P is any k-place predicate in the language of $\mathfrak{B}$, then for every $1 \leqq i_1, \ldots, i_k \leqq m$ we have a formula $\varphi_{P,i_1,\ldots,i_k}$ in the language of $\mathfrak{A}$ such that the following holds (where for simplicity we take $k = 2$):

$$\mathfrak{B} \models P(I_{i_1}(\boldsymbol{x}_1), I_{i_2}(\boldsymbol{x}_2)) \Leftrightarrow \mathfrak{A} \models \varphi_{P,i_1,i_2}(\boldsymbol{x}_1, \boldsymbol{x}_2)$$

(where $\boldsymbol{x}_j = (x_{j,1}, \ldots, x_{j,n}), j = 1, 2$).

This includes the case where P is the equality symbol (that is, different tuples might yield the same member of $\mathfrak{B}$, but their "equality" is defined in $\mathfrak{A}$). Distinguished members of $\mathfrak{B}$ are treated along the same lines; if b is a distinguished member then, for each i ($i = 1, \ldots, m$), $\{\boldsymbol{x} : I_i(\boldsymbol{x}) = b\}$ is defined in $\mathfrak{A}$. One can, of course, use tuples of varying length, but by adding dummy coordinates to the shorter tuples this reduces to the case where the length is uniform.

The defining schema is the system of all the formulas which are used. Given $\mathfrak{A}$ and a defining schema, one can construct $\mathfrak{B}$ as follows. Take the set of all tuples $\langle i, x_1, \ldots, x_n \rangle$, $i = 1, \ldots, m$, such that $\mathfrak{A} \models \sigma_i(x_1, \ldots, x_n)$. Using the formulas which define in $\mathfrak{A}$ the equalities $I_i(\boldsymbol{x}) = I_j(\boldsymbol{y})$, one gets an equivalence relation for this set. Divide by this equivalence, i.e., choose representatives of the equivalence classes—these will be the members of $\mathfrak{B}$. ($I_i(\boldsymbol{x})$ is defined as the representative of $\langle i, \boldsymbol{x} \rangle$.) The relations and distinguished members of $\mathfrak{B}$ are defined in the obvious way using the corresponding defining formulas in $\mathfrak{A}$. In order for the defining schema to be applicable, $\mathfrak{A}$ must, of course, satisfy certain sentences (e.g., sentences asserting that the relation by which $\langle i, \boldsymbol{x} \rangle$ and $\langle j, \boldsymbol{y} \rangle$ are identified is indeed a congruence relation).

This concept of definability has been introduced in [**5**]. The idea of using tuples is mentioned explicitly by Vaught in [**22**]. Many special cases of this technique had been a common mathematical practice long before the discovery of model theory, or mathematical logic. Recently the author has shown that natural semantical characterizations of this definability concept, which do not use the notion of a defining schema, can be given which amount to a generalization of Beth's theorem; in a certain natural sense this is the most general concept of definability.

If $\mathfrak{A}$ and $\mathfrak{B}$ have the same language an embedding $f : \mathfrak{A} \to \mathfrak{B}$ is defined in $\mathfrak{A}$ if there are m formulas $\psi_i(u, \boldsymbol{v})$, $i = 1, \ldots, m$, which define, respectively, the relations $f(x) = I_i(x_1, \ldots, x_n)$. A self-extension schema (in the general sense) is obtained by adding such formulas to the defining schema.

All of our claims extend to this general concept (the proven generalized version of (15) is a little weaker; (18)(ii) should then be changed to: $J_1 \subset J_2 \Rightarrow \mathfrak{A}^{J_1} \subset \mathfrak{A}^{J_2}$).

We use the simpler special concept, first for the sake of convenience, second, because the structures with which we are most concerned are models of ZF or rich enough subtheories of it; consequently they include the natural numbers and tuples of fixed finite length are represented in them. If $x_1, \ldots, x_n \in |\mathfrak{A}|$ then $\langle i, x_1, \ldots, x_n \rangle$ is again a member of $\mathfrak{A}$, or, if the $\in$-relation of $\mathfrak{A}$ is not the "real" membership, there is such a tuple, in the sense of $\mathfrak{A}$, which will do the same job. Furthermore, the choosing of representatives from equivalence classes can be done in most cases inside $\mathfrak{A}$, using some fixed customary procedure. In the case when this is not possible, which is exceptional in this work, we will let the equality symbol be interpreted in $\mathfrak{B}$ as a congruence relation. It makes no difference, and causes no difficulties.

Since $O_e\mathfrak{A}$ and $\mathfrak{A} \to O_e\mathfrak{A}$ are defined in $\mathfrak{A}$, properties which can be expressed by first order formulas, concerning $O_e\mathfrak{A}$ and the embedding, can be stated by first order formulas inside $\mathfrak{A}$. If φ is any formula, and if φ^* is obtained by substituting for each predicate in the language of $\mathfrak{A}$ the formula which defines in $\mathfrak{A}$ the corresponding relation of $O_e\mathfrak{A}$, by eliminating the distinguished members via their definitions in $\mathfrak{A}$ and by restricting all quantifiers to $\sigma(u)$ (which defines $|O_e\mathfrak{A}|$), then the satisfaction of φ in $O_e\mathfrak{A}$ is equivalent to the satisfaction of φ^* in $\mathfrak{A}$. If we put $\hat{\varphi} = \forall \boldsymbol{u} \forall \boldsymbol{v}[\bigwedge_{i=1}^{n} \psi(u_i, v_i) \to (\varphi(\boldsymbol{u}) \leftrightarrow \varphi^*(\boldsymbol{v}))]$, then $\hat{\varphi}$ asserts that $\mathfrak{A} \to O_e\mathfrak{A}^*$ is φ-elementary. We get

(8) $O_e\mathfrak{A} \models \varphi(\boldsymbol{x}) \Leftrightarrow \mathfrak{A} \models \varphi^*(\boldsymbol{x})$; *$f_{e,\mathfrak{A}}$ is φ-elementary* $\Leftrightarrow \mathfrak{A} \models \hat{\varphi}$.

It follows that, given $h: \mathfrak{A} \to \mathfrak{B}$ in $\boldsymbol{C}$, $O_e h$ is φ-elementary iff, for every $x_1, \ldots, x_n \in |\mathfrak{A}|$, $\mathfrak{A} \models \bigwedge_{i=1}^{n} \sigma(x_i) \wedge \varphi^*(x_1, \ldots, x_n) \Leftrightarrow \mathfrak{B} \models \bigwedge_{i=1}^{n} \sigma(hx_i) \wedge \varphi^*(hx_1, \ldots, hx_n)$; consequently, putting $\varphi^{**} = \bigwedge_{i=1}^{n} \sigma(v_i) \wedge \varphi^*(v_1, \ldots, v_n)$ we get

(9) *$O_e h$ is φ-elementary $\Leftrightarrow$ h is φ^{**}-elementary.*

So far all the definitions and statements can be made equally well for structures that are proper classes. The situation is different when the iteration of a self-extension operator is to be defined.

Iterations of a self-extension operator. We let e be a fixed self-extension schema and put $\boldsymbol{O} = \boldsymbol{O}_e, f = f_e, O = O_e$.

If $O\mathfrak{A} \in \boldsymbol{C}_e$ then O can be iterated once more and we get $O(O\mathfrak{A})$, which we denote as "$O^2\mathfrak{A}$"; if $O^2\mathfrak{A} \in \boldsymbol{C}_e$ we can get $O^3\mathfrak{A}$, and so on. The same applies to embeddings. We shall want also to apply O to $f: \mathfrak{A} \to O\mathfrak{A}$ and get $Of: O\mathfrak{A} \to O^2\mathfrak{A}$, and so on. All such iterations can be performed provided that $\mathfrak{A}$ is in some subcategory $\boldsymbol{C}'$ such that the application of $\boldsymbol{O}$ to its structures and embeddings always yields structures and embeddings in $\boldsymbol{C}'$.

Let $\boldsymbol{C}'$ be a subcategory of $\boldsymbol{C}_e$. We say that *$\boldsymbol{O}$ acts within $\boldsymbol{C}'$* if, for all $\mathfrak{A} \in \boldsymbol{C}'$, $O\mathfrak{A}$ and $f_{\mathfrak{A}}$ are in $\boldsymbol{C}'$ and, for every $h: \mathfrak{A} \to \mathfrak{B} \in \boldsymbol{C}'$, $Oh \in \boldsymbol{C}'$.

There is a maximal subcategory of $\boldsymbol{C}_e$, which will be denoted as "$\boldsymbol{C}_{e,*}$", such that $\boldsymbol{O}$ acts within it. $\boldsymbol{C}_{e,*}$ is defined as follows:

$\mathfrak{A} \in \boldsymbol{C}_{e,*}$ iff the following two conditions hold:

(i) For every n, O, $O^2\mathfrak{A}, \ldots, O^n\mathfrak{A}$ are defined (implying that, for every n, $O^n\mathfrak{A} \in \boldsymbol{C}_e$).

(ii) For every n and every m, if $\mathfrak{B} = O^n\mathfrak{A}$, then $Of_{\mathfrak{B}}, O^2f_{\mathfrak{B}}, \ldots, O^mf_{\mathfrak{B}}$ are defined (implying that, for every m, $O^mf_{\mathfrak{B}} \in \boldsymbol{C}_e$).

$h: \mathfrak{A} \to \mathfrak{B} \in \boldsymbol{C}_{e,*}$ iff the following holds:

(iii) $\mathfrak{A}, \mathfrak{B} \in \boldsymbol{C}_{e,*}$ and, for every n, $Oh, \ldots, O^nh$ are defined (implying that, for every n, $O^nh \in \boldsymbol{C}_e$).

It is not difficult to verify that $\boldsymbol{C}_{e,*}$ is a category, that $\boldsymbol{O}$ acts within $\boldsymbol{C}_{e,*}$ and, if $\boldsymbol{C}' \subset \boldsymbol{C}_e$ and $\boldsymbol{O}$ acts within $\boldsymbol{C}'$, then $\boldsymbol{C}' \subset \boldsymbol{C}_{e,*}$.

Using (8) and (9) one can construct a set $S_{e,*}$ of sentences and a set $\Gamma_{e,*}$ of formulas such that

$$\mathfrak{A} \in \boldsymbol{C}_{e,*} \Leftrightarrow \mathfrak{A} \models S_{e,*}.$$

$$h: \mathfrak{A} \to \mathfrak{B} \in \boldsymbol{C}_{e,*} \Leftrightarrow \mathfrak{A}, \mathfrak{B} \in \boldsymbol{C}_{e,*} \quad \text{and} \quad h \text{ is } \Gamma_{e,*}\text{-elementary}.$$

(The construction is as follows: $O\mathfrak{A} \in \boldsymbol{C}_e$ iff $\mathfrak{A} \models \{\varphi^* : \varphi \in S_e\}$. Similarly, the satisfaction of S_e in $O^2\mathfrak{A}$ is equivalent to the satisfaction of $\{\varphi^* : \varphi \in S_e\}$ in $O\mathfrak{A}$, which again reduces to the satisfaction of a set of sentences in $\mathfrak{A}$, and so on. Condition (i) can be therefore expressed as the satisfaction of a certain set of sentences in $\mathfrak{A}$. Using (9) in a similar way it follows that the condition that $O^m h$ is Γ_e-elementary is equivalent to the condition that h is $\Gamma_e^{(m)}$-elementary, for some set $\Gamma_e^{(m)}$, and by (8) the condition that $f_{O^n\mathfrak{A}}$ is $\Gamma_e^{(m)}$-elementary is again equivalent to the satisfaction of a certain set of sentences in $\mathfrak{A}$.)

From (8) and (9) it follows that if $\mathfrak{A} \equiv \mathfrak{B}$ and $f_{\mathfrak{A}}$ is elementary, so is $f_{\mathfrak{B}}$, and if $h : \mathfrak{A}_1 \to \mathfrak{A}_2$ is elementary so is Oh. Consequently, we have:

(10) *If $\mathfrak{A} \in \boldsymbol{C}_e$ and $f_{\mathfrak{A}}$ is elementary then $\mathfrak{A} \in \boldsymbol{C}_{e,*}$. The same is true with "elementary" replaced by "Φ-elementary", provided that Φ has the property: If $\mathfrak{A} \in \boldsymbol{C}_e$ and $f_{\mathfrak{A}}$ is Φ-elementary then $O\mathfrak{A} \in \boldsymbol{C}_e$ and $f_{O\mathfrak{A}}$ is Φ-elementary, and if $h \in \boldsymbol{C}_e$ and h is Φ-elementary then $Oh \in \boldsymbol{C}_e$ and is Φ-elementary.*

From (6) and (7) it follows that $\boldsymbol{C}_{e,*}$ is *closed under direct limits.*

(I.e., if $X = (X, \leqq)$ is a directed set and for every $x, y \in X$ such that $x \leqq y$, $\mathfrak{A}_x \subset \mathfrak{A}_y$ and the inclusion is in $\boldsymbol{C}_{e,*}$ then $\bigcup_{x\in X} \mathfrak{A}_x \in \boldsymbol{C}_{e,*}$ and whenever Y is a directed subset of X then $\bigcup_{x\in Y} \mathfrak{A}_x \subset \bigcup_{x\in X} \mathfrak{A}_x$ is in $\boldsymbol{C}_{e,*}$. The argument runs as follows: Put $\mathfrak{A} = \bigcup_{x\in X} \mathfrak{A}_x$, then $\mathfrak{A} \in \boldsymbol{C}_e$ and $O\mathfrak{A} = \bigcup_{x\in X} O\mathfrak{A}_x$. The system $\{O\mathfrak{A}_x\}_{x\in X}$ has by our assumptions the same properties as those of $\{\mathfrak{A}_x\}_{x\in X}$, hence $O\mathfrak{A} \in \boldsymbol{C}_e$ and $O^2\mathfrak{A} = \bigcup_{x\in X} O^2\mathfrak{A}_x$, and so on: $O^n\mathfrak{A} = \bigcup_x O^n\mathfrak{A}_x$. If $\mathfrak{B} = \bigcup_{x\in X} \mathfrak{B}_x$ where $\{\mathfrak{B}_x\}_{x\in X}$ has the same properties as $\{\mathfrak{A}_x\}_{x\in X}$ and if for each $x \in X$, $h_x : \mathfrak{A}_x \to \mathfrak{B}_x$ is an embedding in $\boldsymbol{C}_{e,*}$ such that, whenever $x \leqq y$, $\mathfrak{A}_x \to \mathfrak{B}_x \subset \mathfrak{B}_y = \mathfrak{A}_x \subset \mathfrak{A}_y \to \mathfrak{B}_y$, then, putting $h = \bigcup_{x\in X} h_x$, it follows that h is an embedding $\mathfrak{A} \to \mathfrak{B}$ and, because, for all x, $\mathfrak{A}_x \subset \mathfrak{A}$, $\mathfrak{B}_x \subset \mathfrak{B}$ and $\mathfrak{A}_x \to \mathfrak{B}_x$ are Γ_e-elementary, h is Γ_e-elementary. Thus $h \in \boldsymbol{C}_e$. It follows that $\{O\mathfrak{A}_x\}_{x\in X}$, $\{O\mathfrak{B}_x\}_{x\in X}$ and $\{Oh_x : O\mathfrak{A}_x \to O\mathfrak{B}_x\}_{x\in X}$ have the same properties as the original families and $Oh = \bigcup_x Oh_x$. Consequently $Oh \in \boldsymbol{C}_e$, and so on: $O^n h \in \boldsymbol{C}_e$. Putting $\mathfrak{B}_x = O\mathfrak{A}_x$, $h_x = f_{\mathfrak{A}_x}$, $h = f_{\mathfrak{A}}$ it follows that $O^m f_{\mathfrak{A}}$ is defined for all m. The same argument works with $\mathfrak{A}$ and $\{\mathfrak{A}_x\}_{x\in X}$ replaced by $O^n\mathfrak{A}$ and $\{O^n\mathfrak{A}_x\}_{x\in X}$. Consequently $O^m f_{O^n\mathfrak{A}}$ is defined for all m and n. Hence $\mathfrak{A} \in \boldsymbol{C}_{e,*}$. The other claim is established by the same kind of arguments.)

The finite iterations of $\boldsymbol{O}$, $\boldsymbol{O}^n$ are defined as operators acting on the structures and embeddings of $\boldsymbol{C}_{e,*}$ as follows:

$$
\begin{aligned}
O^0\mathfrak{A} &= \mathfrak{A}, & O^{n+1}\mathfrak{A} &= O(O^n\mathfrak{A}),\\
O^0 h &= h, & O^{n+1}h &= O(O^n h),\\
f^0_{\mathfrak{A}} &= 1_{\mathfrak{A}}, & f^{(n+1)}_{\mathfrak{A}} : \mathfrak{A} \to O^{n+1}\mathfrak{A} &= \mathfrak{A} \to O^n\mathfrak{A} \to O^{n+1}\mathfrak{A},
\end{aligned}
$$

where $\mathfrak{A} \to O^n\mathfrak{A} = f^{(n)}_{\mathfrak{A}}$ and $O^n\mathfrak{A} \to O^{n+1}\mathfrak{A} = f_{O^n\mathfrak{A}}$. (We write here $f^{(n)}$ rather than f^n to avoid confusing $f^{(n)}$ with $f \circ f \circ \cdots \circ f$, n times. When we replace later n by any ordered set the parentheses will be omitted.)

$\boldsymbol{O}^n$ is easily shown to have the same essential properties ((1)–(6)) of $\boldsymbol{O}$. (Actually, for each fixed n, $\boldsymbol{O}^n$ is a self-extension operator, but we do not have to use this.)

Let $I, J, \ldots$ be finite ordered sets. If I and J have m and n elements, respectively, we shall correlate with every embedding $j:I \to J$ an induced embedding $O^m\mathfrak{A} \to O^n\mathfrak{A}$. This correlation is the key to the definition of the infinite iterating process.

DEFINITION 1. $O^{\varnothing}\mathfrak{A} = \mathfrak{A}$, $O^{\varnothing}h = h$ (here $\varnothing$ = empty ordered set).

$O^I\mathfrak{A} = O(O^{I'}\mathfrak{A})$, $O^I h = O(O^{I'}h)$, where I' is obtained from I by omitting its last member ($\mathfrak{A}, h \in \boldsymbol{C}_{e,*}$).

If $I \subset J$ (i.e., I is a subset of J and $<_I = <_J \cap (I \times I)$) define $f_{I,J,\mathfrak{A}}:O^I\mathfrak{A} \to O^J\mathfrak{A}$ by induction on the number of elements of J:

(0) $J = \varnothing$. Then $I = \varnothing$ and $O^I\mathfrak{A} \to O^J\mathfrak{A} = 1_{\mathfrak{A}}$.

Assume that J has $n+1$ elements. Let j = maximal member of J (in the ordering $<_J$) and let $J' = J - \{j\}$, with the induced ordering. There are two cases:

(i) $j \notin I$. Then $I \subset J'$ and we put

$$f_{I,J,\mathfrak{A}}:O^I\mathfrak{A} \to O^J\mathfrak{A} =_{\mathrm{Df}} O^I\mathfrak{A} \to O^{J'}\mathfrak{A} \to O^J\mathfrak{A},$$

where $O^I\mathfrak{A} \to O^{J'}\mathfrak{A} = f_{I,J',\mathfrak{A}}$ and $O^{J'}\mathfrak{A} \to O^J\mathfrak{A} = f_{O^{J'}\mathfrak{A}}$.

(ii) $j \in I$. Let $I' = I - \{j\}$ and put

$$f_{I,J,\mathfrak{A}} =_{\mathrm{Df}} O(f_{I',J',\mathfrak{A}}).$$

The two cases can be described schematically as follows:

(i)

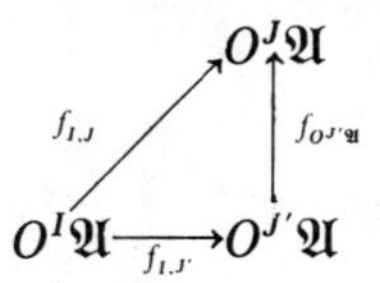

(ii)

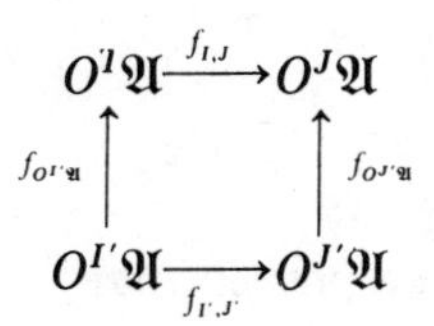

If $j:I \to J$ is an embedding of I into J (i.e., an order preserving mapping) let $I^* = j''I$, then I^* is isomorphic to I, $O^{I^*}\mathfrak{A} = O^I\mathfrak{A}$ and we put: $f_{j,\mathfrak{A}} : O^I\mathfrak{A} \to O^J\mathfrak{A} = f_{I^*,J,\mathfrak{A}} : O^{I^*}\mathfrak{A} \to O^J\mathfrak{A}$. This concludes the definition.

It is easily seen that if J has n elements then $f_{\varnothing,J} = f^{(n)}$.

The following can now be proved:

(11) *For every $\mathfrak{A} \in \boldsymbol{C}_{e,*}$ the assignment which correlates with every $j:I \to J$ the embedding $f_{j,\mathfrak{A}}:O^I\mathfrak{A} \to O^J\mathfrak{A}$ is a functor from the category of all finite ordered sets into $\boldsymbol{C}_{e,*}$; that is,*

$$I \to K = I \to J \to K \Rightarrow O^I\mathfrak{A} \to O^K\mathfrak{A} = O^I\mathfrak{A} \to O^J\mathfrak{A} \to O^K\mathfrak{A}.$$

(12) *For every $h:\mathfrak{A} \to \mathfrak{B} \in \boldsymbol{C}_{e,*}$ the following diagram is commutative:*

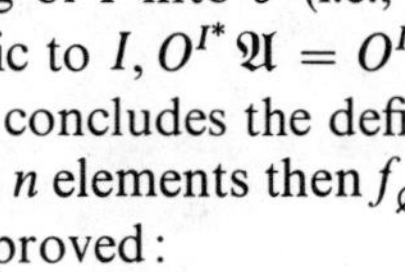

To prove (11) it is enough to assume that $I \to J$ and $J \to K$ are inclusions. The proof is by induction on the number of elements of K and relies directly on the definition of $f_{I,J,\mathfrak{A}}$. The proof of (12) is by induction on J. For details see [**6**].

If I and J are disjoint, let $I + J$ be the ordered set obtained by placing J after I (i.e., $i < j$ if $i \in I, j \in J$, and $<|I = <_I$, $<|J = <_J$, where $< = <_{I+J}$ and $<|I$ is the restriction of $<$ to I). The following is easy to prove. We assume there that I is disjoint from J, J_1 and J_2, that J is disjoint from I_1 and I_2, and that $I_1 \subset I_2$, $J_1 \subset J_2$ (all are finite ordered sets).

(13) (i) $O^J(O^I\mathfrak{A}) = O^{I+J}\mathfrak{A}$.

(ii) $f_{O^I\mathfrak{A},J_1,J_2}: O^{J_1}(O^I\mathfrak{A}) \to O^{J_2}(O^I\mathfrak{A}) = f_{\mathfrak{A},I+J_1,I+J_2}: O^{I+J_1}\mathfrak{A} \to O^{I+J_2}\mathfrak{A}$.

(iii) $O^J(f_{\mathfrak{A},I_1,I_2}): O^J(O^{I_1}\mathfrak{A}) \to O^J(O^{I_2}\mathfrak{A}) = f_{\mathfrak{A},I_1+J,I_2+J}: O^{I_1+J}\mathfrak{A} \to O^{I_2+J}\mathfrak{A}$.

(11) and (12) make the following definition possible.

DEFINITION 2. (i) Let I be any ordered class. For $\mathfrak{A} \in \boldsymbol{C}_{e,*}$ define $O^I\mathfrak{A}$ as the direct limit of $\{O^{I'}\mathfrak{A}: I' \subset I \text{ and } I' \text{ is finite}\}$ with respect to the embeddings $O^{I'}\mathfrak{A} \to O^{I''}\mathfrak{A}$, $I' \subset I''$.

(ii) If $h: \mathfrak{A} \to \mathfrak{B}$ is in $\boldsymbol{C}_{e,*}$ define $O^I h: O^I\mathfrak{A} \to O^I\mathfrak{B}$ as the embedding which is induced by the family of embeddings $O^{I'}h: O^{I'}\mathfrak{A} \to O^{I'}\mathfrak{B}$, $I' \subset I$ and I' finite.

(iii) If $I \subset J$, then $\{O^{I'}\mathfrak{A}: I' \subset I \text{ and } I' \text{ finite}\}$ is a subfamily of $\{O^{I'}\mathfrak{A}: I' \subset J$ and I' finite$\}$, and $f_{I,J,\mathfrak{A}}: O^I\mathfrak{A} \to O^J\mathfrak{A}$ is defined as the embedding of the direct limit of the subfamily into that of the family.

(iv) An isomorphism $j: I \to I^*$, of I onto I^*, maps every finite subset of I onto a corresponding subset of I^*, and induces, in this way, an isomorphism of the direct limit of $\{O^{I'}\mathfrak{A}: I' \subset I \text{ and } I' \text{ finite}\}$ onto the direct limit of $\{O^{I'}\mathfrak{A}: I' \subset I^*$ and I' finite$\}$. This isomorphism is defined to be $f_{j,\mathfrak{A}}: O^I\mathfrak{A} \to O^{I^*}\mathfrak{A}$.

(v) If $j: I \to J$ is an embedding, put $I^* = j''I$ and let j^* be the isomorphism $I \to I^*$. Define $f_{j,\mathfrak{A}}$ as $O^I\mathfrak{A} \to O^{I^*}\mathfrak{A} \to O^J\mathfrak{A}$, where the first embedding is the isomorphism $f_{j^*,\mathfrak{A}}$, and the second is $f_{I^*,J,\mathfrak{A}}$.

With these definitions it can be proved that (11) and (12) hold for all ordered classes I, J, K. They follow from the basic properties of direct limits.

We shall omit sometimes "$\mathfrak{A}$" and write "$f_{I,J}$" instead of "$f_{I,J,\mathfrak{A}}$".

Concerning direct limits. By a *directed family of structures* we mean an indexed family $\{\mathfrak{A}_x\}_{x \in X}$, where $X = (X, \leqq)$ is a directed class, together with a family of embeddings $h_{x,y}: \mathfrak{A}_x \to \mathfrak{A}_y$, $x \leqq y$, such that $\mathfrak{A}_x \to \mathfrak{A}_y \to \mathfrak{A}_z = \mathfrak{A}_x \to \mathfrak{A}_z$ for all x, y, z for which $x \leqq y \leqq z$.

If $\boldsymbol{C}$ is any category we say that the *family lies in* $\boldsymbol{C}$ if $h_{x,y} \in \boldsymbol{C}$ for all $x \leqq y$.

A direct limit of $\{\mathfrak{A}_x\}_{x \in X}$ is a structure, $\mathfrak{A}$, and a family of embeddings $h_x: \mathfrak{A}_x \to \mathfrak{A}$, $x \in X$, such that $\mathfrak{A} = \bigcup_x h_x''\mathfrak{A}_x$ and $h_y \circ h_{x,y} = h_x$ for all $x \leqq y$.

We shall use "direct limit" to denote ambiguously both the structure $\mathfrak{A}$ and the system consisting of $\mathfrak{A}$ and the embeddings $\mathfrak{A}_x \to \mathfrak{A}$, $x \in X$. Explicit mention of the embeddings of the directed family, or the direct limit, is often omitted and we put

$$\mathfrak{A} = \operatorname*{dir\,lim}_{x \in X} \mathfrak{A}_x \quad \text{or} \quad \mathfrak{A} = \operatorname{dir\,lim} \mathfrak{A}_x.$$

A direct limit is unique up to an isomorphism over the given family. If $\mathfrak{A}'$ with the embeddings $\mathfrak{A}_x \to \mathfrak{A}', x \in X$, is another direct limit then there is a unique isomorphism $\mathfrak{A} \simeq \mathfrak{A}'$ such that, for all $x \in X$, $\mathfrak{A}_x \to \mathfrak{A}' = \mathfrak{A}_x \to \mathfrak{A} \simeq \mathfrak{A}'$. This justifies the term "the direct limit". If a better justification is sought, a procedure for defining a specific direct limit can be given. For example, given the family $\{\mathfrak{A}_x\}_{x \in X}$ consider $\{\langle a, x\rangle : a \in |\mathfrak{A}_x|, x \in X\}$, and put $\langle a, x\rangle \equiv \langle b, y\rangle$ if, for some z, such that $x, y \leqq z$, we have $h_{x,z}(a) = h_{y,z}(b)$. Choose representatives of the equivalence classes, for example the sets of equivalent members of minimal rank. These will be the members of $|\mathfrak{A}|$. The relations and distinguished members of $\mathfrak{A}$ and the embeddings $h_x : \mathfrak{A}_x \to \mathfrak{A}$ are defined in the obvious way. Cases will arise where by such a procedure two direct limits which one tends to regard as identical become only "naturally isomorphic." For example, if x^* is the maximal member of X, one tends to identify the direct limit with $\mathfrak{A}_{x^*}$, whereas the given procedure yields only an isomorphic replica of $\mathfrak{A}_{x^*}$. This can be avoided if it is agreed that the procedure is to be employed only if X has no maximal element and, otherwise, $\mathfrak{A}_{x^*}$ is to be the direct limit. By adopting such a convention the general definition of $O^I\mathfrak{A}$ yields the same structure as the previous definition of $O^I\mathfrak{A}$ for the case where I is finite. Another modification is to agree that, in the case where all the embeddings $\mathfrak{A}_x \to \mathfrak{A}_y$, $x \leqq y$, are inclusions, the direct limit is to be simply $\bigcup_{x \in X} \mathfrak{A}_x$. There are, however, other cases where "naturally isomorphic" has to replace the desired "identical"; additional provisions can be made to take care of some of them but others are bound to arise.

Note that the definitions of $O^I\mathfrak{A}$ and $O^I\mathfrak{A} \to O^I\mathfrak{B}$ apply equally well to ordered classes as to ordered sets. If I is a proper class and $f_{\mathfrak{A}} : \mathfrak{A} \to O\,\mathfrak{A}$ is not an isomorphism onto $O\mathfrak{A}$ then, as we shall see, $O^I\mathfrak{A}$ is a proper class.

Put $f^I_{\mathfrak{A}} = f_{\varnothing,I,\mathfrak{A}} : O^{\varnothing}\,\mathfrak{A} \to O^I\,\mathfrak{A}$, and let $\boldsymbol{O}^I$ be (O^I, f^I). $\boldsymbol{O}^I$ is an operator acting on structures and embeddings in $\boldsymbol{C}_{e,*}$, and it is not difficult to verify that it satisfies (1)–(5), as well as (6).

Other properties of the iterated operation are the following:

(14) *If* $I = \bigcup \mathscr{I}$, *where* $\mathscr{I}$ *is a family of subclasses of* I *directed under inclusion then* $O^I\mathfrak{A} = \text{dir lim}\,\{O^J\mathfrak{A} : J \in \mathscr{I}\}$ (*the embeddings of the directed family are* $f_{J,J',\mathfrak{A}}$, $J \subset J'$, *and those of the direct limit are* $f_{J,I,\mathfrak{A}}$).

DEFINITION 3. Say that an element x of $O^I\mathfrak{A}$ is *contributed by* $O^{I'}\mathfrak{A}$, where $I' \subset I$, if it is an image of an element of $O^{I'}\mathfrak{A}$ under the embedding $f_{I',I,\mathfrak{A}} : O^{I'}\mathfrak{A} \to O^I\mathfrak{A}$. (In stricter formulation one should speak about $(I', I, \mathfrak{A})$-contributed.)

Obviously x is contributed by $O^{I'}\mathfrak{A}$ iff it is contributed by $O^{I''}\mathfrak{A}$ for some finite $I'' \subset I'$.

By repeated applications of (5) one can show that if x is contributed by $O^{I_1}\mathfrak{A}$ and by $O^{I_2}\mathfrak{A}$, where I_1 and I_2 are finite, and if $I_1 \cap I_2 = \varnothing$, then x is contributed by $\mathfrak{A}$; furthermore, if $I_1 \cap I_2 \neq \varnothing$ and $i^* = \max I_1 \cap I_2$, then x is contributed by $O^{I'_j}\mathfrak{A}$, $j = 1, 2$, where $I'_j = \{i \in I_j : i \leqq i^*\}$. Now self-extension operators have also the following property:

If $\mathfrak{A} \subset \mathfrak{A}_j \subset \mathfrak{B}, j = 1, 2$, *and all these inclusions are in* $\boldsymbol{C}_e$ *then* $\mathfrak{A} = \mathfrak{A}_1 \cap \mathfrak{A}_2 \Rightarrow O\,\mathfrak{A} = O\,\mathfrak{A}_1 \cap O\,\mathfrak{A}_2$. (This condition is also meaningful for any natural extension operator which need not carry inclusions to inclusions, provided that $O\,\mathfrak{A}$, $O\,\mathfrak{A}_1$, $O\,\mathfrak{A}_2$ are replaced by their images under $O\,\mathfrak{A} \to O\mathfrak{B}$, $O\,\mathfrak{A}_1 \to O\mathfrak{B}$ and $O\,\mathfrak{A}_2 \to O\mathfrak{B}$.)

From this one can deduce:

(15) *For each* $x \in |O^I\mathfrak{A}|$ *there exists a unique* $I_x \subset I$ *such that, for all* $J \subset I$, $O^J\mathfrak{A}$ *contributes* x *iff* $J \supset I_x$.

In some cases I_x can tell one something important about the properties of x in $O^I\mathfrak{A}$. Let $i(x) = \max I_x$. Certain natural extension operators act on models of Peano's arithmetic and yield elementary end-extensions of these models. They are not self-extension operators but the machinery described here applies to them. For these operators we have $i(x_1) <_I i(x_2) \Rightarrow x_1 < x_2$, where "$<$" on the right-hand side denotes the natural ordering of the model of Peano's arithmetic. No such simple relationship holds for the self-extension operators to be discussed in Parts II and III of this work. Theorem III.1 relates the ordering of two ordinals x_1, x_2 in $O^I\mathfrak{A}$ to the ordering of $i(x_1)$ and $i(x_2)$ in I; but this is done in a special case, under additional assumptions. There are good reasons to suppose that there are no general simple connections between the ordering of two ordinals x_1, x_2 of $O^I\mathfrak{A}$ and the finite subsets I_1, I_2 of I, such that $O^{I_i}\mathfrak{A}$ contributes x_i (they will be outlined in Part III).

$\boldsymbol{C}_{e,*}$ is closed under direct limits; consequently if $\mathfrak{A} \in \boldsymbol{C}_{e,*}$ and I, I' are ordered sets such that $I' \subset I$, then $O^I\mathfrak{A}$ and $O^{I'}\mathfrak{A} \to O^I\mathfrak{A}$ are in $\boldsymbol{C}_{e,*}$. It can also be shown that $h:\mathfrak{A} \to \mathfrak{B} \in \boldsymbol{C}_{e,*} \Rightarrow O^I h : O^I\mathfrak{A} \to O^I\mathfrak{B} \in \boldsymbol{C}_{e,*}$.

Since $O^I\mathfrak{A} \in \boldsymbol{C}_{e,*}$ one can apply to it $\boldsymbol{O}^J$ and get $O^J(O^I\mathfrak{A})$. (13) can be now generalized to all ordered sets, whether finite or infinite. There is, however, one small difference: strictly speaking, $O^J(O^J\mathfrak{A})$ need not be identical to $O^{I+J}\mathfrak{A}$ but only naturally isomorphic. The reason is that when infinite ordered sets are involved the structures in question are obtained as direct limits. As we remarked, the direct limit is determined up to an isomorphism; if a precise determination is to be made, representatives should be chosen in some uniform way, and this can make two direct limits which we tend to identify only naturally isomorphic. For all practical purposes one can identify $O^J(O^I\mathfrak{A})$ with $O^{I+J}\mathfrak{A}$. The generalization of (13) is as follows, where, as in (13), I is disjoint from J, J_1, J_2 and J is disjoint from I_1, I_2 and $I_1 \subset I_2, J_1 \subset J_2$.

(16) *There is a natural isomorphism* $O^J(O^I\mathfrak{A}) \simeq O^{I+J}\mathfrak{A}$, *such that the following diagrams are commutative*:

$$
\begin{array}{ccc}
O^{J_2}(O^I\mathfrak{A}) & \simeq & O^{I+J_2}\mathfrak{A} \\
{\scriptstyle f_{O^I\mathfrak{A},J_1,J_2}} \uparrow & & \uparrow {\scriptstyle f_{\mathfrak{A},I+J_1,I+J_2}} \\
O^{J_1}(O^I\mathfrak{A}) & \simeq & O^{I+J_1}\mathfrak{A}
\end{array}
\qquad\qquad
\begin{array}{ccc}
O^{J}(O^{I_2}\mathfrak{A}) & \simeq & O^{I_2+J}\mathfrak{A} \\
{\scriptstyle O^J(f_{\mathfrak{A},I_1,I_2})} \uparrow & & \uparrow {\scriptstyle f_{\mathfrak{A},I_1+J,I_2+J}} \\
O^{J}(O^{I_1}\mathfrak{A}) & \simeq & O^{I_1+J}\mathfrak{A}
\end{array}
$$

(The natural isomorphism $O^J(O^I\mathfrak{A}) \simeq O^{I+J}\mathfrak{A}$ is first established for finite J's using the continuity property, (6) of $\boldsymbol{O}$, then one uses properties of direct limits.)

If I is any ordered class, and if we put $I|i = \{j \in I : j < i\}$, with the induced ordering, then (13) and (14) yield the following:

(17) $O^{I|i}\mathfrak{A} \simeq O(O^{I|i'}\mathfrak{A})$ *if i' is the immediate predecessor of i.*
$O^{I|i}\mathfrak{A} = \operatorname{dir\,lim}_{j<i} O^{I|j}\mathfrak{A}$ *if i is a limit member of I.*

Taking I to be the class of ordinals this gives $O^{\alpha+1}\mathfrak{A} \simeq O(O^{\alpha}\mathfrak{A})$, $O^{\alpha}\mathfrak{A} = \operatorname{dir\,lim}_{\beta<\alpha} O^{\beta}\mathfrak{A}$, for $\alpha = \bigcup \alpha > 0$.

Now, keep I fixed, take $\mathfrak{A} \in \boldsymbol{C}_{e,*}$ and, for every $J \subset I$, let $\mathfrak{A}^J = f''_{J,I}(O^J\mathfrak{A})$ (i.e., $|\mathfrak{A}^J|$ = set of all members of $O^I\mathfrak{A}$ which are contributed by $O^J\mathfrak{A}$). Given $J_1, J_2 \subset I$ and an embedding $q: J_1 \to J_2$ we have an induced embedding $f_{q,\mathfrak{A}}: O^{J_1}\mathfrak{A} \to O^{J_2}\mathfrak{A}$ and consequently we have an induced embedding $\mathfrak{A}^{J_1} \to \mathfrak{A}^{J_2}$, which we will call "$\bar{q}$". The following now follows easily from (11) and (15).

(18) (A) *For any ordered class I, the substructures $\mathfrak{A}^J$, $J \subset I$, and the embeddings $\bar{q}: \mathfrak{A}^{J_1} \to \mathfrak{A}^{J_2}$, induced by the embeddings $q: J_1 \to J_2$, have the following properties:*

(i) $\mathfrak{A}^I = O^I\mathfrak{A}$ *and $\mathfrak{A}^{\varnothing}$ is isomorphic to $\mathfrak{A}$.*

(ii) $J = J_1 \cap J_2 \Rightarrow \mathfrak{A}^J = \mathfrak{A}^{J_1} \cap \mathfrak{A}^{J_2}$.

(iii) *If q is the inclusion $J_1 \subset J_2$ then $\bar{q}$ is the inclusion $\mathfrak{A}^{J_1} \subset \mathfrak{A}^{J_2}$.*

(iv) $q = q_1 \circ q_2 \Rightarrow \bar{q} = \bar{q}_1 \circ \bar{q}_2$.

(v) *If $J \subset J_1$ then* $[\forall i \in J\ q(i) = i] \Rightarrow [\forall x \in |\mathfrak{A}^J|\ \bar{q}(x) = x]$. ((v) *follows from* (iii) *and* (iv).)

(B) *If $f_{\mathfrak{A}}: \mathfrak{A} \to O\mathfrak{A}$ is not onto and if $c \in |O\mathfrak{A}| - f''(|\mathfrak{A}|)$, then, putting $c_i = f_{\{i\},I}(c)$, the embedding $I \to O^I\mathfrak{A}$ which sends i to c_i has the following properties:*

(vi) $c_i \in \mathfrak{A}^J \Leftrightarrow i \in J$ *(implying $i \neq j \Rightarrow c_i \neq c_j$).*

(vii) *If $i \in J_1$ then for every $q: J_1 \to J_2$, $c_{q(i)} = \bar{q}(c_i)$.*

It is easy to see that if $h: \mathfrak{A} \to \mathfrak{B}$ is in $\boldsymbol{C}_{e,*}$, and if we choose c to define $I \to O^I\mathfrak{A}$ and $h(c)$ to define $I \to O^I\mathfrak{B}$ then the following diagram is commutative:

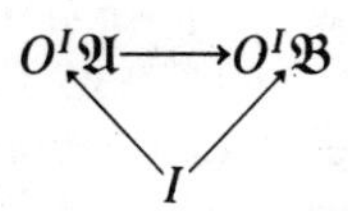

The main interest of (18) for Part III is in the case where $f_{\mathfrak{A}}: \mathfrak{A} \to O\mathfrak{A}$ is elementary, in which case all the structures $\mathfrak{A}^J$, $J \subset I$, are elementary substructures of $\mathfrak{A}^I$ and the c_i's are indiscernible with respect to the ordering of I. This is established as follows.

If $f_{\mathfrak{A}}$ is elementary then $\mathfrak{A} \in \boldsymbol{C}_{e,*}$ and $f_{O\mathfrak{A}}: O\mathfrak{A} \to O^2\mathfrak{A}$ is elementary as well. Moreover $Og: O\mathfrak{A} \to O\mathfrak{B}$ is elementary whenever $g: \mathfrak{A} \to \mathfrak{B}$ is. It now follows easily by induction that $f_{I,J}: O^I\mathfrak{A} \to O^J\mathfrak{A}$ is elementary, for all finite I, J such that $I \subset J$. By a well-known property of elementary embeddings concerning direct limits, this is also true for any ordered I, J. Consequently,

(19) *If $f_{\mathfrak{A}}: \mathfrak{A} \to O\mathfrak{A}$ is elementary then, for every embedding $j: I \to J$ of the ordered class I into the ordered class J, the induced $f_j: O^I\mathfrak{A} \to O^J\mathfrak{A}$ is elementary. The same is true with "elementary" replaced everywhere by "Φ-elementary" where Φ is a set of formulas satisfying the property of* (10) *and closed under subformulas.*

(18) and (19) immediately imply

(20) *If $f_{\mathfrak{A}}: \mathfrak{A} \to O\mathfrak{A}$ is elementary then*

(i) *all the structures* $\mathfrak{A}^J$ *of* (18) *are elementary substructures of* $\mathfrak{A}^I$ *and all* $\bar{q}$ *are elementary embeddings, and*

(ii) *the* c_i*'s are indiscernible with respect to the ordering* $<$ *of* I; *that is to say, if* $i_1 < \cdots < i_n$ *and* $k_1 < \cdots < k_n$, *then*

$$\mathfrak{A}^I \models \varphi(c_{i_1}, \ldots, c_{i_n}) \leftrightarrow \varphi(c_{k_1}, \ldots, c_{k_n}).$$

(To get (ii) apply the isomorphism $q: \{i_1, \ldots, i_n\} \to \{k_1, \ldots, k_n\}$ and use (18)(ii). Note that (20) is true if "elementary" is replaced everywhere by "Φ-elementary", where Φ is any set of formulas satisfying the property of (10) and closed under subformulas, provided that in (ii) "indiscernibles" is replaced by "Φ-indiscernibles", i.e., φ ranges over Φ only.)

Getting additional information about the embeddings $\bar{q}: \mathfrak{A}^{J_1} \to \mathfrak{A}^{J_2}$ can be extremely useful. For example, one can ask: Which x's in the domain of $\bar{q}$ satisfy $\bar{q}(x) = x$? In models of set theory $\bar{q}$ can be represented as a class of ordered pairs and one can ask: Which subsets of $\bar{q}$ are sets inside the model $\mathfrak{A}^I$? One can get much more than indiscernibles, as we shall see in Part III, where the analysis of these embeddings plays a crucial role.

This concludes the outline of iterated self-extensions for structures which are sets. It should be noted that the concept $\boldsymbol{O}^I$ generalizes in the obvious way to that of $\prod_{i \in I} \boldsymbol{O}_i$, where I is an ordered class and, with each $i \in I$, an operator $\boldsymbol{O}_i$ is associated.

Structures which are proper classes. If $\mathfrak{A}$ is a proper class and $\mathfrak{A} \models S_{e,*}$ then, for every fixed n, one can define $O^n\mathfrak{A}$, but the definition will depend on n and there is no direct way of defining $O^n\mathfrak{A}$ by induction.

DEFINITION 4. Say that $\boldsymbol{O}$ *can be iterated on* $\mathfrak{A}$ if definitions of $O^I\mathfrak{A}$ and $f_{I,J}: O^I\mathfrak{A} \to O^J\mathfrak{A}$ can be given, in which I and J are parameters which range over finite ordered sets, such that the recursive conditions of Definitions 1 can be proved. (This involves also the provability of: For all finite I, J, $O^I\mathfrak{A} \models S_e$ and $O^I\mathfrak{A} \to O^J\mathfrak{A}$ is Γ_e-elementary. Without this the recursive clauses of Definition 1 are meaningless, for we apply $\boldsymbol{O}$ only to models of S_e and to Γ_e-elementary embeddings.) In a similar way, we say that $\boldsymbol{O}$ *can be iterated on* $h: \mathfrak{A} \to \mathfrak{B}$, if it can be iterated on $\mathfrak{A}$ and $\mathfrak{B}$ and, furthermore, a definition of $O^I h$, with I a parameter ranging over finite ordered sets, can be given.

Being able to iterate $\boldsymbol{O}$ on $\mathfrak{A}$ (or h) is a metaconcept which depends on the set-theory in question and on the definition of $\mathfrak{A}$ (or h). In this definition parameters might occur, in which case they will occur also in the definition of $O^I\mathfrak{A}$ ($O^I h$).

If $\boldsymbol{O}$ can be iterated on $h_1: \mathfrak{A} \to \mathfrak{B}$ and on $h_2: \mathfrak{B} \to \mathfrak{C}$ then it can be iterated on $h_2 \circ h_1: \mathfrak{A} \to \mathfrak{C}$, for one can define: $O^I(h_2 \circ h_1) =_{\mathrm{Df}} O^I h_2 \circ O^I h_1$. The recursive conditions are easily seen to hold. Thus, the structures and embeddings on which $\boldsymbol{O}$ can be iterated form a "category".

Suppose that $\{\mathfrak{A}_x\}_{x \in X}$ is a given family of structures parametrized by x, $x \in X$, such that $\boldsymbol{O}$ can be iterated on $\mathfrak{A}_x$ (i.e., a definition of the iteration can be given in which x occurs as a parameter). Assume moreover that X is a directed class and with every $x, y \in X$ such that $x \leqq y$ an embedding $h_{x,y}: \mathfrak{A}_x \to \mathfrak{A}_y$ is associated, so that we get a directed family and $\boldsymbol{O}$ can be iterated on $h_{x,y}$. For simplicity, assume that all the $h_{x,y}$'s are inclusions, and put $\mathfrak{A} = \bigcup_{x\in X} \mathfrak{A}_x$. Then $\boldsymbol{O}$ can be iterated on $\mathfrak{A}$ and on the inclusions $h_x : \mathfrak{A}_x \subset \mathfrak{A}$. If Y is a directed subclass of X and $\mathfrak{A}' = \bigcup_{x\in Y} \mathfrak{A}_x$ then $\boldsymbol{O}$ can be iterated on the inclusion $h : \mathfrak{A}' \subset \mathfrak{A}$. The definitions of the iterations will be as follows, where I, J range over finite sets:

$O^I\mathfrak{A} =_{\mathrm{Df}} \bigcup_x O^I\mathfrak{A}_x$; $f_{I,J,\mathfrak{A}} =_{\mathrm{Df}} \bigcup_x f_{I,J,\mathfrak{A}_x}$ (that is, for $a \in |O^I\mathfrak{A}_x|$, $f_{I,J,\mathfrak{A}}(a) =_{\mathrm{Df}} f_{I,J,\mathfrak{A}_x}(a)$); $O^I h_x =_{\mathrm{Df}}$ the inclusion $O^I\mathfrak{A}_x \subset O^I\mathfrak{A}$; $O^I h =_{\mathrm{Df}}$ the inclusion $O^I\mathfrak{A}' \subset O^I\mathfrak{A}$.

The proof that these definitions satisfy the recursive conditions of Definition 1 is straightforward but involves the checking of several details. It rests on the following observation: For every finite I we know that $O^I\mathfrak{A}_x \models S_e$ for all $x \in X$, and, whenever $x \leqq y$, $O^I\mathfrak{A}_x \prec_{\Gamma_e} O^I\mathfrak{A}_y$. Using the same argument as in the proof of (7) it follows that $\bigcup_x O^I\mathfrak{A}_x \models S_e$ and $\mathfrak{A}_x \prec_{\Gamma_e} \bigcup_x O^I\mathfrak{A}_x$. (Remember that Γ_e and S_e are finite and the argument applies equally well to structures which are proper classes.) Consequently $O(\bigcup_x O^I\mathfrak{A}_x) = \bigcup_x O(O^I\mathfrak{A}_x)$; thus $O(O^I\mathfrak{A}) = O^{I+1}\mathfrak{A}$, where $I + 1$ is obtained by adding a largest member to I. The other induction clauses are proved in a similar way.

The same is true in the general case, where the $h_{x,y}$ are not necessarily inclusions and $\mathfrak{A} = \operatorname{dir\,lim} \mathfrak{A}_x$. The easiest way to be convinced of this is to replace the family $\{\mathfrak{A}_x\}_{x\in X}$ by a replica $\{\mathfrak{A}^*_x\}_{x\in X}$, where $\mathfrak{A}^*_x$ is the image of $\mathfrak{A}_x$ under $\mathfrak{A}_x \to \mathfrak{A}$. One has to observe that if $\boldsymbol{O}$ can be iterated on $\mathfrak{A}_x$ and if we are given an isomorphism $i_x : \mathfrak{A}_x \to \mathfrak{A}^*_x$ then $\boldsymbol{O}$ can be iterated on $\mathfrak{A}^*_x$, and a similar claim is true concerning the embeddings $\mathfrak{A}_x \to \mathfrak{A}_y$, $\mathfrak{A}^*_x \to \mathfrak{A}^*_y$. Now the embeddings of $\{\mathfrak{A}^*_x\}_{x\in X}$ are inclusions; hence one can apply the previous result, and deduce that $\boldsymbol{O}$ can be iterated on $\mathfrak{A}$. To show that it can be iterated on $\mathfrak{A}_x \to \mathfrak{A}$ use again the isomorphism $i_x : \mathfrak{A}_x \to \mathfrak{A}^*_x$, and the fact that it can be iterated on $\mathfrak{A}^*_x \to \mathfrak{A}$.

This is the analog of the closeness of $\boldsymbol{C}_{e,*}$ under direct limits. In an appropriate sense those structures and embeddings on which $\boldsymbol{O}$ can be iterated form a "category" which is "closed" under direct limits.

If $\boldsymbol{O}$ can be iterated on $\mathfrak{A}$ one can define

$$O^I\mathfrak{A} =_{\mathrm{Df}} \operatorname{dir\,lim} \{O^{I'}\mathfrak{A} : I' \subset I \text{ and } I' \text{ is finite}\}$$

where I is an ordered class. Similarly, the rest of Definition 2 can be used without any changes. There are no difficulties in constructing direct limits of directed families whose "members" are classes. (The use of "classes of classes" is only apparent; what we mean is that the structures depend on a parameter which ranges over some directed class.) All the properties of direct limits which were used for structures which are sets can be proved in the same way for structures which are classes. (11)–(18) *hold without exception for structures and embeddings on which* $\boldsymbol{O}$ *can be iterated.* (To get (16) observe that $\boldsymbol{O}$ can be iterated on $O^I\mathfrak{A}$. This is so because $O^I\mathfrak{A}$ is a direct limit of a family on whose structures and embeddings $\boldsymbol{O}$ can be iterated; if $\boldsymbol{O}$ can be iterated on $\mathfrak{A}$ then it can be iterated on $O^{I'}\mathfrak{A}$ where I' ranges over finite ordered sets: Put $O^J(O^{I'}\mathfrak{A}) = O^{I'+J}\mathfrak{A}$ where J is a finite ordered set, and similarly for embeddings.)

Note that the recursive conditions of Definition 1 determine $O^I\mathfrak{A} \to O^J\mathfrak{A}$ and $O^I\mathfrak{A} \to O^I\mathfrak{B}$ uniquely; hence any two definitions of the iteration will turn out to be equivalent.

To get a useful sufficient condition by which $\boldsymbol{O}$ can be iterated on $\mathfrak{A}$ we proceed as follows. Let $\boldsymbol{C}$ be any category.

DEFINITION 5. (I) A family $\{\mathfrak{A}_\alpha\}_{\alpha\in\mathrm{Ord}}$ is a *$\boldsymbol{C}$-representation* of $\mathfrak{A}$ if the following hold:

(i) $\mathfrak{A} = \bigcup_\alpha \mathfrak{A}_\alpha$.

(ii) If $\alpha \leqq \beta$ then $\mathfrak{A}_\alpha \subset \mathfrak{A}_\beta$ and the inclusion is in $\boldsymbol{C}$.

(iii) There is a limit ordinal λ such that, for every set of ordinals X of order type λ, $\bigcup_{\alpha\in X} \mathfrak{A}_\alpha = \mathfrak{A}_\beta$, where $\beta = \sup X$.

(iii) implies the weaker second order condition:

(iii′) $\forall X \subset \mathrm{Ord}$ if X is unbounded then there is a limit ordinal β such that $\bigcup_{\alpha\in X\cap\beta} \mathfrak{A}_\alpha = \mathfrak{A}_\beta$.

(iii′) is another way of saying that $\{\beta : \mathfrak{A}_\beta = \bigcup_{\alpha<\beta} \mathfrak{A}_\alpha\}$ is stationary, i.e., intersects every closed class. For all purposes, any first-order condition which implies (iii′) can be substituted for (iii).

(II) $\{h_\alpha : \mathfrak{A}_\alpha \to \mathfrak{B}_\alpha\}_{\alpha\in\mathrm{Ord}}$ is a *$\boldsymbol{C}$-representation* of $h : \mathfrak{A} \to \mathfrak{B}$ if $\{\mathfrak{A}_\alpha\}_\alpha$ and $\{\mathfrak{B}_\alpha\}_\alpha$ satisfy (i) and (ii) with respect to $\mathfrak{A}$ and $\mathfrak{B}$ and, for some limit ordinal λ, (iii) holds for both families with the same λ (or any first-order condition which implies that $\{\beta : \mathfrak{A}_\beta = \bigcup_{\alpha<\beta} \mathfrak{A}_\alpha$ and $\mathfrak{B}_\beta = \bigcup_{\alpha<\beta} \mathfrak{B}_\alpha\}$ is stationary), and $h = \bigcup_\alpha h_\alpha$ (i.e., for $x \in |\mathfrak{A}_\alpha|$, $h(x) = h_\alpha(x)$).

(21) *If $\{\mathfrak{A}_\alpha\}_\alpha$ is a $\boldsymbol{C}_{e,*}$-representation of $\mathfrak{A}$ then the following is a good definition of the iteration of $\boldsymbol{O}$ on $\mathfrak{A}$ (where I, J range over finite ordered sets)*:

$$O^I\mathfrak{A} =_{\mathrm{Df}} \bigcup_\alpha O^I\mathfrak{A}_\alpha, \qquad f_{I,J,\mathfrak{A}} = \bigcup_\alpha f_{I,J,\mathfrak{A}_\alpha}.$$

If $\{h_\alpha : \mathfrak{A}_\alpha \to \mathfrak{B}_\alpha\}_\alpha$ is a $\boldsymbol{C}_{e,*}$-representation of h then the following is a good definition of $O^I h$:

$$O^I h = \bigcup_\alpha O^I h_\alpha.$$

(OUTLINE OF THE PROOF. By the assumptions we have $\alpha \leqq \beta \Rightarrow O\mathfrak{A}_\alpha \subset O\mathfrak{A}_\beta$. Put $\mathfrak{B} = \bigcup_\alpha O\mathfrak{A}_\alpha$. The equality $\mathfrak{B} = O\mathfrak{A}$ means that $\mathfrak{B}$ is the structure which is defined in $\mathfrak{A}$ by the self-extending schema e. To show it, note that, for all α, $\mathfrak{A}_\alpha \models S_e$, and $\mathfrak{A}_\alpha \prec_{\Gamma_e} \mathfrak{A}_\beta$ for $\alpha \leqq \beta$; it follows that $\mathfrak{A} \models S_e$; thus O is applicable to $\mathfrak{A}$. Using (iii) or (iii′), one can get arbitrarily large α's such that $\mathfrak{A}_\alpha \prec_{\Gamma_e} \mathfrak{A}$; the method is the same which proves the reflection principle. Hence, for arbitrarily large α's, $O\mathfrak{A}_\alpha \subset O\mathfrak{A}$, which implies $\bigcup_\alpha O\mathfrak{A}_\alpha = O\mathfrak{A}$. Since $\{O^n\mathfrak{A}_\alpha\}_\alpha$ is easily seen to be a $\boldsymbol{C}_{e,*}$-representation of $\bigcup_\alpha O^n\mathfrak{A}_\alpha$, the same argument proves that $O\bigcup_\alpha O^n\mathfrak{A}_\alpha = \bigcup_\alpha O^{n+1}\mathfrak{A}_\alpha$. The other clauses of Definition 1 are proved along the same lines. When the two families $\{\mathfrak{A}_\alpha\}_\alpha$ and $\{\mathfrak{B}_\alpha\}_\alpha$ satisfy (II), one can get, for any finite Φ, arbitrarily large α's such that both $\mathfrak{A}_\alpha \prec_\Phi \mathfrak{A}$ and $\mathfrak{B}_\alpha \prec_\Phi \mathfrak{B}$ and this is used to show that $O^I h$ satisfies the required conditions.)

If $\boldsymbol{O}$ can be iterated on $\mathfrak{A}$ and $f_{\mathfrak{A}} : \mathfrak{A} \to \boldsymbol{O}\,\mathfrak{A}$ is elementary then for every fixed finite I and J such that $I \subset J$ and for every φ it can be proved that $\boldsymbol{O}^I\mathfrak{A} \to \boldsymbol{O}^J\mathfrak{A}$ is φ-elementary. The proof will depend, however, on the given I, J. In order to prove (19) and (20) for structures which are classes one has to prove:

For all finite I, J, $\boldsymbol{O}^I\mathfrak{A} \to \boldsymbol{O}^J\mathfrak{A}$ is φ-elementary,

that is to say, one has to prove that $\boldsymbol{O}^I\mathfrak{A} \to \boldsymbol{O}^J\mathfrak{A}$ is φ-elementary, with I, J occurring as parameters in the proof (the proof will depend, of course, on φ). Once this is done (19), and therefore (20), can be proved as follows: Given any finite set Φ of formulas, which is closed under subformulas, one proves that $\boldsymbol{O}^{I'}\mathfrak{A} \to \boldsymbol{O}^{J'}\mathfrak{A}$ is Φ-elementary for all finite I', J' such that $I' \subset J'$, $I' \subset I$, $J' \subset J$. From this it can be easily deduced that $\boldsymbol{O}^I\mathfrak{A} \to \boldsymbol{O}^J\mathfrak{A}$ is Φ-elementary.

(22) *If* $\{\mathfrak{A}_\alpha\}_\alpha$ *is a* $\boldsymbol{C}_{e,*}$*-representation of* $\mathfrak{A}$ *and, for every* α, $f_{\mathfrak{A}_\alpha} : \mathfrak{A}_\alpha \to \boldsymbol{O}\mathfrak{A}_\alpha$ *is elementary, then for all finite* I, J *such that* $I \subset J$, $\boldsymbol{O}^I\mathfrak{A} \to \boldsymbol{O}^J\mathfrak{A}$ *is elementary* (*i.e., we have the schema: for all finite* I, J, $\boldsymbol{O}^I\mathfrak{A} \to \boldsymbol{O}^J\mathfrak{A}$ *is* φ*-elementary*). *Consequently,* $\boldsymbol{O}^I\mathfrak{A} \to \boldsymbol{O}^J\mathfrak{A}$ *is elementary, where* I, J *are any ordered classes. The same is true with "elementary" replaced by "*Φ*-elementary" where* Φ *is as in* (19).

(PROOF OF (22). We have $\boldsymbol{O}^I\mathfrak{A} = \bigcup_\alpha \boldsymbol{O}^I\mathfrak{A}_\alpha$, $\boldsymbol{O}^J\mathfrak{A} = \bigcup_\alpha \boldsymbol{O}^J\mathfrak{A}_\alpha$, the diagrams

$$\begin{array}{ccc} \boldsymbol{O}^I\mathfrak{A} & \to & \boldsymbol{O}^J\mathfrak{A} \\ \cup & & \cup \\ \boldsymbol{O}^I\mathfrak{A}_\alpha & \to & \boldsymbol{O}^J\mathfrak{A}_\alpha \end{array}$$

are commutative and, for all α, $\boldsymbol{O}^I\mathfrak{A}_\alpha \to \boldsymbol{O}^J\mathfrak{A}_\alpha$ is elementary.

Moreover, if $\mathfrak{A}_\beta = \bigcup_{\alpha<\beta}\mathfrak{A}_\alpha$ then $\boldsymbol{O}^I\mathfrak{A}_\beta = \bigcup_{\alpha<\beta}\boldsymbol{O}^I\mathfrak{A}_\alpha$ and the same is true for J. It follows that for some limit ordinal λ both $\{\boldsymbol{O}^I\mathfrak{A}_\alpha\}_\alpha$ and $\{\boldsymbol{O}^J\mathfrak{A}_\alpha\}_\alpha$ are closed under unions of subsets of order type λ. Given any φ one can get arbitrarily large α's such that $\boldsymbol{O}^I\mathfrak{A}_\alpha \prec_\varphi \boldsymbol{O}^I\mathfrak{A}$ and $\boldsymbol{O}^J\mathfrak{A}_\varphi \prec_\varphi \boldsymbol{O}^J\mathfrak{A}$ hence $\boldsymbol{O}^I\mathfrak{A} \to \boldsymbol{O}^J\mathfrak{A}$ is φ-elementary.)

The condition of (22) can be phrased as: $\{\mathfrak{A}_\alpha\}_\alpha$ is a $\boldsymbol{C}_{e,\#}$-representation of $\mathfrak{A}$, where $\boldsymbol{C}_{e,\#}$ is defined as follows:

$$\mathfrak{A} \in \boldsymbol{C}_{e,\#} \Leftrightarrow_{\mathrm{Df}} \mathfrak{A} \in \boldsymbol{C}_{e,*} \text{ and } f_{\mathfrak{A}} : \mathfrak{A} \to \boldsymbol{O}\mathfrak{A} \text{ is elementary,}$$

$$h : \mathfrak{A}_1 \to \mathfrak{A}_2 \in \boldsymbol{C}_{e,*} \Leftrightarrow_{\mathrm{Df}} \mathfrak{A}_1, \mathfrak{A}_2 \in \boldsymbol{C}_{e,\#} \text{ and } h \in \boldsymbol{C}_{e,*}.$$

(Note that if $\mathfrak{A}$ has a $\boldsymbol{C}_{e,\#}$-representation, $\{\mathfrak{A}_\alpha\}_\alpha$ say, then $\boldsymbol{O}^I\mathfrak{A}$, where I is any ordered class, has a $\boldsymbol{C}_{e,\#}$-representation: Put $I_\alpha = I \cap R\alpha$, then $\boldsymbol{O}^I\mathfrak{A} = \operatorname{dir\,lim} \{\boldsymbol{O}^{I_\alpha}\mathfrak{A}_\alpha\}_\alpha$ where the embeddings of the directed family are $\boldsymbol{O}^{I_\alpha}\mathfrak{A}_\alpha \to \boldsymbol{O}^{I_\beta}\mathfrak{A}_\beta = \boldsymbol{O}^{I_\alpha}\mathfrak{A}_\alpha \to \boldsymbol{O}^{I_\alpha}\mathfrak{A}_\beta \to \boldsymbol{O}^{I_\beta}\mathfrak{A}_\beta$, for $\alpha \leqq \beta$. It is easily verified that this family yields a $\boldsymbol{C}_{e,\#}$-representation of $\boldsymbol{O}^I\mathfrak{A}$; by using replicas one can make all the embeddings inclusions.)

The following observations will be used later. If $\{\mathfrak{B}_\alpha\}_\alpha$ is a $\boldsymbol{C}_{e,*}$-representation of $\mathfrak{B}$ and if $\mathfrak{A}$ is a set and $h_0 : \mathfrak{A} \to \mathfrak{B}_0$ is in $\boldsymbol{C}_{e,*}$ then the embedding $h : \mathfrak{A} \to \mathfrak{B}$, which is defined as $\mathfrak{A} \overset{h_0}{\to} \mathfrak{B}_C \subset \mathfrak{B}$, is easily represented as $\bigcup_\alpha h_\alpha$ where $h_\alpha : \mathfrak{A}_\alpha \to \mathfrak{B}_\alpha$, $\mathfrak{A}_\alpha = \mathfrak{A}$ for all α, and $h_\alpha(x) = h_0(x)$ for all $x \in |\mathfrak{A}|$. Consequently, $\boldsymbol{O}$ can be iterated on h, and we get $\boldsymbol{O}^I h$ which embeds $\boldsymbol{O}^I\,\mathfrak{A}$ into $\boldsymbol{O}^I\mathfrak{B}$. The same is true if $h_0 = h_0 : \mathfrak{A}$

$\to \mathfrak{B}_\gamma$, for some γ; for if we put $\mathfrak{B}'_\alpha = \mathfrak{B}_{\gamma+\alpha}$ then $\{\mathfrak{B}'_\alpha\}_\alpha$ is obviously a $\boldsymbol{C}_{e,*}$-representation of $\mathfrak{B}$ and $h_0 = h_0 : \mathfrak{A} \to \mathfrak{B}'_0$. Consequently,

(23) *If* $\{\mathfrak{B}_\alpha\}_\alpha$ *is a* $\boldsymbol{C}_{e,*}$*-representation of* $\mathfrak{B}$, $\mathfrak{A}$ *is a set and, for some* γ, $h_0 : \mathfrak{A} \to \mathfrak{B}_\gamma$ *is in* $\boldsymbol{C}_{e,*}$, *then* $\boldsymbol{O}$ *can be iterated on* h, *where* $h : \mathfrak{A} \to \mathfrak{B}$ *is the composition* $\mathfrak{A} \to \mathfrak{B}_\gamma \subset \mathfrak{B}$.

PART II

Cofinal embeddings and self-extensions of models of Z^+. From now on, the structures to be considered are for the language of set theory, where, besides $=$ and $\in$, finitely many additional predicates are allowed. ZF is Zermelo-Fraenkel's theory (with the replacement schema including all the formulas of the language).

Z is Zermelo's theory (i.e., instead of the replacement schema we have the separation schema).

OR = class of all ordinals.

We let $\alpha, \beta, \gamma, \ldots$ range over ordinals. They will be used both in our exposition and as variables of the given language. Thus, "$\forall\alpha\varphi$" is to be read $\forall\alpha[\alpha \in \mathrm{OR} \to \varphi]$. Moreover, when discussing a certain structure, $\mathfrak{A}$ say, we will let $\alpha, \beta, \ldots$ denote ordinals in the sense of $\mathfrak{A}$, although they need not be "real" ordinals ($\mathfrak{A}$ can even be not well-founded).

$R\alpha$ = set of all sets of rank $< \alpha$.

'$R\alpha$ exists' is the formula $\exists f\{f$ is a function $\wedge$ Domain $(f) = \alpha + 1 \wedge \forall\beta \leqq \alpha\,\forall x[x \in f(\beta) \leftrightarrow \exists\gamma < \beta\,(x \subset f(\gamma))]\}$.

$Z^+ = Z \cup \{\forall v \exists\alpha[R\alpha \text{ exists} \wedge v \in R\alpha]\}$.

If α is an ordinal in $\mathfrak{A}$ and $\mathfrak{A} \models$ '$R\alpha$ exists' we let $R^{\mathfrak{A}}\alpha$ be the member of $\mathfrak{A}$ which is $R\alpha$ in the sense of $\mathfrak{A}$ (we assume that $\mathfrak{A}$ satisfies certain axioms of Z, such as extensionality and others, which guarantee that "$R\alpha$" is meaningful in it).

V = Cantor's universe.

If $\mathfrak{A} \models$ '$R\alpha$ exists,' then we define $\mathfrak{A}_\alpha$ = the substructure of $\mathfrak{A}$ whose universe is $\{x : \mathfrak{A} \models x \in R\alpha\}$.

In particular, V_α = the substructure of V whose universe is $R\alpha$.

Σ_0 = set of all Σ_0-formulas in Lévy's hierarchy (i.e., the smallest set containing the atomic formulas, closed under sentential connectives and under bounded quantification—$\forall x \in y\, \varphi$).

$j : \mathfrak{A} \to \mathfrak{B}$ is *cofinal* if $\forall y \in |\mathfrak{B}|\, \exists x[\mathfrak{B} \models y = jx \vee y \in jx]$.

(In general, the weaker $\forall y \in |\mathfrak{B}|\, \exists n, x_1, \ldots, x_n[\mathfrak{B} \models (y = jx_n \vee y \in x_1 \in x_2 \in \ldots \in jx_n)]$ is the more natural condition. This, however, will be equivalent to the given definition in all the cases to be discussed. If $\varphi(x, y) = \forall v_1, v_2[v_1 \in v_2 \in x \to v_1 \in y]$, then the equivalence holds if $\mathfrak{A} \models \forall x \exists y \varphi(x, y)$ and $\mathfrak{A} \models \varphi(x, y) \Rightarrow \mathfrak{B} \models \varphi(jx, jy)$.)

In all the following theorems the structures can be either sets or proper classes.

THEOREM 1. *If* $\mathfrak{A} \models Z^-$ *and* $j : \mathfrak{A} \to \mathfrak{B}$ *is cofinal, then* j *is elementary iff* j *is* Σ_0*-elementary and, for every formula* φ, *the following holds*:

$(*_{j,\varphi})$ *For all* $x \in |\mathfrak{A}|$ *which in* $\mathfrak{A}$ *(i.e., in the sense of* $\mathfrak{A}$*) is an ordered set and, for all* $a_1, \ldots, a_k \in |\mathfrak{A}|$, *if* $\varphi(a_1, \ldots, a_k, v_1, v_2)$ *defines in* $\mathfrak{A}$ *a function* F *which is a cofinal mapping of* x *into the ordinals of* $\mathfrak{A}$, *then* j *maps* x *cofinally into* jx.

In a more precise phrasing $(*_{j,\varphi})$ should be read as follows: For all $x_1, x_2 \in |\mathfrak{A}|$ if the following holds in $\mathfrak{A}$:
[x_2 is an ordering of x_1]

$$\wedge \forall v_1, v_2 \in x_1[(Fv_1 \in \mathrm{OR}) \wedge (\langle v_1, v_2\rangle \in x_2 \leftrightarrow Fv_1 \in Fv_2)] \wedge \forall\alpha\exists v \in x_1[\alpha \in Fv]$$

then, for every y such that $\mathfrak{B} \models y \in jx_1$ there is x such that $\mathfrak{A} \models x \in x_1$ and $\mathfrak{B} \models \langle y, jx\rangle \in jx_2$. (Here "$Fv$" is to be eliminated, via φ, as "the unique v^* such that $\varphi(a_1, \ldots, a_k, v, v^*)$".)

From now on the phrase "j maps x cofinally into jx" has to be understood in this sense.

Note that we do not assume that $\mathfrak{B}$ satisfies any axioms.

REMARK 1. From the proof it will be obvious that a weaker assumption than $\mathfrak{A} \models Z^+$ is sufficient. All that is needed is that some formula, τ say, defines in $\mathfrak{A}$, from parameters, a function G which correlates with every ordinal α of $\mathfrak{A}$ an element $G\alpha$ such that

$$\mathfrak{A} \models \forall\alpha, \beta[\alpha \leqq \beta \to G\alpha \subset G\beta] \wedge \forall x\exists\alpha[x \in G\alpha].$$

If Z^τ is obtained from Z by adding the axiom which states that there are parameters from which τ defines such a function, then $\mathfrak{A} \models Z^\tau$ is sufficient.

Neither do we need the full power set axiom to hold in $\mathfrak{A}$. It is sufficient to require first, the existence of cartesian products and, second, that whenever a formula correlates with every ordinal a subset of a fixed set x, then the family of all such subsets constitutes a set; this second requirement is not needed in the proof if the replacement schema holds in $\mathfrak{A}$.

The claim is also true for $\mathfrak{A} \models Z_{\text{finite}}$, where Z_{finite} is obtained from Z by replacing the infinity axiom by its negation. (In that case $(*_{j,\varphi})$ holds trivially.)

REMARK 2. $(*_{j,\varphi})$ holds trivially if $\mathfrak{A} \models \mathrm{ZF}$ (for no φ defines a cofinal embedding of a set into the ordinals). In that case, the claim is known and can be proved by a very easy argument, using reflection. (It is usually stated for a transitive $\mathfrak{B}$, but the argument can be generalized.)

REMARK 3. If x and y are ordered sets in $\mathfrak{A}$ and if for each there is a formula which defines from parameters a cofinal embedding of it into the ordinals of $\mathfrak{A}$, then there is $z \in |\mathfrak{A}|$ such that $\mathfrak{A} \models$ 'z is a function $\wedge$ z embeds x cofinally into y' (and vice versa). This implies that if j maps x cofinally into jx then it maps y cofinally into jy (assuming that j is Σ_0-elementary). Hence if the claim of $(*_{j,\varphi})$ holds for one ordered set x which is mapped, via φ, cofinally into the ordinals of $\mathfrak{A}$, then it holds always. Using this observation, and some of the preceding remarks, it is easy to deduce the following corollary.

COROLLARY. *If $j: V_\alpha \to \mathfrak{B}$ is cofinal and Σ_0-elementary, where $\alpha = \bigcup\alpha > 0$, then j is elementary iff one of the following cases takes place*:

(i) *$V_\alpha \models ZF$ or $\alpha = \omega$.*

(ii) *$V_\alpha \not\models ZF$, α is singular and j maps β cofinally into $j\beta$, where $\beta = \mathrm{cof}(\alpha)$.*

(iii) *$V_\alpha \not\models ZF$, α is regular, and for some ordered set $x \in V_\alpha$ whose order type is α, j maps x cofinally into jx.*

Note that $\mathfrak{B}$ can be a proper class. Note also that the structure V_α can have any subsets of $(R\alpha)^n$, $n = 1, 2, \ldots$, as relations, in addition to $\in|R\alpha$.

PROOF of THEOREM 1. The proof that $(*_{j,\varphi})$ is necessary is easy and is left to the reader.

In the other direction, let ψ be a formula in which a block of unbounded quantifiers precedes a Σ_0-formula. We show, by induction on the number of unbounded existential quantifiers, that $\mathfrak{A} \models \psi(x_1, \ldots, x_m) \Rightarrow \mathfrak{B} \models \psi(jx_1, \ldots, jx_m)$.

If $\psi = \forall v_1, \ldots, v_n\psi^*$ where ψ^* is Σ_0, and $z_1, \ldots, z_n \in |\mathfrak{B}|$ take $y_1, \ldots, y_n \in |\mathfrak{A}|$ such that $\mathfrak{B} \models$ '$z_i \in jy_i$', $i = 1, \ldots, n$. (This is always possible, since if $z_i = jy_i^*$ take y_i such that $\mathfrak{A} \models y_i = \{y_i^*\}$.) Then $\mathfrak{A} \models \forall u_1 \in y_1 \cdots \forall u_n \in y_n\psi^*(u_1, \ldots, u_n, x_1, \ldots, x_m)$; hence $\mathfrak{B} \models \forall u_1 \in jy_1, \ldots, \psi^*(u_1, \ldots, u_n, jx_1, \ldots)$ implying that $\mathfrak{B} \models \psi^*(z_1, \ldots, z_n, jx_1, \ldots, jx_m)$.

The parameters $x_1, \ldots, x_m$ play no role in the argument, and we shall not refer, from now on, to them or to $jx_1, \ldots, jx_m$. It is obvious that if the claim is true for ψ it is true for $\exists v\psi$. It remains to prove the claim for $\forall u_1, \ldots, u_n\exists v\psi$, assuming that it is true for formulas with fewer unbounded existential quantifiers. For the sake of simplicity take $n = 1$. Assuming $\mathfrak{A} \models \forall u\exists v\psi(u, v)$ and letting $b \in |\mathfrak{B}|$, we show $\mathfrak{B} \models \exists v\psi(b, v)$. Take $a \in |\mathfrak{A}|$ such that $\mathfrak{B} \models$ '$b \in ja$'. We have $\mathfrak{A} \models \forall u \in a\exists v\psi(u, v)$; hence $\mathfrak{A} \models \forall u \in a\exists\alpha\exists v \in R\alpha\, \psi(u, v)$. There are two cases:

(i) For some ordinal of $\mathfrak{A}$, α, we have $\mathfrak{A} \models \forall u \in a\exists v \in R\alpha\, \psi(u, v)$.

(ii) Otherwise.

In case (i), using separation in $\mathfrak{A}$, we get $a^* \in |\mathfrak{A}|$ such that the following hold in $\mathfrak{A}$:

$$\forall u \in a\exists v \in R\alpha[\langle u, v\rangle \in a^*],$$

$$\forall u, v[\langle u, v\rangle \in a^* \to \psi(u, v)].$$

The first is a Σ_0-formula and the second is logically equivalent to one with fewer unbounded $\exists$'s. Hence, both hold in $\mathfrak{B}$ (with every parameter replaced by its image under j). Hence

$$\mathfrak{B} \models \exists v \in j(R\alpha)[\langle b, v\rangle \in ja^*] \wedge \forall v[\langle b, v\rangle \in ja^* \to \psi(u, v)],$$

implying $\mathfrak{B} \models \exists v\psi(b, v)$.

(ii) Put

$$\sigma(x, \alpha, a) = \forall u[u \in x \leftrightarrow (u \in a \wedge \exists v \in R\alpha\, \psi(u, v))].$$

Using twice the separation axiom and once the power set axiom, in $\mathfrak{A}$, we get $a_1 \in |\mathfrak{A}|$ such that

$$\mathfrak{A} \models \forall x[x \in a_1 \leftrightarrow \exists\alpha\, \sigma(x, \alpha, a)].$$

Roughly speaking, a_1 consists of those subsets of a which, for some α, are of the form $\{u : \exists v \in R\alpha\, \psi(u, v)\}$. Let

$$\varphi(x, \alpha) = x \in a_1 \wedge \sigma(x, \alpha, a) \wedge \forall\beta < \alpha \neg\, \sigma(x, \beta, a),$$

then $\mathfrak{A} \models \forall x \in a_1\exists!\alpha\varphi(x, \alpha)$. ($\varphi(x, \alpha)$ asserts that x is the set of all u's in a for which there is $v \in R\alpha$ such that $\psi(u, v)$, and α is the minimal ordinal which bears this relation to x.) Putting $F(x)$ = the unique α such that $\varphi(x, \alpha)$, we obviously have

$$\mathfrak{A} \models \forall x, y \in a_1[Fx \leqq Fy \leftrightarrow x \subset y].$$

Hence, if $a_2 \in |\mathfrak{A}|$ is the inclusion relation on a_1 (i.e., $\mathfrak{A} \models a_2 = \{\langle x, y\rangle : x, y \in a_1 \wedge x \subset y\}$) then a_2 orders a_1 and F embeds this ordering into the ordinals of $\mathfrak{A}$. Because (i) does not hold, $F(x)$, where x ranges over a_1, becomes arbitrarily large. Applying $(*_{j,\varphi})$ it follows that j embeds a_1 (ordered by a_2) cofinally into ja_1 (ordered by ja_2). Now $\mathfrak{A} \models \forall u \in a \exists x \in a_1[u \in x]$; this, being a Σ_0-statement, holds in $\mathfrak{B}$ (with ja, ja_1 replacing a, a_1). Hence $\mathfrak{B} \models \exists x \in ja_1[b \in x]$. Let $y \in |\mathfrak{B}|$ be such that $\mathfrak{B} \models b \in y \in ja_1$. Get $x \in |\mathfrak{A}|$ such that $\mathfrak{B} \models \langle y, jx\rangle \in ja_2$. Since a_2 is (in $\mathfrak{A}$) the inclusion relation over a_1, and this is expressible as a Σ_0-statement, ja_2 is (in $\mathfrak{B}$) the inclusion relation over ja_1:

$$\mathfrak{B} \models \forall v_1, v_2 \in ja_1[\langle v_1, v_2\rangle \in ja_2 \leftrightarrow v_1 \subset v_2].$$

Hence, $\mathfrak{B} \models y \subset jx$, implying $\mathfrak{B} \models b \in jx$. If $\mathfrak{A} \models$ '$F(x) = \alpha$,' then $\mathfrak{A} \models \forall u \in x \exists v \in R\alpha\, \psi(u, v)$. Now we proceed as in case (i). Q.E.D.

We shall now construct a finite set of formulas Ψ such that the following will be true:

THEOREM 2. *If $\mathfrak{A}^1 \models Z$ and $\mathfrak{A}^1 \models \forall x \exists y[x \subset y \wedge y$ is transitive], then $j : \mathfrak{A}^1 \to \mathfrak{A}^2$ is Ψ-elementary $\Rightarrow j : \mathfrak{A}^1 \to \mathfrak{A}^2$ is Σ_0-elementary.*

The requirement that $\mathfrak{A}^1 \models Z$ can be weakened by replacing the power set axiom by the axiom asserting the existence of the cartesian products of any two sets. Neither do we need the infinity axiom.

The argument rests on the fact that the relation that the model x satisfies the formula φ under the assignment z (where z is supposed to be a function defined on the free variables of φ, with values in x) can be defined in Z, or even in the weaker theory, so that the usual recursion conditions are provable. The definition in Z is the standard one. Some modifications are used in the case of the weaker theory; they are left to the reader.

Choose Ψ such that the following hold:

(i) All subformulas of the following formulas are in Ψ:

$$\exists v \forall u[\neg u \in v], \quad \forall v_1, v_2 \exists v \forall u[u \in v \leftrightarrow u = v_1 \vee u \in v_2],$$

$$\forall v_1 \exists v \forall u[u \in v \leftrightarrow \exists v_2(u \in v_2 \in v_1)], \quad \forall v_1, v_2[v_1 \in v_2 \in v \to v_1 \in v].$$

(ii) The formula '$x \models \varphi[z]$' and all the sentences expressing the usual recursion conditions for satisfaction are in Ψ. Here '$x \models \varphi[z]$' is to be read as 'φ is satisfied in x under the assignment z,' where x is the submodel of the whole universe whose domain is the set x. [To be more specific, let $\sigma(x, \varphi, z)$ be 'φ is a wff $\wedge$ z is a function $\wedge \forall y_1 \exists y_2$ [y_1 is a free variable in $\varphi \to \langle y_1, y_2\rangle \in z \wedge y_2 \in x$]', the recursion conditions are the universal generalizations of the formulas:

$$x \models \langle P, t_1, \ldots, t_k\rangle[z] \leftrightarrow \sigma(x, \langle P, t_1, \ldots, t_k\rangle, z) \wedge P(z(t_1), \ldots, z(t_k))$$

(where P is any k-place predicate in the language of $\mathfrak{A}^1$; there are finitely many altogether),

$$x \models \langle \vee, \varphi, \psi\rangle[z] \leftrightarrow \sigma(x, \langle \vee, \varphi, \psi\rangle, z) \wedge (x \models \varphi[z] \vee x \models \psi[z]),$$

$$x \models \langle \neg, \varphi\rangle[z] \leftrightarrow \sigma(x, \langle \neg, \varphi\rangle, z) \wedge \neg(x \models \varphi[z]),$$

$$x \models \langle \exists, v, \varphi\rangle[z] \leftrightarrow \sigma(x, \langle \exists, v, \varphi\rangle, z) \wedge \exists y \in x \exists z' \, (z' = z^v_y \wedge x \models \varphi[z'])$$

(where $z' = z^v_y$ asserts that z' is a function and, for all v_1, v_2, $\langle v_1, v_2\rangle \in z' \leftrightarrow (\langle v_1, v_2\rangle \in z \wedge v_1 \neq v) \vee (v_1 = v \wedge v_2 = y))$.]

PROOF OF THEOREM 2. With no loss of generality we can assume that the hereditary finite sets are in $\mathfrak{A}^1$ and that, restricted to them, the membership relation of $\mathfrak{A}^1$ is the real membership. ($\mathfrak{A}^1$ may have nonstandard natural numbers, in which case it will have additional nonstandard hereditary finite sets.)

(i) implies that j maps the standard hereditary finite sets of $\mathfrak{A}^1$ onto an isomorphic replica in $\mathfrak{A}^2$. Hence with no loss of generality it can be assumed that $jx = x$ for every standard hereditary finite set of $\mathfrak{A}^1$. Formulas are construed as hereditary finite sets; hence we have $j\varphi = \varphi$ for every formula φ.

For $x \in |\mathfrak{A}^i|$ put $(x)^i = \{y : \mathfrak{A}^i \models y \in x\}$; thus $x^i \subset |\mathfrak{A}^i|$. We shall also use "$(x)^i$" to denote the submodel of $\mathfrak{A}^i$ whose universe is $(x)^i$. Because satisfaction is absolute in $\mathfrak{A}^1$ we have

$$(x)^1 \models \varphi[y_1, \ldots, y_k] \Leftrightarrow \mathfrak{A}^1 \models \text{`}x \models \varphi[y_1, \ldots, y_k]\text{'}.$$

Since, by (ii), the recursion conditions for satisfaction hold in $\mathfrak{A}^2$ the same is true for $\mathfrak{A}^2$. Furthermore, by (ii) we have $\mathfrak{A}^1 \models \text{`}x \models \varphi[y_1, \ldots, y_k]\text{'} \Leftrightarrow \mathfrak{A}^2 \models \text{`}jx \models \varphi[jy_1, \ldots, jy_k]\text{'}$, implying that

$$(x)^1 \models \varphi[y_1, \ldots, y_k] \Leftrightarrow (jx)^2 \models \varphi[jy_1, \ldots, jy_k].$$

If $\varphi(v_1, \ldots, v_k)$ is a Σ_0-formula then the universal generalization of the following formula holds in both $\mathfrak{A}^i$, $i = 1, 2$:

v is transitive $\wedge \bigwedge_{1 \leq i \leq k} v_i \in v \rightarrow (\varphi(v_1, \ldots, v_k) \leftrightarrow v \models \varphi[v_1, \ldots, v_k])$ (where 'v is transitive' $= \forall v_1, v_2[v_1 \in v_2 \in v \rightarrow v_1 \in v]$). This is easy to verify.

Given $x_1, \ldots, x_k \in |\mathfrak{A}^1|$, we get $x \in |\mathfrak{A}^1|$ such that $\mathfrak{A}^1 \models$ 'x is transitive $\wedge \bigwedge_{1 \leq i \leq k} x_i \in x$.' By (i), $\mathfrak{A}^2 \models$ 'jx is transitive $\wedge \bigwedge_{1 \leq i \leq k} jx_i \in jx$'. We get

$$\mathfrak{A}^1 \models \varphi(x_1, \ldots, x_k) \Leftrightarrow (x)^1 \models \varphi(x_1, \ldots, x_k) \Leftrightarrow (jx)^2 \models \varphi(jx_1, \ldots, jx_k)$$

$$\Leftrightarrow \mathfrak{A}^2 \models \varphi(jx_1, \ldots, jx_k). \quad \text{Q.E.D.}$$

If $\mathfrak{A} \models Z^+$, $j : \mathfrak{A} \rightarrow \mathfrak{B}$ is Ψ-elementary and cofinal and $(*_{j,\varphi})$ for all φ, j then is elementary. This follows from Theorems 1 and 2. On the other hand, if j is elementary and it is a self-extension then we can prove that it is cofinal (implying also $(*_{j,\varphi})$).

THEOREM 3. *If* $\mathfrak{A} \models Z$ *and* $j : \mathfrak{A} \rightarrow \mathfrak{B}$ *is an elementary self-extension then*

(I) *there is no* $b \in |\mathfrak{B}|$ *such that, for all* $x \in |\mathfrak{A}|$, $\mathfrak{B} \models jx \in b$; *and*

(II) *if* $\mathfrak{A} \models Z^+$, j *is cofinal.*

REMARK 1. The claim is that from the infinite set of all the sentences $\forall x_1, \ldots, x_n$ $[\mathfrak{A} \models \varphi(x_1, \ldots, x_n) \rightarrow \mathfrak{B} \models \varphi(jx_1, \ldots, jx_n)]$ and the infinite set $\{\mathfrak{A} \models \psi : \psi$ is an axiom of $Z\}$ one can deduce the sentence asserting (I), and similarly for (II). Given a self-extension, there will be a finite subset of these sentences which will imply (I) (or (II)). In particular, there is a finite set, Δ say, such that $\mathfrak{A} \models Z \cup \{j$ is Δ-elementary$\} \Rightarrow$ (I), and $\mathfrak{A} \models Z^+ \cup \{j$ is Δ-elementary$\} \Rightarrow j$ is cofinal. Δ depends on the self-extension schema e which defines $\mathfrak{B}$ and j. The proof will indicate how it can be constructed.

REMARK 2. In the proof of (I) only the satisfaction of the separation axiom in $\mathfrak{A}$ is used. To get (II) it is enough that, for some τ, $\mathfrak{A}$ satisfies Z^τ (see Remark 1 to Theorem 1) without the power set axiom. Some of the other axioms can also be omitted or weakened, and we leave the details to the reader.

REMARK 3. The claim is also true (with the same proof) if $\mathfrak{A} \models Z_{\text{finite}}$ (i.e., the infinity axiom is replaced by its negation). Obviously, an elementary self-extension of $\mathfrak{A}$ cannot introduce new members of "finite" rank. Hence the only elementary self-extension of a model of Z_{finite} is the trivial one. This is essentially known (see [**1**]) for the case where $\mathfrak{A}$ = the standard model of Z_{finite}. The proof as given here is, however, shorter and does not go through Tarski's undefinability theorem.

PROOF OF THEOREM 3. (I) Since $j : \mathfrak{A} \rightarrow \mathfrak{B}$ is a self-extension, every member of $|\mathfrak{B}|$ is a member of $|\mathfrak{A}|$ and there is a formula $\varphi(v_1, v_2)$ such that, for all $x_1 \in |\mathfrak{A}|$ and all $x_2 \in |\mathfrak{B}|$, $\mathfrak{A} \models \varphi(x_1, x_2) \Leftrightarrow \mathfrak{B} \models jx_1 \in x_2$. If $b \in |\mathfrak{B}|$ and $\mathfrak{B} \models jx \in b$ for all $x \in |\mathfrak{A}|$, use separation in $\mathfrak{B}$ and get $c \in |\mathfrak{B}|$ such that $\mathfrak{B} \models \forall x[x \in c \leftrightarrow x \in b \wedge \neg\varphi(x, x)]$. We get $\mathfrak{A} \models \varphi(c, c) \Leftrightarrow \mathfrak{B} \models jc \in c \Leftrightarrow \mathfrak{B} \models jc \in b \wedge \neg\varphi(jc, jc)$. Since $\mathfrak{B} \models jc \in b$ and since j is φ-elementary we get $\mathfrak{B} \models jc \in b \wedge \neg\varphi(jc, jc) \Leftrightarrow \mathfrak{B} \models \neg\varphi(jc, jc) \Leftrightarrow \mathfrak{A} = \neg\varphi(c, c)$; hence $\mathfrak{A} \models \varphi(c, c) \leftrightarrow \neg\varphi(c, c)$. Contradiction.

(II) follows from (I) by observing that if there is an ordinal β of $\mathfrak{B}$ such that, for every ordinal α of $\mathfrak{A}$, $\mathfrak{B} \models j\alpha \in \beta$, then for every ordinal α of $\mathfrak{A}$ we have $\mathfrak{B} \models j(R^{\mathfrak{A}}\alpha) \in R\beta$; hence for every $x \in |\mathfrak{A}|$, $\mathfrak{B} \models jx \in R\beta$, contradicting (I). Consequently, for every ordinal β of $\mathfrak{B}$ there is α in $\mathfrak{A}$ such that $\mathfrak{B} \models \beta \in j\alpha$, implying $\mathfrak{B} \models R\beta \in j(R^{\mathfrak{A}}\alpha)$, from which the cofinality follows. Q.E.D.

[Analogous results can be obtained if the assumption that j is elementary is replaced by the weaker assumption that j is Ψ-elementary (where Ψ is as in Theorem 2) and $(*_{j,\varphi})$ for every φ. In that case, assuming also $\mathfrak{A} \models Z^+$, we get the following:

Let $B_0 = \{y \in |\mathfrak{B}| : \exists x \in |\mathfrak{A}|$ such that $\mathfrak{B} \models y \in jx \vee y = jx\}$; then B_0 is not a set in $\mathfrak{B}$, i.e., there is no $b \in |\mathfrak{B}|$ such that $B_0 = \{y : \mathfrak{B} \models y \in b\}$. From this one concludes that if $\mathfrak{B} \models \forall\alpha\ [R\alpha$ exists$]$, then $\{\beta : \mathfrak{B} \models$ '$\beta \in \text{OR} \wedge j\alpha \in \beta$', for all ordinals α in $\mathfrak{A}\}$ is either empty, and then j is cofinal, or has no first member (with respect to the $\in$ of $\mathfrak{B}$). In particular, if $\mathfrak{B}$ is well-founded, j is cofinal. The proof uses the submodel of $\mathfrak{B}$, $\mathfrak{B}_0$, whose universe is B_0. Using Theorems 1 and 2 one shows that j embeds $\mathfrak{A}$ elementarily in $\mathfrak{B}_0$. If B_0 is a set in $\mathfrak{B}$, one can give in $\mathfrak{B}$ a truth definition for $\mathfrak{B}_0$. This can be translated into a definition in $\mathfrak{A}$, from which one gets a truth definition for $\mathfrak{A}$ inside $\mathfrak{A}$, leading to the well-known contradiction.]

Theorems 1, 2, 3 imply that for self-extensions $j:\mathfrak{A} \to \mathfrak{B}$ of models of ZF the following are equivalent:

(i) j is elementary.

(ii) j is cofinal and Σ_0-elementary.

(iii) j is cofinal and Ψ-elementary.

(ZF can be replaced here by any theory obtained from a theory Z^τ by omitting the power set axiom and adding the replacement schema and the existence of the cartesian products of any two sets.)

Iterating self-extensions of models of ZF.

THEOREM 4. *If $j:V \to M$ is an elementary self-extension which is obtained by applying the self-extending schema e, and if for some limit ordinal λ, j embeds λ cofinally into $j\lambda$, then one can define a $\boldsymbol{C}_{e,*}$-representation $\{\mathfrak{A}_\beta\}_\beta$ of V such that $\mathfrak{A}_\beta \to O_e\mathfrak{A}_\beta$ is elementary for all β. Consequently, $\boldsymbol{O}_e$ can be iterated on V and, for all ordered classes I, J, $O_e^I V \to O_e^J V$ is elementary.*

The $\boldsymbol{C}_{e,*}$-representation is obtained by putting $\mathfrak{A}_\beta = V_{\alpha_\beta}$ where $\alpha_\beta =$ the βth ordinal of Ω, and Ω is a certain, definable unbounded class of ordinals satisfying the following:

(i) Ω is closed under limits of subsets of order type λ.

(ii) $\alpha \in \Omega \Rightarrow V_\alpha \in \boldsymbol{C}_{e,*}$ and $V_\alpha \to O_e V_\alpha$ is elementary.

(iii) $\alpha, \beta \in \Omega$ and $\alpha \leqq \beta \Rightarrow V_\alpha \subset V_\beta$ is in $\boldsymbol{C}_{e,*}$.

REMARK 1. If M is well-founded then j embeds ω cofinally into $j\omega$; hence, the assumption of the theorem holds with $\lambda = \omega$.

REMARK 2. M is not assumed to be a transitive class. It can also be well-founded without being isomorphic to a transitive class, namely when $\{y:M \models y \in x\}$ is not a set for some $x \in M$.

REMARK 3. The assumption that $j: V \to M$ is elementary is the schema $\varphi(x_1, \ldots, x_n) \Leftrightarrow \varphi^M(jx_1, \ldots, jx_n)$. On the other hand the class Ω of ordinals is defined by some fixed formula (which depends on the schema e), and the claims (i)–(iii) concerning Ω amount to a conjunction of finitely many sentences. Consequently, these claims follow from the assumption that j is Δ-elementary for some finite set Δ; the proof shows how Δ can be chosen. Furthermore, to establish (i)–(iii) only the satisfaction in V of finitely many axioms of ZF is needed. The iterated $O_e^I V$ is defined, for finite I's, by I.(21) and then it is extended to arbitrary I's using the direct limit construction of Definition I.2. We use I.(21) to show that this satisfies the recursion conditions of the iteration. Again, all this is expressed by finitely many sentences and finitely many axioms of ZF are needed. The only claim which amounts to an infinite schema is the claim that $O_e^I V \to O_e^J V$ is elementary. If we drop it, the theorem can be applied within any structure $\mathfrak{A}$ satisfying certain finitely many axioms of ZF; hence

THEOREM. *There is a finite set Γ of axioms of ZF and a finite set of formulas Δ such that, whenever $\mathfrak{A} \models \Gamma$ and $\mathfrak{A} \to O_e\mathfrak{A}$ is Δ-elementary, a good definition of the iteration $O_e^I\mathfrak{A}$ and the embeddings $O_e^I\mathfrak{A} \to O_e^J\mathfrak{A}$ can be given inside $\mathfrak{A}$. Here I, J* *range over the ordered classes of $\mathfrak{A}$. Since all finite ordered sets (or isomorphic replicas) will be in $\mathfrak{A}$, we get a definition of $O_e^I\mathfrak{A} \to O_e^J\mathfrak{A}$ for all finite I, J and we can extend it to all I, J (not necessarily in $\mathfrak{A}$) by the usual direct limit construction. If, moreover, $\mathfrak{A} \models$ ZF and $\mathfrak{A} \to O_e\mathfrak{A}$ is elementary all the embeddings $O_e^I\mathfrak{A} \to O_e^J\mathfrak{A}$ will be elementary.*

(Note that this implies that $\mathfrak{A} \in \boldsymbol{C}_{e,*}$, whenever $\mathfrak{A}$ is a set such that $\mathfrak{A} \models \Gamma$ and $\mathfrak{A} \to O_e\mathfrak{A}$ is Δ-elementary, with Γ and Δ finite.)

REMARK 4. If the set theory is such that the whole of V can be well-ordered then one can define a self-extension $j: V \to M$ such that, for every limit ordinal γ, j does not embed γ cofinally in $j\gamma$. This is accomplished by successive ultrapowers, using in the process, for each infinite cardinal κ, an ultrafilter over κ which contains no set of cardinality $<\kappa$. It can be shown that in such a case there is no Ω satisfying (i)–(iii). Yet, in this example $\boldsymbol{O}_e$ can be iterated on V. It is the author's conjecture that, in any set theory (extending ZF), any elementary self-extension of V can be iterated on V. Once the iteration is defined, one can show that it gives rise to elementary embeddings by the following theorem:

THEOREM. *If $f_{e,V}: V \to O_eV$ is elementary and $\boldsymbol{O}_e$ can be iterated on V then, for all ordered I and J, $O_e^IV \to O_e^JV$ is elementary (i.e., we have the schema: $O_e^I \to O_e^JV$ is φ-elementary).*

(The proof uses Theorems 1, 2, 3. Since "elementary" is equivalent to "cofinal and Ψ-elementary" and Ψ is finite, it is enough to prove by induction on J that, for finite J, $O_e^IV \to O_e^JV$ is cofinal and Ψ-elementary, and this can be done.)

PROOF OF THEOREM 4. If $\lambda^* = \operatorname{cof}(\lambda)$ then it is not difficult to show that j embeds λ^* cofinally into $j\lambda^*$. Hence without loss of generality assume that λ is regular.

Let Ψ be as in Theorem 2.

The assertion that $\mathfrak{A} \to O_e\mathfrak{A}$ is cofinal can be expressed by a first-order sentence in $\mathfrak{A}$. Since Ψ is finite, the same is true of the assertion that $\mathfrak{A} \to O_e\mathfrak{A}$ is Ψ-elementary. Let Φ be the finite set consisting of all subformulas of these two sentences and all the formulas of Γ_e (i.e., those which figure in the extension schema). Put

$$\Omega = \{\alpha : V_\alpha \prec_\Phi V, \alpha > \lambda \text{ and } \operatorname{cof}(\alpha) = \lambda\}.$$

By the reflection principle, Ω is unbounded and closed under limits of λ-sequences.

For $\alpha \in \Omega$, put $j_\alpha = f_{e,V_\alpha}: V_\alpha \to O_eV_\alpha$. Since j_α is Ψ-elementary, Theorem 2 implies that it is Σ_0-elementary. We also have $O_eV_\alpha \subset O_eV$ and, for every $x \in V_\alpha$, $j_\alpha(x) = j(x)$. It is not difficult to deduce from this that j_α embeds λ cofinally in $j_\alpha(\lambda)$. By the corollary to Theorem 1 it follows that j_α is an elementary embedding for all $\alpha \in \Omega$.

The main point is to show that the inclusions $V_\alpha \subset V_\beta$ are in $\boldsymbol{C}_{e,*}$ (where $\alpha \leqq \beta$, $\alpha, \beta \in \Omega$); that is to say, for all n, $O_e^nV_\alpha \subset O_e^nV_\beta$ and the inclusion is in $\boldsymbol{C}_e$, i.e., Γ_e-elementary. It is enough to prove that for all φ, $V_\alpha \prec_\varphi V_\beta \Rightarrow O_eV_\alpha \prec_\varphi O_eV_\beta$, for then one shows, by induction on n, that $O_e^nV_\alpha \subset O_e^nV_\beta$ and, for all φ, $V_\alpha \prec_\varphi V_\beta$

$\Rightarrow O_e^n V_\alpha \prec_\varphi O_e^n V_\beta$. Assume it true for n, then, since $V_\alpha \prec_{\Gamma_e} V_\beta$, we have $O_e^n V_\alpha \prec_{\Gamma_e} O_e^n V_\beta$; hence $O_e^{n+1} V_\alpha \subset O_e^{n+1} V_\beta$. By I.(9), whenever $\mathfrak{A} \subset \mathfrak{B}$ is in C_e, $O_e \mathfrak{A} \prec_\varphi O_e \mathfrak{B} \Leftrightarrow \mathfrak{A} \prec_{\varphi**} \mathfrak{B}$. Assume $V_\alpha \prec_\varphi V_\beta$, then $O_e V_\alpha \prec_\varphi O_e V_\beta$, hence $V_\alpha \prec_{\varphi**} V_\beta$, implying, by the induction hypothesis, $O_e^n V_\alpha \prec_{\varphi**} O_e^n V_\beta$. This by I.(9) implies $O_e^{n+1} V_\alpha \prec_\varphi O_e^{n+1} V_\beta$.

To prove $V_\alpha \prec_\varphi V_\beta \Rightarrow O_e V_\alpha \prec_\varphi O_e V_\beta$ we proceed as follows:

For $x \in M$ put $(x) = \{y : M \models y \in x\}$, and let also (x) denote the submodel of M whose universe is (x). Using the formula which in M defines satisfaction in a structure, we get a definition of satisfaction for the substructures (x), $x \in M$; that is to say, we can use '$(x) \models \varphi[y_1, \ldots, y_k]$' with x varying over M, φ over the formulas of the language and $y_1, \ldots, y_k$ over (x). (Note that (x) might be a proper class.)

For every $\alpha \in \Omega$ let $(j_\alpha): V_\alpha \to (j(R\alpha))$ be the embedding such that $(j_\alpha)x = jx$ for all $x \in R\alpha$. For every $\varphi(v)$ we get

$$V_\alpha \models \varphi(x) \Leftrightarrow V \models \text{`}R\alpha \models \varphi(x)\text{'} \Leftrightarrow M \models \text{`}jR\alpha \models (j\varphi)(jx)\text{'} \Leftrightarrow (jR\alpha) \models \varphi(jx)$$

and similarly if $\varphi = \varphi(v_1, \ldots, v_k)$. Consequently, (j_α) is elementary.

If $y \in |O_e V_\alpha|$ there is $x \in R\alpha$ such that $O_e V_\alpha \models y \in jx$ (by the cofinality of j_α) and there is $\gamma < \alpha$ such that $x \subset R\gamma$, implying $O_e V_\alpha \models y \in j(R\gamma)$. Hence, $M \models y \in j(R\gamma)$, implying $M \models y \in j(R\alpha)$. This shows that every member of $|O_e V_\alpha|$ is in $j(R\alpha)$; thus, $O_e V_\alpha$ is a substructure of $(j(R\alpha))$ and $j'' V_\alpha$ is an elementary submodel of both.

Claim. $O_e V_\alpha \prec (j(R\alpha))$.

Take, for example, a formula $\varphi(v)$ with one free variable (the argument is the same for n free variables) and let $x \in |O_e V_\alpha|$. Take $\gamma < \alpha$ such that $M \models x \in j(R\gamma)$. Let $z = \{x \in R\gamma : V_\alpha \models \varphi(x)\}$; then $V_\alpha \models \forall v \in R\gamma[\varphi(v) \leftrightarrow v \in z]$. Since j embeds V_α elementarily in $O_e V_\alpha$ as well as in $(j(R\alpha))$, the same holds in either structure with $R\gamma$ and z replaced by $j(R\gamma)$ and jz. Since in both structures '$x \in j(R\gamma)$' is true, we get

$$O_e V_\alpha \models \varphi(x) \leftrightarrow x \in jz \quad \text{and} \quad (j(R\alpha)) \models \varphi(x) \leftrightarrow x \in jz.$$

Since $O_e V_\alpha \subset (j(R\alpha))$, $O_e V_\alpha \models x \in jz \Leftrightarrow (j(R\alpha)) \models x \in jz$; hence $O_e V_\alpha \models \varphi(x) \Leftrightarrow (j(R\alpha)) \models \varphi(x)$. This proves the claim.

Now assume $V_\alpha \prec_\varphi V_\beta$ where $\alpha, \beta \in \Omega$. We have

$$V_\beta \models \forall x_1, \ldots, x_n \in R\alpha[V_\alpha \models \varphi(x_1, \ldots, x_n) \leftrightarrow \varphi(x_1, \ldots, x_n)].$$

Since (j_β) is elementary, the same holds in $(j(R\beta))$, with $R\alpha$, V_α replaced, respectively, by $j(R\alpha)$ and $(j(R\alpha))$. Hence, $(j(R\alpha)) \prec_\varphi (j(R\beta))$. By the claim, $O_e V_\alpha \prec (j(R\alpha))$ and $O_e V_\beta \prec (j(R\beta))$ from which it follows easily that $O_e V_\alpha \prec_\varphi O_e V_\beta$.

Q.E.D.

THEOREM 5 (WELL-FOUNDEDNESS THEOREM). *If $j: V \to M$ is an elementary self-extension of V, with the defining schema e $(M = O_e V)$ and M is well-founded, then, for every well-ordered class I, $O_e^I V$ is well-founded.*

The proof uses these two observations:

(A) If $O_e^\alpha\mathfrak{A}$ is well-founded for all countable α's, then $O_e^I\mathfrak{A}$ is well-founded for every well-ordered class I. This is so because every member of $|O_e^I\mathfrak{A}|$ is contributed by some $O_e^{I'}\mathfrak{A}$, $I' \subset I$, where I' is finite. If ω members of $O_e^I\mathfrak{A}$ make a descending chain with respect to the $\in$ of $O_e^I\mathfrak{A}$, then this is also true for some $O_e^{I^*}\mathfrak{A}$, with I^* a countable subset of I.

(B) If $\mathfrak{A}$ is a well-founded model of ZF whose ordinals are not cofinal with ω, then $\mathfrak{A} \models$ 'X is a well-ordering' $\Rightarrow X$ is a well-ordering, where X is any class of $\mathfrak{A}$. (To be precise one should replace in the consequent "X" by "$\{\langle x, y\rangle : \mathfrak{A} \models$ '$\langle x, y\rangle \in X$'$\}$.") For if X has a descending ω-chain, there will be an ordinal α of $\mathfrak{A}$ such that $\mathfrak{A} \models$ '$x \in R\alpha$' for all x in this chain. But, inside $\mathfrak{A}$, $X \cap R\alpha$, is isomorphic to some ordinal, which is a contradiction.

Sketch of the Proof. Put $V^{(\alpha)} = O_e^\alpha V$, let $\in^{(\alpha)}$ be the $\in$-relation of $V^{(\alpha)}$ and let $j_{\alpha,\beta} = f_{e,\alpha,\beta}: O_e^\alpha V \to O_e^\beta V$, for $\alpha \leqq \beta$. In particular, $j_{0,\alpha}$ is an elementary cofinal embedding (by Theorem 4) of V into $V^{(\alpha)}$. Hence, the ordinals of $V^{(\alpha)}$ are not cofinal with ω. Now, $V \models$ 'the structure defined by the schema e is well-founded', hence the same sentence holds in $V^{(\alpha)}$. The application of e in $V^{(\alpha)}$ yields $O_e V^{(\alpha)}$ which is $V^{(\alpha+1)}$ (or more precisely, naturally isomorphic to it—see I.(15)). By (B), if $V^{(\alpha)}$ is well-founded, then the ordinals of $V^{(\alpha+1)}$ are well-ordered, implying that $V^{(\alpha+1)}$ is also well-founded.

Hence, if γ is the first ordinal such that $V^{(\gamma)}$ is not well-founded, then $\gamma = \bigcup\gamma > 0$; hence $V^{(\gamma)} = \text{dir lim}\,\{V^{(\alpha)}: \alpha < \gamma\}$. We get a contradiction as follows. The ordinals of $V^{(\gamma)}$ are not well-ordered and, since $j_{0,\gamma}$ is cofinal, there is a first ordinal ξ such that $\{\lambda : V^{(\gamma)} \models \lambda \in j_{0,\gamma}\xi\}$ is not well-ordered by $\in^{(\gamma)}$. We get a descending chain $\ldots \in^{(\gamma)} \eta_2 \in^{(\gamma)} \eta_1 \in^{(\gamma)} j_{0,\gamma}\xi$. Now, η_1 is contributed by some V^β, $\beta < \gamma$, so let $\eta_1 = j_{\beta,\gamma}\eta$, where η is an ordinal in $V^{(\beta)}$. Let $\bar{\xi} = j_{0,\beta}\xi$; then $j_{\beta,\gamma}\bar{\xi} = j_{0,\gamma}\xi$; hence, $\eta \in^{(\beta)} \bar{\xi}$. Let $\bar{\gamma} = j_{0,\beta}\gamma$.

The following sentence is true in V:

$$\forall\alpha \leqq \gamma \forall\zeta < \xi[\{\lambda : \lambda \in^{(\alpha)} j_{0,\alpha}\zeta\} \text{ is well-ordered by } \in^{(\alpha)}].$$

Consequently the same sentence holds in $V^{(\beta)}$ with ξ and γ replaced, respectively, by $\bar{\xi}$ and $\bar{\gamma}$.

$\{\alpha : V^{(\beta)} \models \alpha \leqq \bar{\gamma}\}$ is well-ordered by $\in^{(\beta)}$ and $j_{0,\beta}$ embeds $\gamma + 1$ into it. Let γ' be such that $\beta + \gamma' = \gamma$. There is an ordinal γ^* of $V^{(\beta)}$ such that $V^{(\beta)} \models \gamma^* \leqq \bar{\gamma}$ and $\{\alpha : \alpha \in^{(\beta)} \gamma^*\}$, ordered by $\in^{(\beta)}$, is isomorphic to γ' (ordered by $\in$). Since $V^{(\beta)} \models \eta < \bar{\xi}$, we have

$$V^{(\beta)} \models \{\lambda : \lambda \in^{(\gamma^*)} j_{0,\gamma^*}\eta\} \text{ is well-ordered by } \in^{(\gamma^*)}.$$

Read in greater detail the sentence says: If $\boldsymbol{O}_e$ is iterated γ^* times then in the resulting universe the $\in$ relation well-orders the class of all elements which are members (in that universe) of the image of the ordinal η.

By the absoluteness of the definitions which determine $\boldsymbol{O}_e$ and its iterations, it follows that if we apply, inside $V^{(\beta)}$, the γ^*th iteration of $\boldsymbol{O}_e$ to the whole universe (i.e., to $V^{(\beta)}$) we will get a structure naturally isomorphic to $O_e^{\gamma'}(O_0^{(\beta)}V)$, which, by

I.(14), is naturally isomorphic to $V^{(\beta+\gamma')}$. The image of η under the embedding $f_{0,\gamma*}$, as defined inside $V^{(\beta)}$, will correspond by the natural isomorphism to $j_{\beta,\beta+\gamma'}(\eta)$, that is to say to η_1. Using (B) again we find that $\{\lambda : O_e^{(\gamma)} V : \lambda \in \eta_1\}$ is well-ordered by $\in^{(\gamma)}$ which is a contradiction.

REMARK 1. The theorem can be applied inside every model $\mathfrak{A}$ of ZF such that $\mathfrak{A} \models$ '$O_e \mathfrak{A}$ is well-founded $\wedge$ $\mathfrak{A} \to O_e \mathfrak{A}$ is elementary'. If $\mathfrak{A}$ is well-founded and its ordinals are not cofinal with ω, then, by (B), it follows that $O_e^I \mathfrak{A}$ is indeed well-founded for every well-ordered I which is isomorphic to an ordinal of $\mathfrak{A}$. By (A), $O_e^I \mathfrak{A}$ is well-founded for all well-ordered I.

An inspection of the proof, taken in conjunction with Remark 3 to Theorem 4, reveals that in order to derive inside $\mathfrak{A}$ the statement 'I is well-ordered $\to O_e^I \mathfrak{A}$ is well-founded', it is enough to have $\mathfrak{A} \models \Gamma$ and $\mathfrak{A} \to O_e \mathfrak{A}$ is Δ-elementary, where ZF $\models \Gamma$ and Γ and Δ are finite. Γ and Δ depend on the self-extension schema e (but not on the parameter I). Since Γ and Δ do not arise naturally in situations where one wishes to apply Theorems 4 and 5 inside $\mathfrak{A}$, this observation is however of little importance.

REMARK 2. If

(S*) $$\mathfrak{A} \models \forall x[\{y : O_e \mathfrak{A} \models y \in x\} \text{ is a set}]$$

where $\mathfrak{A} \models$ ZF and $\mathfrak{A} \models$ '$O_e \mathfrak{A}$ is well-founded', then inside $\mathfrak{A}$ one can define an isomorphism of the ordinals of $O_e \mathfrak{A}$ onto those of $\mathfrak{A}$. If $\mathfrak{A} \to O_e \mathfrak{A}$ is elementary then, inside $\mathfrak{A}$, one can prove (using Theorem 5)

$$\forall \alpha \,[\text{the ordinals of } O_e^{\alpha} \mathfrak{A} \text{ are order isomorphic to OR}].$$

If $\mathfrak{A}$ is well-founded, then $O_e^I \mathfrak{A}$ is well-founded for every I which is order isomorphic to an ordinal of $\mathfrak{A}$. If, moreover, $\mathfrak{A}$ has uncountably many ordinals, then, by (A), $O_e^I \mathfrak{A}$ is well-founded for all well-ordered I.

All the elementary self-extensions of V which arise in the study of large cardinals satisfy (S*).

The only known method of getting self-extensions (which are elementary, or Φ-elementary, for some nontrivial Φ) which do not satisfy (S*) is by iterating many times the ones that do, where "many" means a proper class. For a well-founded elementary self-extension $V \to O_e V$ which is not an isomorphism, one can, using I.(18)B and I.(20), embed I into a set of indiscernible ordinals of $O_e^I V$. It will be shown that this embedding is order-preserving and is into a proper initial segment of these ordinals. Thus, if I is a well-ordered proper class, $O_e^I V$ is a very "long" universe—i.e., its ordinals are well-ordered and have a larger order type than the real ordinals. Choosing "longer" I's yields "longer" universes.

By assuming (S*) one can considerably simplify the proof of Theorem 4. Although all the natural known examples satisfy (S*) the author preferred to present a more general theorem.

PART III

General properties of $O^I \mathfrak{A}$. In this section the assumptions are that $\mathfrak{A} \models Z^+$ (where, as before, $Z^+ = Z \cup \{\forall x \exists \alpha \,[R\alpha \text{ exists} \wedge x \in R\alpha]\}$) and that

$f_{\mathfrak{A}}:\mathfrak{A} \to O\mathfrak{A}$ is an elementary self-extension of $\mathfrak{A}$. If $\mathfrak{A}$ is a set, then $\boldsymbol{O}$ can be iterated on $\mathfrak{A}$ and, for all ordered I, J, such that $I \subset J, f_{I,J,\mathfrak{A}}:O^I\mathfrak{A} \to O^J\mathfrak{A}$ is elementary. This is also true if $\mathfrak{A}$ is a proper class provided that $\mathfrak{A} \models$ ZF and $f_{\mathfrak{A}}$ maps the natural numbers of $\mathfrak{A}$ onto those of $O\mathfrak{A}$ (i.e., ω is not moved by $f_{\mathfrak{A}}$; actually, a weaker requirement is sufficient—see Theorem II.4). For cases not covered by these two, let the assumption that $\boldsymbol{O}$ can be iterated on $\mathfrak{A}$ and that $O^I\mathfrak{A} \to O^J\mathfrak{A}$ is elementary be an additional assumption.

The theorems and corollaries remain true if $\mathfrak{A} \to O\mathfrak{A}$ is the ultrapower extension based on a $\mathfrak{A}$-ultrafilter as defined by Kunen [**11**], but the proofs are different.

We put $\mathfrak{A}^{(I)} = O^I\mathfrak{A}$, $x^{(I)} = f_{\varnothing,I}(x)$ (where $x \in |\mathfrak{A}|$) and if X is a class in $\mathfrak{A}$ then $X^{(I)}$ is the corresponding class in $O^I\mathfrak{A}$. (*Note.* $y \in X^{(I)} \Leftrightarrow \exists x[\mathfrak{A} \models x \subset X$ and $\mathfrak{A}^{(I)} \models y \in x^{(I)}]$, because $\mathfrak{A} \to O\mathfrak{A}$ must be cofinal.)

If $x \in |\mathfrak{A}^{(I)}|$ and $J \subset I$ then we shall say that x *is contributed* by J if it is contributed by $\mathfrak{A}^{(J)}$ (i.e., $x \in f''_{J,I}(|\mathfrak{A}^{(J)}|)$).

Unless otherwise stated, $f_{J,I} = f_{J,I,\mathfrak{A}}$ and $\mathfrak{A}^{(J)} \to \mathfrak{A}^{(I)} = f_{J,I}:\mathfrak{A}^{(J)} \to \mathfrak{A}^{(I)}$.

The easiest way to visualize the situation is for the case where $\mathfrak{A} = V, OV$ is a transitive class and I is a well-ordered set, in which case $\mathfrak{A}^{(I)}$ can be identified with a transitive class. The theorems and proofs are, however, formulated without any of these assumptions. The only price which one pays here for generality is the minor inconvenience in formulation caused by the fact that the $\in$ relation of the structures in question is not the real $\in$. Later, when this becomes too bothersome, we restrict ourselves to the special case and leave the generalizations to the reader.

We know that $\mathfrak{A} \to O\mathfrak{A}$ is cofinal (Theorem II.3). Moreover, every $\mathfrak{A}^{(J)} \to \mathfrak{A}^{(I)}$ is elementary and cofinal (see Remark 4 to Theorem II.4). Given any ordinal θ of $\mathfrak{A}^{(I)}$ there is a smallest ordinal α of $\mathfrak{A}$ such that $\mathfrak{A}^{(I)} \models \theta \leqq \alpha^{(I)}$. This is obvious if $\mathfrak{A}$ is well-founded. It is true in general because θ is contributed by some finite $J \subset I$, say $\theta = f_{J,I}(\theta')$; since $\mathfrak{A} \to \mathfrak{A}^{(J)}$ is definable inside $\mathfrak{A}$, we can establish inside $\mathfrak{A}$ the existence of a minimal α for which $\mathfrak{A}^{(J)} \models \theta' \leqq \alpha^{(J)}$, and, using the equality $\mathfrak{A} \to \mathfrak{A}^{(I)} = \mathfrak{A} \to \mathfrak{A}^{(J)} \to \mathfrak{A}^{(I)}$, this implies that α has the same property with respect to θ and $\mathfrak{A}^{(I)}$.

From now on we use "$\leqq$" to denote also the ordering of the ordinals in the models under consideration; e.g., "$\beta^{(I)} \leqq \theta$" means that $\mathfrak{A}^{(I)} \models \beta^{(I)} \leqq \theta$.

THEOREM 1. *If, for every ordinal β of $\mathfrak{A}$, $\beta^{(I)} \leqq \theta_1 \Rightarrow \beta^{(I)} \leqq \theta_2$, and if θ_1, θ_2 are contributed, respectively, by I_1, I_2 and every member of I_1 is $<$ (in the ordering of I) every member of I_2, then $\theta_1 \leqq \theta_2$.*

PROOF. Since θ_k is contributed by some finite subset of $I_k, k = 1, 2$, we can assume with no loss of generality that I_1 and I_2 are finite.

It is enough to prove the claim for the case $I = I_1 + I_2$, for the general case will follow from this using the well-known facts about the embeddings $\mathfrak{A}^{(J)} \to \mathfrak{A}^{(J')}$, $J' \subset J$ (e.g., $\mathfrak{A}^{(I_k)} \to \mathfrak{A}^{(I)} = \mathfrak{A}^{(I_k)} \to \mathfrak{A}^{(I_1+I_2)} \to \mathfrak{A}^{(I)}, k = 1, 2$, and $\mathfrak{A} \to \mathfrak{A}^{(I)} = \mathfrak{A} \to \mathfrak{A}^{(I_1+I_2)} \to \mathfrak{A}^{(I)}$).

Let α be the smallest ordinal in $\mathfrak{A}$ such that $\theta_1 < \alpha^{(I)}$. Then for every ordinal β in $\mathfrak{A}$ such that $\beta < \alpha$ we have $\beta^{(I)} \leqq \theta_1$ and, therefore, by our assumption $\beta^{(I)} \leqq \theta_2$.

Let θ_2^* be the ordinal in $\mathfrak{A}^{(I_2)}$ such that $\theta_2 = f_{I_2,I}(\theta_2^*)$. It follows that, for every ordinal β in $\mathfrak{A}$ which is (in $\mathfrak{A}$) $< \alpha$, we have in $\mathfrak{A}^{(I_2)}: \beta^{(I_2)} \leqq \theta_2^*$. This fact is expressible as a statement inside $\mathfrak{A}$ because $\mathfrak{A} \to \mathfrak{A}^{(I_2)}$ is definable in $\mathfrak{A}$. Thus, the following holds in $\mathfrak{A}$:

For every $\beta < \alpha$ the image of β in the universe obtained by applying $\boldsymbol{O}^{I_2}$ to the whole universe is (in that universe) $\leqq \theta_2^*$.

Now employ the elementary embedding $\mathfrak{A} \to \mathfrak{A}^{(I_1)}$. The same statement holds in $\mathfrak{A}^{(I_1)}$, with α replaced by $\alpha^{(I_1)}$, θ_2^* by $\theta_2^{*(I_1)}$ and "the whole universe" referring now to $\mathfrak{A}^{(I_1)}$. It is obviously absolute. Consequently, for every β such that $\mathfrak{A}^{(I_1)} \models \beta < \alpha^{(I_1)}$ we have, in $O^{I_2}(\mathfrak{A}^{(I_1)})$,

$$f_{\varnothing,I_2,\mathfrak{A}^{(I_1)}}(\beta) \leqq (\theta_2^*)^{(I_1)}.$$

Recall that $O^{I_2}(\mathfrak{A}^{(I_1)}) = O^{I_2}(O^{I_1}\mathfrak{A}) = O^{I_1+I_2}\mathfrak{A}$ and that we have the following commutative diagram:

$$\begin{array}{ccc} O^{I_2}\mathfrak{A} & \xrightarrow{f_{I_2,I_1+I_2}} & O^{I_2}(O^{I_1}\mathfrak{A}) = O^{I_1+I_2}\mathfrak{A} \\ \uparrow & & \uparrow \\ \mathfrak{A} & \xrightarrow[f_{\varnothing,I_1}]{} & O^{I_1}\mathfrak{A} \end{array}$$

and that $f_{I_2,I_1+I_2} = O^{I_2}(f_{\varnothing,I_1})$, i.e., $f_{I_2,I_1+I_2}(x) = f_{\varnothing,I_1}(x)$ for all $x \in |O^{I_2}\mathfrak{A}|$ (see I.(13)).

Consequently $(\theta_2^*)^{(I_1)} = f_{\varnothing,I_1}(\theta_2^*) = f_{I_2,I_1+I_2}(\theta_2^*) = \theta_2$ and we have the following: For all β, if $\mathfrak{A}^{(I_1)} \models \beta < \alpha^{(I_1)}$ then in $\mathfrak{A}^{(I)}$ we have $f_{I_1,I}(\beta) \leqq \theta_2$.

Let θ_1^* be such that $\theta_1 = f_{I_1,I}(\theta_1^*)$. Since $\mathfrak{A}^{(I)} \models \theta_1^* < \alpha^{(I)}$ we have $\mathfrak{A}^{(I_1)} \models \theta_1^* < \alpha^{(I_1)}$; hence (in $\mathfrak{A}^{(I)}$) $f_{I_1,I}(\theta_1^*) < \theta_2$. Q.E.D.

In the following two corollaries we put $\mathfrak{A}^1 = O\mathfrak{A}$, $x^1 = f_{\mathfrak{A}}(x)$ (for all $x \in |\mathfrak{A}|$). Given any $x \in |\mathfrak{A}^1|$ we put $x_i = x_i(I) = f_{\{i\},I}(x)$.

COROLLARY 1. *If θ is any ordinal in $\mathfrak{A}^1$ not contributed by $\varnothing$ (i.e., by $\mathfrak{A}$) then $i < i' \Rightarrow \theta_i < \theta_{i'}$ ("$<$" on the left-hand side denotes the ordering of I and on the right-hand side the ordering of the ordinals of $\mathfrak{A}^{(I)}$).*

PROOF. Let $\alpha =$ smallest ordinal of $\mathfrak{A}$ such that $\theta < \alpha^1$. Then $\beta^1 < \theta < \alpha^1$ for all $\beta < \alpha$. Employing $f_{\{i\},I}$ and the fact that $f_{\{i\},I}(x^1) = x^{(I)}$ we get $\mathfrak{A}^{(I)} \models \beta^{(I)} < \theta_i < \alpha^{(I)}$, for all $\beta < \alpha$. Consequently for every $i, i' \in I$, $\beta^{(I)} \leqq \theta_i \Leftrightarrow \beta^{(I)} \leqq \theta_{i'}$. Now use the theorem and the fact that $i \neq i' \Rightarrow \theta_i \neq \theta_{i'}$.

(It is a general fact that if we have some definable ordered subclass, $X = (X, <)$, of $|\mathfrak{A}|$ and if we choose $x \in |O\mathfrak{A}| - f''(|\mathfrak{A}|)$ such that $x \in X^1$ then the embedding $i \to x_i(I)$ will embed I into $X^{(I)}$ and will be either order preserving ($i < i' \Rightarrow x_i <^{(I)} x_{i'}$) or ordering reversing. This holds for arbitrary structures and can be generalized to natural extension operators which are not self-extensions. If there are predicates in the language which denote $(X, <)$ then one need not even assume that $\mathfrak{A} \to O\mathfrak{A}$ is elementary. The claim follows from I.(18). Corollary 1 is trivial if one assumes $\mathfrak{A} \models$ ZF and $\mathfrak{A} \models$ '$O\,\mathfrak{A}$ is well-founded', for then, if $i < i'$ and $\theta_i > \theta_{i'}$, one can argue inside $\mathfrak{A}$ that $\omega \to O^{\omega}\mathfrak{A}$ maps ω into a descending chain of ordinals of $O^{\omega}\mathfrak{A}$, contradicting the well-foundedness theorem.)

COROLLARY 2. *If, for a given* $\alpha, \mathfrak{A} \models \beta < \alpha \Rightarrow \mathfrak{A}^1 \models \beta^1 < \theta < \alpha^1$, *for all ordinals* β *of* $\mathfrak{A}$, *and if* I *has no last element then* $\alpha^{(I)} = \sup\{\theta_i : i \in I\}$, *where the supremum is with respect to the ordering of the ordinals in* $\mathfrak{A}^{(I)}$.

PROOF. Since, in $\mathfrak{A}^1, \theta < \alpha^1$, it is obvious that, in $\mathfrak{A}^{(I)}, \theta_i < \alpha^{(I)}$. If $\xi < \alpha^{(I)}$, then choose a finite subset $J \subset I$ such that ξ is contributed by J and take i such that $i' < i$ for all $i' \in J$. By the theorem, $\xi < \theta_i$.

DEFINITION 1. Let $j = j : \mathfrak{A} \to \mathfrak{B}$ be an elementary embedding. An ordinal α of $\mathfrak{A}$ is *critical* (for j) if there is an ordinal θ of $\mathfrak{B}$ such that $\mathfrak{B} \models j\beta < \theta < \alpha$ for all $\beta < \alpha$ (in $\mathfrak{A}$).

If $j : \mathfrak{A} \to \mathfrak{B}$ is definable in $\mathfrak{A}$ then α is *moved* (by j) if $\mathfrak{A} \models$ 'α, ordered by $\in$, is not isomorphic to $\{\beta : \mathfrak{B} \models \beta \in j\alpha\}$, ordered by $\in^{\mathfrak{B}}$'. This concept is also defined in the case where j is not definable in $\mathfrak{A}$, provided that $\{\beta : \mathfrak{A} \models \beta \in \alpha\}$, ordered by $\in^{\mathfrak{A}}$, is isomorphic to some ordinal. In that case the well-known definition is: α is moved if $\{\beta : \mathfrak{A} \models \beta \in \alpha\}$, ordered by $\in^{\mathfrak{A}}$, is not isomorphic to $\{\beta : \mathfrak{B} \models \beta \in j\alpha\}$, ordered by $\in^{\mathfrak{B}}$. (If j is not definable in $\mathfrak{A}$ and $\mathfrak{A}$ is not well-founded there seems to be no way to generalize this concept, for the ordered classes in question can be isomorphic for the wrong reasons.)

The following claims are obvious or almost obvious. (We assume that the concepts are meaningful.)

(1) *Every critical ordinal of* $\mathfrak{A}$ *is moved by* j.

(2) *The first critical ordinal of* $\mathfrak{A}$ *is the first ordinal moved by* j.

(3) *If* $\mathfrak{A} \models$ '$\mathrm{cof}(\alpha) = \mathrm{cof}(\beta)$' *then* α *is critical* $\Leftrightarrow \beta$ *is critical* ($\mathrm{cof}(\alpha) =$ *cofinality of* α; $\mathrm{cof}(\alpha + 1) = 1$).

Assuming that $j : \mathfrak{A} \to \mathfrak{B}$ is a self-extension we also have the following:

(4) *If every* $\beta \leqq \alpha$ *is unmoved by* j *then* j *maps* $\mathfrak{A}_\alpha$ *isomorphically onto* $\mathfrak{B}_{j\alpha}$. (The proof is by induction on α. For non-self-extensions this need not be true in general.)

DEFINITION 2. *An initial segment* of $\mathfrak{A}$ is any substructure whose universe is of the form $\bigcup_{\alpha \in \Xi} \{x : \mathfrak{A} \models x \equiv R\alpha\}$. An initial segment of $\mathfrak{A}$ is *proper* if it is $\neq \mathfrak{A}$. If $\mathfrak{A}'$ is a proper initial segment of $\mathfrak{A}$ then $\mathfrak{A}$ is said to be an *end extension* of $\mathfrak{A}'$. If $B \subset |\mathfrak{A}|$, then by an initial segment of B (with respect to $\mathfrak{A}$) we understand any subclass B' of the form $|\mathfrak{A}'| \cap B$, where $\mathfrak{A}'$ is an initial segment of $\mathfrak{A}$. B' is a proper initial segment of B if it is $\neq B$.

THEOREM 2. *Let* μ *be the first ordinal moved by* $\mathfrak{A} \to O\mathfrak{A}$. *If* I *is a (proper) initial segment of* J *then* $f_{I,J} : \mathfrak{A}^{(I)} \to \mathfrak{A}^{(J)}$ *maps* $(\mathfrak{A}^{(I)})_{\mu^{(I)}}$ *onto a (proper) initial segment and an elementary submodel of* $(\mathfrak{A}^{(J)})_{\mu^{(J)}}$.

PROOF. Since $f_{I,J}$ is elementary and $f_{I,J}(\mu^{(I)}) = \mu^{(J)}$, it maps $(\mathfrak{A}^{(I)})_{\mu^{(I)}}$ onto an elementary submodel of $(\mathfrak{A}^{(J)})_{\mu^{(J)}}$. It remains to show that if I is a proper initial segment of J, then (i) if $\mathfrak{A}^{(J)} \models$ '$\mathrm{rank}(x) \leqq \mathrm{rank}(y) < \mu^{(J)}$' and y is contributed by I then x is contributed by I, and (ii) there are in $\mathfrak{A}^{(J)}$ ordinals $< \mu^{(J)}$ not contributed by I. Since every member of $|\mathfrak{A}^{(J)}|$ is contributed by some finite subset of J, and similarly for $|\mathfrak{A}^{(I)}|$, it suffices to prove the claim assuming that J is finite.

For the case $I = 0 = \varnothing, J = 1 = \{0\}$, (i) is true, for μ is a limit ordinal in $\mathfrak{A}$ and, by (4), $f_{\mathfrak{A}}: \mathfrak{A} \to O\mathfrak{A}$ maps every $(\mathfrak{A})_\beta$, where $\mathfrak{A} \models \beta < \mu$, isomorphically onto $(O\mathfrak{A})_{\beta^{(1)}}$. Since μ is critical for $\mathfrak{A} \to O\mathfrak{A}$, (ii) is also true. $\mathfrak{A} \models$ 'μ is the first ordinal moved by $\mathfrak{A} \to O\mathfrak{A}$', therefore $\mathfrak{A}^{(n)} \models$ '$\mu^{(n)}$ is the first ordinal moved by $\mathfrak{A}^{(n)} \to O\mathfrak{A}^{(n)}$' (where we put $n = \{0, 1, \ldots, n-1\}$ with the natural ordering). Hence, by the same argument, applied to $\mathfrak{A}^{(n)}$, we have $f_{n,n+1}$ maps $(\mathfrak{A}^{(n)})_{\mu^{(n)}}$ onto a proper initial segment of $(\mathfrak{A}^{(n+1)})_{\mu^{(n+1)}}$. Now $f_{n,n+k}: \mathfrak{A}^{(n)} \to \mathfrak{A}^{(n+k)} = \mathfrak{A}^{(n)} \to \mathfrak{A}^{(n+1)} \to \cdots \to \mathfrak{A}^{(n+k)}$; hence the general claim follows by induction. Q.E.D.

From now on we assume that $f_{\mathfrak{A}}$ is not an isomorphism of $\mathfrak{A}$ onto $O\mathfrak{A}$ and we let μ be the first ordinal moved by it.

If $\mathfrak{A} \models \omega < \mu$ then μ is very large in $\mathfrak{A}$, in fact

THEOREM. $\mathfrak{A} \models$ *'there is a $<\mu$-complete nonprincipal ultrafilter over μ.'*

This is an easy generalization of an idea which goes back to Keisler [**9**] (see also Keisler–Tarski [**10**]). Take any θ such that $O\mathfrak{A} \models \beta^{(1)} < \theta < \mu^{(1)}$, for all β which, in $\mathfrak{A}$, are $<\mu$ (where $x^{(1)} = f_{\mathfrak{A}}(x)$). Working inside $\mathfrak{A}$, put $\mathfrak{F} = \{X \subset \mu : O\mathfrak{A} \models \theta \in X^{(1)}\}$, then $\mathfrak{F}$ is the required ultrafilter in $\mathfrak{A}$.

If $\mathfrak{A} \models$ '$O\mathfrak{A}$ has a μth ordinal', then letting $\bar{\mu}$ be such that $\mathfrak{A} \models$ '$\bar{\mu}$ is the μth ordinal of $O\mathfrak{A}$', it follows easily that the initial segment of $O\mathfrak{A}$ onto which $f_{\mathfrak{A}}$ maps $\mathfrak{A}_\mu$ is $(O\mathfrak{A})_{\bar{\mu}}$. Without loss of generality we can put $\bar{\mu} = \mu$ and assume that $f_{\mathfrak{A}}(x) = x$ for all $x \in |\mathfrak{A}_\mu|$. For if it were otherwise, one could modify in the obvious way the self-extension schema so as to get an isomorphic copy, $O'\mathfrak{A}$, of $O\mathfrak{A}$, with an embedding $f'_{\mathfrak{A}}: \mathfrak{A} \to O'\mathfrak{A}$, having this property, such that $\mathfrak{A} \to O\mathfrak{A} = \mathfrak{A} \to O'\mathfrak{A} \simeq O\mathfrak{A}$.

We shall therefore always assume that, for any $\mathfrak{A}$, if $\mathfrak{A} \models$ '$O\mathfrak{A}$ has a μth ordinal' then μ is this ordinal and, on $\mathfrak{A}_\mu$, $f_{\mathfrak{A}}$ is the identity.

Obviously the μth ordinal of $O\mathfrak{A}$ is also the first ordinal θ in it satisfying: for all β, $\mathfrak{A} \models \beta < \mu \Rightarrow O\mathfrak{A} \models \beta^{(1)} < \theta < \mu^{(1)}$.

Hence, the μth ordinal of $O\mathfrak{A}$ is the first ordinal in it which is not contributed by $\mathfrak{A}$; the existence of each implies that of the other and their equality. If we put $\theta = \mu$ in the definition of the above-mentioned $\mathfrak{F}$ (which, in $\mathfrak{A}$, is an ultrafilter over μ), we get, in $\mathfrak{A}$, a normal ultrafilter.

We also assume that, unless otherwise stated, $\mathfrak{A} \models$ '$O\mathfrak{A}$ has a μth ordinal'.

Since μ is not contributed by $\mathfrak{A}$, we can use it to produce indiscernibles in $\mathfrak{A}^{(I)}$. As before, we put $\mu_i = \mu_i(I) = f_{\{i\},I}(\mu)$. All the interesting results in which the iteration method is used to produce indiscernibles are based on this choice. As was noted, however, in Part I, any member of $O\mathfrak{A}$ not contributed by $\mathfrak{A}$ can be used to produce indiscernibles.

Notation. From now on if $i \in I$ then $I|i = \{k \in I : k < i\}$. We shall always consider $I|i$ as an ordered subclass of I; its ordering is the restriction of the ordering of I.

THEOREM 3. *For every $i \in I$, $f_{I|i,I}$ maps $(\mathfrak{A}^{(I|i)})_{\mu^{(I|i)}}$ isomorphically onto $(\mathfrak{A}^{(I)})_{\mu_i}$.*

REMARK. The story which this theorem tries to tell is very simple. Assume, for simplicity, that $\mathfrak{A}$ is a transitive class and $O\mathfrak{A}$ is well-founded and, in fact, isomorphic to a transitive subclass of $\mathfrak{A}$. The first application of O introduces a block of new ordinals before μ and leaves everything of rank $<\mu$ fixed, so that μ jumps upwards and becomes $\mu^{(1)}$, whereas the "real μ" is now called μ_0. Iterating O once more pushes, in the same manner, $\mu^{(1)}$ to $\mu^{(2)}$, leaves everything below $\mu^{(1)}$ fixed, the former $\mu^{(1)}$ is now called μ_1; and so on. The theorem describes what this "so on" means in the general situation where I is any ordered class, and nothing is assumed on $\mathfrak{A}$ and $O\mathfrak{A}$ besides $\mathfrak{A} \models$ '$O\mathfrak{A}$ has a μth ordinal'.

PROOF OF THEOREM 3. As in other cases which we had, the claim needs establishing only for finite I's. The rest will follow by elementary considerations concerning direct limits (using the routine facts that $\mathfrak{A}^{(I)} = \operatorname{dir\,lim}\{\mathfrak{A}^{(J)}: J \subset I \text{ and } J \text{ finite}\}$, that $\mu_i(I) = f_{J,I}(\mu_i(J))$, etc.).

Now, $f_{\mathfrak{A}}$ maps $\mathfrak{A}_\mu$ isomorphically onto $(O\mathfrak{A})_\mu$ (actually, by our convention, $f_{\mathfrak{A}}(x) = x$, for all $x \in |\mathfrak{A}_\mu|$). Since $\mu^{(n)}$ is the first ordinal moved by $f_{n,n+1}: \mathfrak{A}^{(n)} \to \mathfrak{A}^{(n+1)}$, this embedding maps $(\mathfrak{A}^{(n)})_{\mu^{(n)}}$ onto $(\mathfrak{A}^{(n+1)})_{\mu^{(n)}}$. Recall the commutative diagram

$$\begin{array}{ccc} O\mathfrak{A} & \xrightarrow{f_{\{n\},n+1}} & O(O^n\mathfrak{A}) = O^{n+1}\mathfrak{A} \\ \uparrow & & \uparrow \\ \mathfrak{A} & \xrightarrow[f_{0,n}]{} & O^n\mathfrak{A} \end{array}$$

and the fact that, for all $x \in |O\mathfrak{A}|$, $f_{\{n\},n+1}(x) = f_{0,n}(x)$ (Definition I.1). For $I = n + 1$, we get $\mu_n(n+1) = f_{\{n\},n+1}(\mu) = f_{0,n}(\mu) = \mu^{(n)}$. Consequently $f_{n,n+1}$ maps $(\mathfrak{A}^{(n)})_{\mu^{(n)}}$ onto $(\mathfrak{A}^{(n+1)})_{\mu_n(n+1)}$, and the claim of the theorem is proved if i is the last member of I. Now assume that the claim holds for $I = n$. Let $I = n + 1$, and let $i \in n + 1$ be such that $i < n$. (Since $i = \{0, \ldots, i-1\}$, we have $(n+1)|i = i$.) We have $\mathfrak{A}^{(i)} \to \mathfrak{A}^{(n+1)} = \mathfrak{A}^{(i)} \to \mathfrak{A}^{(n)} \to \mathfrak{A}^{(n+1)}$, and by our assumption $\mathfrak{A}^{(i)} \to \mathfrak{A}^{(n)}$ maps $(\mathfrak{A}^{(i)})_{\mu^{(i)}}$ onto $(\mathfrak{A}^{(n)})_{\mu_i(n)}$. Since $\mathfrak{A}^{(n)} \to \mathfrak{A}^{(n+1)}$ sends $\mu_i(n)$ to $\mu_i(n+1)$ and maps $(\mathfrak{A}^{(n)})_{\mu^{(n)}}$ onto $(\mathfrak{A}^{(n+1)})_{\mu_n(n+1)}$, and since $\mathfrak{A}^{(n)} \models \mu_i(n) < \mu^{(n)}$, our claim for $\mathfrak{A}^{(i)} \to \mathfrak{A}^{(n+1)}$ follows. Q.E.D.

COROLLARY 1. *Any x which, in $\mathfrak{A}^{(I)}$, is of rank $<\mu^{(I)}$ is contributed by $I|i$ iff it is in $\mathfrak{A}^{(I)}$ of rank $<\mu_i$.*

PROOF. If x is in $\mathfrak{A}^{(I)}$ of rank $<\mu^{(I)}$, then x is contributed by $I|i$ iff $x \in f''_{I|i,I}(\mathfrak{A}^{(I|i)})_{\mu^{(I|i)}}$.

COROLLARY 2. *If i is a limit member of I then* $\mu_i = \sup\{\mu_k : k < i\}$.

PROOF. If $\mathfrak{A}^{(I)} \models \theta < \mu_i$, then θ is contributed by $I|i$. Since θ is contributed by some finite subset of $I|i$ and i is a limit member of I, there is $k < i$ such that θ is contributed by $I|k$, and therefore $\theta < \mu_k$.

Keep I fixed and for every $J \subset I$ put $\mathfrak{A}^J = f''_{J,I}(\mathfrak{A}^{(J)})$. Recall that every embedding $q: J \to J'$ induces an elementary embedding $\bar{q}: \mathfrak{A}^J \to \mathfrak{A}^{J'}$, such that the $\bar{q}$'s have the nice properties stated in I.(18). In particular, $\bar{q}(\mu_i) = \mu_{qi}$, and if $x \in |\mathfrak{A}^{J_1}|$ where $J_1 \subset J$ and q keeps fixed every member of J_1, then $\bar{q}(x) = x$. From Corollary 1 we get

COROLLARY 3. *If $J \subset I$ and all $i < i^*$ are in J, and if $q: J \to I$ is order preserving then the domain of the induced elementary embedding $\bar{q}$ includes, among the rest, every x which in $\mathfrak{A}^{(I)}$ is of rank $< \mu_{i^*}$. If, moreover, $q(i) = i$ for all $i < i'$, where $i' \leqq i^*$, then $\bar{q}(x) = x$ for all x which are in $\mathfrak{A}^{(I)}$ of rank $< \mu_{i'}$.*

(This can be generalized as follows: If $J \supset I|i^* \cap I'$ then the domain of $\bar{q}$ includes every $x \in |\mathfrak{A}^{I'}|$ which in $\mathfrak{A}^{(I)}$ is of rank $< \mu_{i^*}$. If $q(i) = i$ for all $i \in I|i' \cap I'$ then $\bar{q}(x) = x$ for all $x \in |\mathfrak{A}^{I'}|$ which is of rank $< \mu_{i'}$.)

Under the assumptions of the last corollary, for any x which in $\mathfrak{A}^{(I)}$ is of rank $< \mu_{i^*}$ there is $x^* = \bar{q}(x)$ such that, for every formula φ, $\mathfrak{A}^{(I)} \models \varphi(\mu_{i_1}, \ldots, \mu_{i_n}, x) \leftrightarrow \varphi(\mu_{q(i_1)}, \ldots, \mu_{q(i_n)}, x^*)$; if $\operatorname{rank}(x) < \mu_{i'}$ then $x^* = x$. Actually one can throw into the formula any additional parameter z which is contributed by J, and replace it on the right-hand side by $\bar{q}(z)$. One can add on both sides any parameter contributed by any subset of J on which q is the identity (for then $\bar{q}(z) = z$); in particular, every member of the form $x^{(I)}$ can be added, where $x \in |\mathfrak{A}|$. Thus, if q maps I into itself, $\bar{q}$ embeds $\mathfrak{A}^{(I)}$ elementarily into itself and, for every ordinal α of $\mathfrak{A}$, it maps $(\mathfrak{A}^{(I)})_{\alpha^{(I)}}$ into itself.

Many properties of the embedding $\mathfrak{A} \to O\mathfrak{A}$ carry over to corresponding properties of $\mathfrak{A}^{(I)}$. For example, consider the following property:

There is a 1-1 mapping $': x \to x'$ of $\{x : \mathfrak{A} \models x \in R(\mu + 1)\}$ onto $\{x : O\mathfrak{A} \models x \in R(\mu + 1)\}$ such that, for every x, y, if $\mathfrak{A} \models x \in R\mu \wedge y \in R(\mu + 1)$ then $x' = x$ and $\mathfrak{A} \models x \in y \Leftrightarrow O\mathfrak{A} \models x \in y'$.

Roughly speaking, it means that $R^{\mathfrak{A}}(\mu + 1)$ and $R^{O\mathfrak{A}}(\mu + 1)$ are the same. This is always so. For $\mathfrak{A} = V$ and $O\mathfrak{A}$ = a well-founded ultrapower extension, it was proved by Scott. Essentially the same idea can be used in general: Let x' be the intersection, in $O\mathfrak{A}$, of $f_{\mathfrak{A}}(x)$ and $R^{O\mathfrak{A}}(\mu)$. (Remember that $f_{\mathfrak{A}}$ is the identity on $\{x : \mathfrak{A} \models x \in R\mu\}$.) The map is onto because $\mathfrak{A} \to O\mathfrak{A}$ is definable in $\mathfrak{A}$ ($O\mathfrak{A}$ cannot have subsets of $R\mu$ which are not in $\mathfrak{A}$). This generalizes as follows:

THEOREM 4. *For every $i \in I$ there is a mapping h_i of $\{x : \mathfrak{A}^{(I|i)} \models x \in R(\mu^{(I|i)} + 1)\}$ onto $\{x : \mathfrak{A}^{(I)} \models x \in R(\mu_i + 1)\}$ such that for every x, y, if $\mathfrak{A}^{(I|i)} \models x \in R(\mu^{(I|i)}) \wedge y \in R(\mu^{(I|i)} + 1)$ then $h_i(x) = f_{I|i,I}(x)$ and $\mathfrak{A}^{(I|i)} \models x \in y \Leftrightarrow \mathfrak{A}^{(I)} \models h_i(x) \in h_i(y)$.*

PROOF. Let $h_i(x)$ be the intersection, in $\mathfrak{A}^{(I)}$, of $f_{I|i,I}(x)$ and $R\mu_i$. To show that h_i is onto, we can assume that I is finite, for the general case will follow from this, using the familiar line of arguments employed in the previous proofs. Let J be such that $I = I|i + J$ then $\mathfrak{A}^{(I|i)} \to \mathfrak{A}^{(I)} = \mathfrak{A}^{(I|i)} \to O^J(\mathfrak{A}^{(I|i)})$ and this last embedding is definable in $\mathfrak{A}^{(I|i)}$. The argument that proves that $'$ is onto can be therefore applied. Q.E.D.

h_i is an isomorphism of $(\mathfrak{A}^{(I|i)})_{\mu^{(I|i)}+1}$ onto $(\mathfrak{A}^{(I)})_{\mu_i+1}$ (where "+1" is to be understood in the sense of the appropriate model) provided that we restrict these structures to the language consisting of $=$ and $\in$. (This provision must be added, for if the set theory with which we started has other predicates, such as say P, we need not have $\mathfrak{A}^{(I|i)} \models P(x) \Leftrightarrow \mathfrak{A}^{(I)} \models P(h_i(x))$, this equivalence is guaranteed only if $x \in (\mathfrak{A}^{(I|i)})_{\mu^{(I|i)}}$, for then h_i coincides with $f_{I|i,I}$.) To avoid additional notation, let us agree that, in the following corollary, a notation of the form "$\mathfrak{B}_\alpha$" denotes the structure with the universe $\{x:\mathfrak{B} \models x \in R\alpha\}$, and no other relations besides equality and the restriction of $\in^{\mathfrak{B}}$ to that universe. Now, $f_{I|i,I}$ embeds $(\mathfrak{A}^{(I|i)})_{\mu^{(I|i)}+1}$ elementarily into $(\mathfrak{A}^{(I)})_{\mu^{(I)}+1}$. Considering the composition $f_{I|i,I} \circ h_i^{-1}$ we get

COROLLARY 1. *There is an elementary embedding of* $(\mathfrak{A}^{(I)})_{\mu_i+1}$ *into* $(\mathfrak{A}^{(I)})_{\mu^{(I)}+1}$ *which leaves fixed every member which in* $\mathfrak{A}^{(I)}$ *is of rank* $<\mu_i$.

(The embedding itself need not be in $\mathfrak{A}^{(I)}$. As we shall see, its being in $\mathfrak{A}^{(I)}$ for the case $I = 1$ implies its being in $\mathfrak{A}^{(I)}$ for all ordered I.)

COROLLARY 2. *The following assertion is true in* $\mathfrak{A}^{(I)}$:
For every $X \subset R\mu$, *there is* $X^* \subset R\mu^{(I)}$ *such that every* Σ_1^2 *sentence which holds in* $(R\mu_i, \in|R\mu_i, X)$ *holds in* $(R\mu^{(I)}, \in|R\mu^{(I)}, X^*)$.

This is an immediate consequence of Corollary 1. (Take $X^* = f_{I|i,I} \circ h_i^{-1}(X)$. The argument is the same as that of Hanf and Scott [7]; see [**20**] for an exposition.)

As we shall see, stronger properties such as, roughly speaking, the identity of $R^{\mathfrak{A}}(\mu + k)$ and $R^{O\mathfrak{A}}(\mu + k)$ where $k > 1$ carry over to $\mathfrak{A}^{(I)}$, and one gets analogous corollaries.

Let $\mathfrak{F}$ be the normal ultrafilter in $\mathfrak{A}$ associated with the embedding $\mathfrak{A} \to \mathfrak{A}^{(1)}$ and let $\mathfrak{A} \models x \subset \mu$. We have

$$\mathfrak{A}^{(1)} \models x^{(1)} \in \mathfrak{F}^{(1)} \Leftrightarrow \mathfrak{A} \models x \in \mathfrak{F} \Leftrightarrow \mathfrak{A}^{(1)} \models \mu \in x^{(1)}.$$

Applying this argument to the embedding $\mathfrak{A} \to \mathfrak{A}^{(n)}$ we get

$$\mathfrak{A}^{(n)} \models x^{(n)} \in \mathfrak{F}^{(n)} \Leftrightarrow \mathfrak{A}^{(n)} \models \mu \in x^{(n)}.$$

By the properties of the $\mu_i(n)$ (recall that $\mu = \mu_0 = \mu_0(n)$), we have $\mathfrak{A}^{(n)} \models \mu \in x^{(n)} \Leftrightarrow \forall i < n\ \mathfrak{A}^{(n)} \models \mu_i(n) \in x^{(n)}$. One can without difficulty extend this to arbitrary ordered I and get

$$\mathfrak{A}^{(I)} \models x^{(I)} \in \mathfrak{F}^{(I)} \Leftrightarrow \forall i \in I\ \mathfrak{A}^{(I)} \models \mu_i \in x^{(I)}.$$

Taking this one further step, one gets the following theorem of which the rest of the proof is left to the reader.

THEOREM 5. *If* $\mathfrak{A}^{(I)} \models x \subset \mu^{(I)}$ *and* x *is contributed by* $I|i$ *then* $\mathfrak{A}^{(I)} \models x \in \mathfrak{F}^{(I)}$ *iff, for all* $i' \geqq i$, $\mathfrak{A}^{(I)} \models \mu_{i'} \in x$.

COROLLARY. *If* I *has no last member and* x *is* (in $\mathfrak{A}^{(I)}$) *a subset of* $\mu^{(I)}$ *then it is in* $\mathfrak{F}^{(I)}$ *iff, for some* $i \in I$, x *contains* (*in* $\mathfrak{A}^{(I)}$), *as members, all* $\mu_{i'}$ *where* $i' \geqq i$.

This, for the case of ultrapower extensions with I well-ordered, has been observed by Kunen [**11**].

All the theorems so far proved are of the upward kind—they relate $\mathfrak{A}^{(J)}$ to $\mathfrak{A}^{(I)}$ where J is an initial segment of I. It is extremely more difficult to get similar information for arbitrary J's. There is no known general description of how the ordinals below $\mu^{(J)}$ are moved by the embedding $\mathfrak{A}^{(J)} \to \mathfrak{A}^{(I)}$ where $J \subset I$ and I

is finite. There is little hope of getting such a description, for that would enable one to prove the well-foundedness of $(\mathfrak{A}^{(\alpha)})_{\mu(\alpha)}$ for all countable α's in cases where such a property would imply the nonexistence of certain large cardinals (see the remarks at the end of the next section).

THEOREM 6. *Keep I fixed. Let θ be an ordinal of $O\mathfrak{A}$ not contributed by $\mathfrak{A}$ and let $\Theta = \{\theta_i : i \in I\}$ be the set of indiscernibles produced from θ, i.e., $\theta_i = f_{\{i\},I}(\theta)$. Then,* (i) *$X$ has the form $\Theta \cap \{z : \mathfrak{A}^I \models z \in x\}$ (where $\mathfrak{A}^I = O^I\mathfrak{A}$), iff X is a finite union of intervals of the forms $\{\theta_i : i > j\}$, $\{\theta_i : i > j\}$, $\{\theta_i : j_1 < i < j_2\}$ and $\{\theta_i\}$;* (ii) *$X = \{z : \mathfrak{A}^I \models z \in x\} \subset \Theta$ for some $x \in |\mathfrak{A}^I|$, iff X is finite.*

PROOF. Intervals of ordinals of $\mathfrak{A}^I$ of the forms $\{\alpha : \mathfrak{A}^I \models \alpha < \beta\}$. etc., are "in" $\mathfrak{A}^I$, hence the "if"-direction of (i) is obvious. Similarly, the "if"-direction of (ii) is trivial. Given $x \in |\mathfrak{A}^I|$ get $J = \{i_1, \ldots, i_n\}$, $i_1 < \ldots < i_n$, such that $x \in |\mathfrak{A}^J|$ (where $\mathfrak{A}^J = f''_{J,I}O^J\mathfrak{A}$). If $i_k < j, j' < i_{k+1}$ let $q : J \cup \{j\} \to J \cup \{j'\}$ be the isomorphism such that $qi = i$ for $i \in J$ and $qj = j'$. Get, by I.(18), an associated isomorphism $\bar{q} : \mathfrak{A}^{J\cup\{j\}} \to \mathfrak{A}^{J\cup\{j'\}}$ such that $\bar{q}x = x$ and $\bar{q}\theta_j = \theta_{j'}$. Hence, $\mathfrak{A}^I \models \theta_j \in x \leftrightarrow \theta_{j'} \in x$. The same works if $j, j' < i_1$, or $j, j' > i_n$. This implies (i). If $X = \{z : \mathfrak{A}^I \models z \in x\} \subset \Theta$ then $\{z : \mathfrak{A}^J \models z \in x\} \subset \Theta \cap |\mathfrak{A}^J| = \{\theta_{i_1}, \ldots, \theta_{i_n}\}$, implying $\mathfrak{A}^J \models$ 'x has $\leqq n$ members' and hence $\mathfrak{A}^I \models$ 'x has $\leqq n$ members'. Q.E.D.

In all the applications of the iteration method in set theory one starts with $\mathfrak{A}$—a transitive class and $O\mathfrak{A}$ isomorphic to a transitive class. There are two kinds of applications. In the first one gets a "small", say countable, submodel $\mathfrak{B}$ of $\mathfrak{A}$, to which $\boldsymbol{O}^I$ can be applied and, by using a large well-ordered class I, blows-up $\mathfrak{B}$ to a large well-founded model $\mathfrak{B}^{(I)}$. This has been the method by which the author originally got his results concerning measurable cardinals and the constructible universe. His results were proven later by Silver [**21**], using partition methods, from weaker assumptions. Still, the iteration method leads in this case to some interesting observations, some of which we will briefly outline.

Applications of the second kind are those in which one keeps iterating the operator on some original "large" model, pushing μ upwards, i.e., pushing μ to $\mu^{(I)}$. The ordinals which $\mu^{(I)}$ can take as values constitute a closed unbounded class, and it is easy to see that after ω iterations any single additional iteration does not increase $\mu^{(I)}$'s cardinality ($\kappa(\mu^{(\gamma+1)}) = \kappa(\mu^{(\gamma)})$ if $\gamma \geqq \omega$; if the operation is the ultrapower extension this is true for $\gamma \geqq 1$). As μ is pushed upwards it leaves behind it a track of its previous values; these are the μ_i's. They have in the resulting structure many properties of μ (e.g., if μ is 2-extendible each μ_i is 1-extendible and much more can be said in this direction). Taking I = OR (= class of all ordinals) we get $\mu^{(I)}$ isomorphic to OR, thus any formula which, in V, describes μ, will in $V^{(\mathrm{OR})}$ describe OR. With larger well-ordered classes μ is pushed still further. The possibility of doing all these is interesting in itself. The more substantial applications of pushing upwards are Kunen's results [**11**].

Blowing up small structures. Assume that $\mathfrak{A}$ is a transitive set and that, for all countable α, $\mathfrak{A}^{(\alpha)}$ is well-founded. As was noted in Part II, this implies that $\mathfrak{A}^{(I)}$ is well founded for all well-ordered I.

If μ is, as before, the first ordinal moved by $\mathfrak{A} \to O\mathfrak{A}$, then we have $\mathfrak{A} \models$ '$\mathfrak{A}_\mu \models$ ZF' and consequently, $\mathfrak{A}_\mu \models$ ZF. This can be proved directly without using the fact that μ is measurable in $\mathfrak{A}$ (the deduction from the measurability of μ uses the axiom of choice which we do not assume, at this stage). The argument is: If h is a function from x into μ and $h \in |\mathfrak{A}|$, $x \in |\mathfrak{A}_\mu|$, then since $f_{\mathfrak{A}}$ is the identity on $\mathfrak{A}_\mu$, $f_{\mathfrak{A}}(x) = x$ and therefore $f_{\mathfrak{A}}(h) = h$. In $O\mathfrak{A}$, h is bounded below $f_{\mathfrak{A}}(\mu)$; hence, in $\mathfrak{A}$, h is bounded below μ.

Put $I =$ OR ($=$ class of all ordinals, ordered by $\in$). Let $\mathfrak{B} = \mathfrak{A}^{(I)}$. Consider $\mathfrak{B}_{\mu(\alpha)}$, $\alpha \in I$. By Theorems 2 and 3, $\alpha < \beta \Rightarrow \mathfrak{B}_{\mu(\beta)}$ is an elementary end-extension of $\mathfrak{B}_{\mu(\alpha)}$. We also have $\alpha = \bigcup \alpha > 0 \Rightarrow \mathfrak{B}_{\mu(\alpha)} = \bigcup_{\beta < \alpha} \mathfrak{B}_{\mu(\beta)}$. Since $\mathfrak{B}_{\mu(0)} = \mathfrak{A}_\mu$, we get a sequence of order type OR of well-founded models of ZF each an elementary end-extension of the preceding ones, the model at a limit ordinal being the union of the preceding ones.

Let $\delta = \sup \{\kappa((\mathfrak{A}^{(n)})_{\mu(n)}) : n < \omega\}$ (where $\kappa(X) =$ cardinality of X and $\kappa(\mathfrak{A}) = \kappa(|\mathfrak{A}|)$). Then we have, for every infinite ordered set J,

$$\kappa((\mathfrak{A}^{(J)})_{\mu(J)}) = \max \{\delta, \kappa(J)\}.$$

This is so because every member of $(\mathfrak{A}^{(J)})_{\mu(J)}$ is of the form $f_{J',J}(x)$, for some finite $J' \subset J$ and some $x \in |(\mathfrak{A}^{(n)})_{\mu(n)}|$. Consequently, $\kappa((\mathfrak{A}^{(J)})_{\mu(J)}) \leqq \sum_{n<\omega} \delta \cdot \kappa(J^n) = \max(\delta, \kappa(J))$. On the other hand, since J is infinite, the cardinality in question is $\geqq \kappa((\mathfrak{A}^{(n)})_{\mu(n)})$ for all n; it is also $\geqq \kappa(J)$, because $i \to \mu_i(J)$ is an embedding of J into $(\mathfrak{A}^{(J)})_{\mu(J)}$.

Since $|\mathfrak{A}^{(n)}|$ is a subset of $|\mathfrak{A}|$, we have $\kappa(|\mathfrak{A}^{(n)}|) = \kappa(|\mathfrak{A}|)$ for all $n < \omega$. Hence $\omega \leqq \delta \leqq \kappa(\mathfrak{A})$.

Suppose we start with a countable $\mathfrak{A}$; then $\delta = \omega$ and we have

$$\kappa(\mathfrak{B}_{\mu(\alpha)}) = \max \{\alpha, \omega\}.$$

Letting M_α be the transitive isomorphic image of $\mathfrak{B}_{\mu(\alpha)}$ and putting $M = \bigcup_{\alpha \in \mathrm{OR}} M_\alpha$ it follows that M is a transitive class and each M_α is an initial segment and elementary submodel of M. (Since $\alpha < \beta \Rightarrow M_\alpha \prec M_\beta$, one has a truth definition for M; namely, define $M \models \varphi(x_1, \ldots, x_n)$ as $M_\alpha \models \varphi(x_1, \ldots, x_n)$ where α is such that $x_1, \ldots, x_n \in M$. We have also an alternative way; we can use the truth definition for $\mathfrak{B}_{\mu(\mathrm{OR})}$ which is available inside $\mathfrak{B}$, where $\mathfrak{B} = \mathfrak{A}^{(\mathrm{OR})}$.)

Let $\xi(\alpha) = \sup \{\beta : \beta \in M_\alpha\}$, then we have the following:

For every α, $\kappa(\xi(\alpha)) = \max \{\kappa(\alpha), \omega\}$ and $\alpha \to \xi(\alpha)$ is a continuous nondecreasing function.

Obviously $M \models$ ZF; let $L =$ constructible universe. By the well-known absoluteness of the constructible universe the formula '$x \in L$' defines inside M the real L, and inside M_α the real $L_{\xi(\alpha)}$. Consequently $L_{\xi(\alpha)} \prec L$, for all α. One can therefore deduce that every definable subset of L is a member of $L_{\xi(0)}$ and hence countable, and other results of similar character which we will not go into. This has been the way in which the author originally got these results in 1963–64.

The only remaining point is to get a countable $\mathfrak{A}$ having the property that, for some elementary self-extension, $\mathfrak{A} \to O\mathfrak{A}$, $O^\alpha\mathfrak{A}$ is well-founded for all countable α. Getting such a model of any cardinality would be sufficient. For let it be $\mathfrak{C}$ such (we assume $\mathfrak{C} \models Z^+$) and take $\mathfrak{A}$ to be a countable elementary submodel of $\mathfrak{C}$. For every α we get an elementary embedding $\mathfrak{A}^{(\alpha)} \to \mathfrak{C}^{(\alpha)}$; hence $\mathfrak{A}^{(\alpha)}$ is well-founded.

The existence of a measurable cardinal implies the existence of such a $\mathfrak{C}$; if κ is measurable, take $\mathfrak{C} = V_\gamma$ where $\gamma > 2^\kappa$ and $\mathrm{cof}(\gamma) > \kappa$ and take the ultrapower extension based on a $<\kappa$-complete nonprincipal ultrafilter over κ. In the original proof an argument similar to that which has been used in the proof of the well-foundedness theorem was used to show that $\mathfrak{C}^{(\alpha)}$ is well-founded for all α. Actually, it can be shown that, for this particular self-extension, any limit ordinal $>\kappa$ can be chosen for γ.

In general, given any elementary self-extension $V \to OV$, with OV well-founded, one can get arbitrarily large α's such that the self-extension operator can be applied to V_γ, $V_\gamma \to OV_\gamma$ is elementary and the inclusion $V_\gamma \subset V$ is in $\boldsymbol{C}_{e,*}$; i.e., one gets an induced embedding $O^I V_\gamma \to O^I V$. This follows from Theorem II.4 and I.(23); the class of such γ's is stationary. Any elementary submodel $\mathfrak{A}$ of V_γ can be used in the way described above.

Taking, as before, a countable $\mathfrak{A}$ one can encode it into a subset Y of ω and then one can carry the whole construction inside $L(Y)$; the sequence of models $\langle M_\alpha : \alpha \in \mathrm{OR}\rangle$ is therefore definable in $L(Y)$. Thus, inside $L(Y)$, L is revealed to be a very "narrow" universe. Now, one can take another countable elementary submodel $\mathfrak{A}_1 \prec V_\gamma$ such that $Y \in \mathfrak{A}_1$. Carrying the construction as before, each of the models M_α will have Y as a member (since $\mathrm{rank}(Y) = \omega < \mu$, Y is not changed by the embeddings $\mathfrak{A}_1^{(\alpha)} \to \mathfrak{A}_1^{(\beta)}$). Consequently we now get $L_{\xi(\alpha)}(Y) \prec L_{\xi(\beta)}(Y)$, and what we concluded concerning L we can now conclude concerning $L(Y)$. If $\mathfrak{A}_1$ is encoded in Y_1 then, inside $L(Y_1)$, $L(Y)$ is revealed to be a very "narrow" universe. The process can be continued on indefinitely, yielding a sequence $\langle Y_\alpha : \alpha < \omega_1\rangle$ such that in $L(Y_\alpha)$ one can represent $L(\langle Y_\beta : \beta < \alpha\rangle)$ as an ascending union of an elementary chain having the above-mentioned properties. One can continue this further and get a sequence $\langle Y_\beta : \beta < \mu\rangle$ having this property, but for $\beta \geqq \omega_1$, Y_β is no longer a subset of ω.

The property that $Y \subset \omega$ encodes a well-founded model satisfying a certain fixed given set of sentences can be expressed by a Π^1_1 formula in second-order arithmetic. (The set of sentences in question is represented by some fixed subset of ω; for our purposes we will need a certain recursive set; the argument works for any set in L.) Consequently the existence of such a Y is expressible by a Σ^1_2 sentence. By Shoenfield's absoluteness theorem [**20**], if there exists such a Y in V there is one also in L.

Now start with an elementary self-extension $V \to OV$ in which OV is a transitive class. It is not difficult to deduce that, for each ordinal α, $O^\alpha V$ is isomorphic to a transitive class. We can, actually, choose the representatives for the direct limit in such a way that $O^\alpha V$ is, by definition, that transitive class. Take γ such that $V_\gamma \to OV_\gamma$ is elementary, and the assertion that for every ordinal α the αth iteration gives a transitive class is true in V_γ. (This is possible since $\{\gamma : V_\gamma \to OV_\gamma \text{ is elementary}\}$ is stationary.) The same assertion will be true inside every elementary submodel of V_γ. Note also that the statement that $\mathfrak{A} \to O\mathfrak{A}$ is elementary amounts to the satisfaction of a certain recursive set of sentences in $\mathfrak{A}$.

By Shoenfield's absoluteness theorem, there will be in L a countable transitive model N such that $N \models Z^+$, $N \to ON$ is elementary and, for every $\alpha \in N$, $N^{(\alpha)}$ is a transitive model. We know, however, that for some β, which is countable in L, we must have a non-well-founded $N^{(\beta)}$, for otherwise we shall get inside $L : N^{(\beta)}$ is well-founded for all β, leading to a contradiction.

What is the first β for which $N^{(\beta)}$ is not well-founded? One can have N's for which that β is very large compared to N. For instance one can show:

There is a transitive $N \in L$ such that for every ordered class I which is definable in N we have $N \models$ 'I is a well-ordered class' $\Rightarrow I$ is a well-ordered class and $N^{(I)}$ is well-founded. This, and much stronger claims in the same direction, can be obtained using the following lemma which follows immediately by an absoluteness argument:

LEMMA. *Let $P(u)$ and $R(u, v)$ be Σ_2^1 formulas in second-order arithmetic such that in V we have* (i) $\exists N[P(N) \wedge \forall\alpha[N^{(\alpha)}$ *is well-founded*]], (ii) $\forall N \exists I[R(N, I) \wedge I$ *is a well-ordering*]. *Then* $L \models \exists N \exists I[P(N) \wedge R(N, I) \wedge N^{(I)}$ *is well-founded*]. (*Here we identified countable models and well-orderings with their encodings as subsets of ω.*)

Let $V_\gamma^{(\alpha)}$ be well-founded for all α, let $\mathfrak{A} \prec V_\gamma$, where $\mathfrak{A}$ is countable, and let N be the transitive model isomorphic to $\mathfrak{A}$ and Y a subset of ω encoding N. Take a countable $\mathfrak{A}_1$ such that $\mathfrak{A} \prec \mathfrak{A}_1 \prec V_\gamma$ and $Y \in |\mathfrak{A}_1|$. Using an absoluteness argument for $L(Y)$ one can show that in $L(Y)$ there is a well-founded elementary extension $\mathfrak{C}$ of N such that $Y \in |\mathfrak{C}|$. As we have seen there is an ordinal α, countable in $L(Y)$, such that $\mathfrak{C}^{(\alpha)}$ is not well-founded. Consequently, we get two well-founded models $\mathfrak{A}, \mathfrak{C}$, with $\mathfrak{A} \prec \mathfrak{C}$ such that $\mathfrak{A}^{(\alpha)}$ is well-founded for all α but $\mathfrak{C}^{(\alpha)}$ is not. Moreover, we can choose $\mathfrak{C}$'s so that the first α for which $\mathfrak{C}^{(\alpha)}$ is not well-founded is very large, as indicated in the preceding lemma (with L replaced by $L(Y)$). This can be done for every countable $\mathfrak{A}$ which is an elementary submodel of some V_γ, where $V_\gamma^{(\omega_1)}$ is well-founded.

It is for this reason that any theorem of the kind of Theorem 1, which will clarify how in general the ordinals contributed by I_1 are related to those contributed by I_2 (where I_1 and I_2 are any finite sets), seems unlikely to be proved. For, using such a version, one would be able to deduce the well-foundedness of the above-mentioned $\mathfrak{C}^{(\alpha)}$ for all countable α and thus contradict the assumption that OV is well-founded. In particular, having such a version for the ultrapower self-extension would imply the nonexistence of measurable cardinals, and, likewise, having it for a well-founded self-extension which is derived from the existence of some other larger cardinals would imply their nonexistence.

Note that in all the constructions carried out in this section we needed only the well-foundedness of $(\mathfrak{A}^{(\alpha)})_{\mu(\alpha)}$ for all countable α, not of the whole $\mathfrak{A}^{(\alpha)}$.

Some properties of $V^{(\alpha)}$ with a very large μ. The richer OV is assumed to be, the larger the first ordinal μ which is moved by $V \to OV$. In the following two theorems we shall see how some properties of OV are passed on to the $V^{(\alpha)}$'s. For the sake of simplicity we assume that OV is a transitive class and we state the results for the case where the well-ordered I is an ordinal. Since $V^{(\alpha)}$ is isomorphic in that case to a transitive class, it can be identified with it. (This can be made part of the definition of representatives in the direct limit.) Note that, for $\beta \in \alpha$, $\alpha|\beta = \{\gamma \in \alpha : \gamma < \beta\} = \beta$; hence $\mu^{(\alpha|\beta)} = \mu^{(\beta)}$. Since $(\mu^{(\alpha|\beta)}, \in|\mu^{(\alpha|\beta)})$ is by Theorem 3 isomorphic to $(\mu_\beta(\alpha), \in|\mu_\beta(\alpha))$ we actually have, for all β and all $\alpha > \beta$, $\mu^{(\beta)} = \mu_\beta(\alpha)$. $\mu_\beta(\alpha)$ does not depend on α and we can omit α altogether; from now on $\mu^{(\beta)} = \mu_\beta$. One can get the whole sequence $\langle \mu_\beta : \beta \in \mathrm{OR}\rangle$ by considering $V^{(\mathrm{OR})}$ and identifying $(V^{(\mathrm{OR})})_{\mu^{(\mathrm{OR})}}$ with the transitive class with which it is isomorphic; $\mu^{(\mathrm{OR})}$ becomes OR.

It is also obvious that the natural isomorphism $O^\gamma(O^\beta V) \simeq O^{\beta+\gamma}V$ becomes an identity: $O^\gamma V^{(\beta)} = V^{(\beta+\gamma)}$.

[The following two theorems can be generalized to the case where OV is not a transitive class—the formulation is then cumbersome because the $\in$ of $V^{(\alpha)}$ is not the real $\in$. Furthermore, not the whole of OV must be well-founded, only large enough initial segments of it, as will be obvious from the formulation. α can be replaced by any well-ordered class I and, finally, V can be replaced by any model $\mathfrak{A}$ of Z^+.]

THEOREM 7. *If* $\nu \leqq \mu_\omega$ *and* $V_\nu \subset V^{(1)}$ *then,* (i) *for all* α, $V_\nu \subset V^{(\alpha)}$ *and,* (ii) *for all* $\beta < \alpha$, $(V^{(\beta)})_{\nu(\beta)} \subset V^{(\alpha)}$.

REMARK 1. Actually there is no need to assume $\nu \leqq \mu_\omega$. Kunen's proof that there is no elementary nontrivial embedding $V_{\alpha+2} \to V_{\alpha+2}$ proves also that there is no elementary nontrivial embedding $V \to V^{(1)}$ with $V_{\mu_\omega+1} \subset V^{(1)}$ ([**12**]; see also the proof of Theorem IV.9).

REMARK 2. The assumption that $V_\nu \subset V^{(\alpha)}$ with $\nu > \mu$ leads to very strong axioms of infinity; the larger the ν the stronger the axiom. More significant than its magnitude is ν's relation to the sequence $\mu_0, \mu_1, \ldots, \mu_n, \ldots, n < \omega$. In view of the contradiction which $\nu = \mu_\omega + 1$ leads to, even $\nu = \mu_\omega$ seems rather risky. It can be shown that $\nu = \mu_2$ implies that, inside V_{μ_2}, μ is γ-extendible for all γ, and more than that. $\nu = \mu + \lambda$, with $\lambda < \mu$ implies that μ is λ-extendible and, for $\lambda \geqq \mu$ this is implied by $\nu = \mu + \lambda^{(1)}$. Vice versa, if μ is $(\lambda + 1)$-extendible one can define an elementary self-extension $V \to V^{(1)}$ with $V_{\mu_1+\lambda^{(1)}} \subset V^{(1)}$ (see Part IV).

REMARK 3. Theorem 4 can be deduced as a special case, where we have $\nu = \mu + 1$. We also get corollaries to Theorem 7 which are analogous to those of Theorem 4. For example, if $V_{\mu+7} \subset V^{(1)}$ then, for every $\beta < \alpha$ the following is true in $V^{(\alpha)}$:

$\forall X \in R(\mu_\beta + 7) \exists X^* \in R(\mu_\alpha + 7)$ such that every Σ_1^8 sentence which holds in (V_β, X) holds in (V_α, X^*).

PROOF OF THEOREM 7. For the case $\nu < \mu_1$, (i) is trivial. (μ_1 is the first ordinal moved by $V^{(1)} \to V^{(\alpha)}$; hence $R\nu \in V^{(1)} \Rightarrow R\nu \in V^{(\alpha)}$. Note that $V_\nu \subset V^{(1)} \Leftrightarrow R^{V^{(1)}}(\nu) = R\nu \Leftrightarrow R\nu \in V^{(1)}$ and similarly for $V_\nu \subset V^{(\alpha)}$ and $(V^{(\beta)})_{\nu(\beta)} \subset V^{(\alpha)}$. This follows from the absoluteness of the concept of rank.)

In general we argue as follows: $V \models R\nu \in |OV|$ implies $V^{(1)} \models R(\nu^{(1)}) \in |OV^{(1)}|$. Since $\nu \leqq \nu^{(1)}$ we have $V^{(1)} \models$ '$R\nu$ is an initial segment of $R(\nu^{(1)})$', implying $V^{(1)} \models$ '$R\nu \in |OV^{(1)}|$'. But $R\nu$ inside $V^{(1)}$ is the real $R\nu$; hence $R\nu \in V^{(2)}$. Continuing so we get $R\nu \in V^{(n)}$ for all $n < \omega$. If $\nu < \mu_\omega$, take n such that $\nu < \mu_n$. The first ordinal moved by $V^{(n)} \to V^{(\alpha)}$ is μ_n; hence $Rv \in V^{(\alpha)}$. If $v = \mu_\omega$ then, for all $n < \omega$, $R\mu_n \in V^{(\alpha)}$, implying $V_{\mu_n} \subset V^{\alpha}$. Hence $V_{\mu_\omega} = \bigcup_{n<\omega} V_{\mu_n} \subset V^{(\alpha)}$.

To get (ii) we apply (i) inside $V^{(\beta)}$. Since $V \models v \leqq \mu_\omega$, we have $V^{(\beta)} \models v^{(\beta)} \leqq (\mu_\omega)^{(\beta)}$. Moreover as, in V, μ_ω is the value of μ under the embedding $V \to O^\omega V$, so, in $V^{(\beta)}$, $(\mu_\omega)^{(\beta)}$ is the value of μ_β under the embedding $V^{(\beta)} \to O^\omega V^{(\beta)}$. Similarly, $V \models Rv \in |OV|$ implies $V^{(\beta)} \models R(v^{(\beta)}) \in |OV^{(\beta)}|$. Consequently, by applying (i) inside $V^{(\beta)}$ we get $V^{(\beta)} \models \forall\gamma[R(v^{(\beta)}) \in |O^\gamma V^{(\beta)}|]$. This is obviously absolute. $O^\gamma(V^{(\beta)}) = V^{(\beta+\gamma)}$ and, obviously, $R^{V^{(\beta)}}(v^{(\beta)}) = (Rv)^{(\beta)}$. Hence, for all $\alpha > \beta$, $(Rv)^{(\beta)} \in V^{(\alpha)}$. Q.E.D.

In the following theorem $R^{(\alpha)} = R^{V^{(\alpha)}}$; i.e., $R^{(\alpha)}(\beta) = \{x : V^{(\alpha)} \models \text{rank}(x) < \beta\}$.

THEOREM 8. *Assume that* $\kappa < \mu_1$. *If* $\{\langle x, f_{0,1}(x)\rangle : x \in R^{(1)}(\kappa)\} \in V^{(1)}$ *(where* $f_{0,1} = V \to V^{(1)}$*) then, for all* α,

(i) *for all* $\beta < \alpha$, $\{\langle x, f_{\beta,\alpha}(x)\rangle : x \in R^{(\alpha)}(\kappa^{(\beta)})\} \in V^{(\alpha)}$, *and*

(ii) *for all* $\beta \leqq \gamma < \alpha$, $\{\langle x, f_{\beta,\gamma}(x)\rangle : x \in R^{(\alpha)}(\kappa^{(\beta)})\} \in V^{(\alpha)}$ *and it is an elementary embedding of* $(V^{(\alpha)})_{\kappa^{(\beta)}}$ *into* $(V^{(\alpha)})_{\kappa^{(\gamma)}}$.

LEMMA. *If* $\kappa < \mu_1$ *then, for all* δ, $\kappa^{(\delta)} < \mu_{\delta+1}$; *for all* $\alpha > \delta$, $(V^{(\alpha)})_{\kappa^{(\delta)}} = (V^{(\delta+1)})_{\kappa^{(\delta)}}$ *and, for all* $\beta \leqq \gamma < \alpha$, $f_{\beta,\gamma}$ *embeds* $(V^{(\alpha)})_{\kappa^{(\beta)}}$ *elementarily into* $(V^{(\alpha)})_{\kappa^{(\gamma)}}$.

PROOF. $V \models \kappa < \mu_1$ implies $V^{(\delta)} \models \kappa^{(\delta)} < \mu_1^{(\delta)}$. In V, μ_1 is the image of μ under $V \to OV$; hence, in $V^{(\delta)}$, $\mu_1^{(\delta)}$ is the image of $\mu^{(\delta)}$ under $V^{(\delta)} \to OV^{(\delta)}$, i.e., under $f_{\delta,\delta+1}$. But $\mu^{(\delta)} = \mu_\delta$ and $f_{\delta,\delta+1}(\mu_\delta) = \mu_{\delta+1}$; hence $\kappa^{(\delta)} < \mu_{\delta+1}$.

Since $\mu_{\delta+1}$ is the first ordinal moved by $f_{\delta+1,\alpha}$, it follows that $f_{\delta+1,\alpha}$ is the identity mapping on $R^{(\delta+1)}(\kappa^{(\delta)})$, implying $(V^{(\alpha)})_{\kappa^{(\delta)}} = (V^{(\delta+1)})_{\kappa^{(\delta)}}$.

Consider $f_{\beta,\gamma} : V^{(\beta)} \to V^{(\gamma)}$. It induces an elementary embedding $Of_{\beta,\gamma} : OV^{(\beta)} \to OV^{(\gamma)}$. $Of_{\beta,\gamma}$ is the restriction of $f_{\beta,\gamma}$ to $OV^{(\beta)}$ i.e., to $V^{(\beta+1)}$ which is a transitive subclass of $V^{(\beta)}$. [By I.(16), $Of_{\beta,\gamma} = f_j : V^{(\beta+1)} \to V^{(\gamma+1)}$ where $j = j : \beta + 1 \to \gamma + 1$ is such that $j|\beta$ = identity and $j(\beta) = \gamma$.] Thus, $f_{\beta,\gamma}$ embeds $V^{(\beta+1)}$ elementarily into $V^{(\gamma+1)}$. Since $f_{\beta,\gamma}(\kappa^{(\beta)}) = \kappa^{(\gamma)}$, it embeds $(V^{(\beta+1)})_{\kappa^{(\beta)}}$ elementarily into $(V^{(\gamma+1)})_{\kappa^{(\gamma)}}$. But $(V^{(\beta+1)})_{\kappa^{(\beta)}} = (V^{(\alpha)})_{\kappa^{(\beta)}}$ and $(V^{(\gamma+1)})_{\kappa^{(\gamma)}} = (V^{(\alpha)})_{\kappa^{(\gamma)}}$. Q.E.D.

Note. A simple checking shows that, for $x \in R^{(\alpha)}(\kappa^{(\beta)})$, $f_{\beta,\gamma}(x) = \bar{q}(x)$ where $\bar{q}$ is the embedding induced by $q : \beta + 1 \to \gamma + 1$ where $q|\beta$ = identity and $q(\beta) = \gamma$. (See I.(18).)

PROOF OF THEOREM 8. (i) Let $p_{0,1} = \{\langle x, f_{0,1}(x)\rangle : x \in R^{(1)}(\kappa)\}$. In $V^{(1)}$, $p_{0,1}$ is a mapping whose domain is $R^{(1)}(\kappa)$; hence, in $V^{(\alpha)}$, $f_{1,\alpha}(p_{0,1})$ is a mapping whose domain is $f_{1,\alpha}(R^{(1)}(\kappa))$.

The first ordinal moved by $f_{1,\alpha}$ is μ_1 and $\kappa < \mu_1$; consequently, $f_{1,\alpha}(R^{(1)}(\kappa)) = R^{(1)}(\kappa) = R^{(\alpha)}(\kappa)$. This easily implies

$$f_{1,\alpha}(p_{0,1}) = \{\langle x, f_{1,\alpha}(f_{0,1}(x))\rangle : x \in R^{(1)}(\kappa)\} = \{\langle x, f_{0,\alpha}(x)\rangle : x \in R^{\alpha}(\kappa)\}$$

and proves

(i_0) $\forall\alpha[\{\langle x, f_{0,\alpha}(x)\rangle : O^\alpha V \models x \in R\kappa\} \in O^\alpha V]$.

To get (i) we replaced V by $V^{(\beta)}$. The assumption of the theorem is true inside $V^{(\beta)}$ provided that we replace every parameter by its value under $f_{0,\beta} : V \to V^{(\beta)}$ and employ all the concepts in the sense of $V^{(\beta)}$. Inside $V^{(\beta)}$ the assertion corresponding to (i_0) is therefore true. It is easily seen to be absolute.

Consequently we have

$$\forall\alpha[\{\langle x, f^{(\beta)}_{0,\alpha}(x)\rangle : O^\alpha(V^{(\beta)}) \models x \in R(\kappa^{(\beta)})\} \in O^\alpha V^{(\beta)}].$$

Now $f^{(\beta)}_{0,\alpha} = f_{0,\alpha,V^{(\beta)}} : V^{(\beta)} \to O^\alpha V^{(\beta)}$ and, by I.(16) it is equal to $f_{\beta,\beta+\alpha} : V^{(\beta)} \to V^{(\beta+\alpha)}$. Also $\{x : V^{(\beta+\alpha)} \models x \in R(\kappa^{(\beta)})\} = R^{(\beta+\alpha)}(\kappa^{(\beta)})$. Hence

$$\{\langle x, f_{\beta,\beta+\alpha}(x)\rangle : x \in R^{(\beta+\alpha)}(\kappa^{(\beta)})\} \in V^{(\beta+\alpha)}.$$

This proves (i) (replace $\beta + \alpha$ by α).

(ii) For every $\delta \leqq \delta' < \alpha$ put $p_{\delta,\delta'} = \{\langle x, f_{\delta,\delta'}(x)\rangle : x \in R^{(\alpha)}(\kappa^{(\delta)})\}$.

By the lemma, $p_{\beta,\gamma}$ embeds $R^{(\alpha)}(\kappa^{(\beta)})$ into $R^{(\alpha)}(\kappa^{(\gamma)})$. Since $f_{\beta,\alpha} = f_{\gamma,\alpha} \circ f_{\beta,\gamma}$ it follows that $p_{\beta,\alpha} = p_{\gamma,\alpha} \circ p_{\beta,\gamma}$. By (i), $p_{\beta,\alpha}, p_{\gamma,\alpha} \in V^{(\alpha)}$. Hence $p_{\beta,\gamma}$ is definable inside $V^{(\alpha)}$ as follows:

$$\langle x, y\rangle \in p_{\beta,\gamma} \Leftrightarrow \exists z[\langle x, z\rangle \in p_{\beta,\alpha} \wedge \langle y, z\rangle \in p_{\gamma,\alpha}]. \quad \text{Q.E.D.}$$

The interesting cases are of course those where $\kappa > \mu$. Assume $\kappa = \mu + \lambda$, and, for simplicity, take $\lambda < \mu$. Then, inside $V^{(\alpha)}$ we have elementary embeddings $(V^{(\alpha)})_{\mu_\beta+\lambda} \to (V^{(\alpha)})_{\mu_\gamma+\lambda}$, which send μ_β to μ_γ, for all $\beta \leqq \gamma < \alpha$ (we can also take $\alpha = \text{OR}$).

Since $V \to V^{(\alpha)}$ sends μ to μ_α, one can derive from the theorem information on what happens below μ. For example it is not difficult to show that there is a subset $X \subset \mu$ such that X is in the normal ultrafilter associated with $V \to OV$ and such that for all ξ, $\eta \in X$, if $\xi < \eta$, there is an elementary embedding $h_{\xi,\eta} : V_{\xi+\lambda} \to V_{\eta+\lambda}$ such that $h_{\xi,\eta}(\xi) = \eta$ and ξ is the first ordinal moved by $h_{\xi,\eta}$. This can, however, be proved directly using standard techniques (see Reinhardt-Solovay [**17**], for similar situations). The additional information which can be deduced from Theorem 8 is: For every $\theta < \omega_1$ there is a subset Y of X of order-type θ and a system of elementary embeddings $\{h_{\xi,\eta} : V_{\xi+\lambda} \to V_{\eta+\lambda}\}_{\xi\leqq\eta,\xi,\eta\in Y}$ such that $h_{\xi,\eta}(\xi) = \eta$, ξ is the first moved by $h_{\xi,\eta}$ and in addition $h_{\xi,\zeta} = h_{\eta,\zeta} \circ h_{\xi,\eta}$ $(\xi, \eta, \zeta \in Y)$. (This is deduced using a certain tree and the fact that if a tree which is defined inside $V^{(\alpha)}$ is well-founded in $V^{(\alpha)}$ then it is well-founded.) Whether the claim is true for uncountable θ's is unknown to the author.

Part IV

Approximations and decompositions of embeddings.

$\text{ZC}^+ =_{\text{Df}} Z^+ \cup \{$*Axiom of Choice*$\}$. All structures to be treated here are, unless otherwise stated, models of ZC^+. As before we put $\kappa(x)$ = cardinality of x. See Postcript A, concerning the interpretation of "cardinality" in models of ZC^+.

Definition 1. Let γ be a cardinal in $\mathfrak{A}$. An embedding $j : \mathfrak{A} \to \mathfrak{B}$ is *γ-cofinal* (*$<\gamma$-cofinal*) if for every x in $\mathfrak{B}$ there is a y of cardinality $\leqq\gamma$ $(<\gamma)$ in $\mathfrak{A}$ such that $\mathfrak{B} \models x \in jy$.

Obviously $<\gamma$-cofinality implies $<\gamma'$-cofinality, if $\gamma \leqq \gamma'$ and γ-cofinality is equivalent to $<\gamma^+$-cofinality if γ^+ = successor cardinal of γ in $\mathfrak{A}$. In this terminology, cofinality is $<\infty$-cofinality where ∞ = "largest ordinal."

In the following theorem M and N are assumed to be transitive structures. The proof can obviously be rephrased for arbitrary structures. For the sake of convenience we shall use the same names for the structures and for their underlying domains.

THEOREM 1. *Let $j:M \to N$ be elementary and cofinal. Let γ be an infinite cardinal in M and let N^* be the submodel of N whose universe is $\{x \in N : \exists y[M \models `\kappa(y) < \gamma'$ and $x \in jy]\}$; then $N^* \prec N$.*

PROOF. Since $jx \in j(\{x\})$ we have $j''M \subset N^*$. Let $j^*: M \to N^*$ be the embedding which sends every x to jx.

Main claim. j^* is elementary.

Obviously j^* is $<\gamma$-cofinal and, a fortiori, cofinal. It is easy to see that if j maps an ordered set x of M (i.e., a pair consisting of x and an ordering $<_x$ on it, both in M) cofinally into jx, so does j^* (see Theorem II.1 for the terminology). By Theorem II.1, since j is elementary and cofinal, it suffices to show that j^* is Σ_0-elementary.

For every universal formula φ we have, since $N^* \subset N$,

(i) $M \models \varphi(x_1, \ldots, x_n) \Rightarrow N \models \varphi(jx_1, \ldots, jx_n) \Rightarrow N^* \models \varphi(jx_1, \ldots, jx_n)$.

Since '$v_1 \subset v_2$' and 'v is transitive' are universal,

(ii) $M \models `x_1 \subset x_2'$ and $M \models `x$ is transitive' $\Rightarrow N^* \models `jx_1 \subset jx_2'$ and $N^* \models `jx$ is transitive'.

Assume that $x_1, x_2 \in N^*$ and let y_1, y_2 be of cardinality $<\gamma$ in M such that $x_i \in jy_i$, $i = 1, 2$. Put $y = \{\{z_1, z_2\} : z_1 \in y_1 \wedge z_2 \in y_2\}$. Then, $N \models jy = \{\{z_1, z_2\} : z_1 \in jy_1 \wedge z_2 \in jy_2\}$, implying $\{x_1, x_2\} \in jy$. Since $M \models \kappa(y) < \gamma$, $\{x_1, x_2\} \in N^*$. Obviously $N^* \models \forall v[v \in \{x_1, x_2\} \leftrightarrow v = x_1 \vee v = x_2]$. Consequently,

(iii) $x_1, x_2 \in N^* \Rightarrow \langle x_1, x_2 \rangle \in N^*$ and the concept of an ordered pair is absolute for N^*. The same holds for n-tuples.

Add to the language a new function symbol $\langle\, ,\, \rangle$ which will be interpreted in our structures as ordered pairs formation. N^* continues to be a substructure of N and (i) continues to hold for universal formulas in the new language. Define φ^v_*, where φ is a Σ_0-formula (in the new language) and v is a new variable as follows:

$$\varphi^v_* = \varphi, \quad \text{if } \varphi \text{ is without quantifiers,} \qquad (\neg\varphi)^v_* = \neg\varphi^v_*,$$

$$(\varphi \vee \psi)^v_* = \varphi^v_* \vee \psi^v_* \quad \text{and} \quad (\exists u_1 \in u_2 \varphi)^v_* = \exists u_1 \in v[u_1 \in u_2 \wedge \varphi^v_*].$$

If φ is a Σ_0-formula with the free variables $v_0, \ldots, v_{n-1}$ then the following is a theorem of logic:

$$\bigwedge_{i<n} (v_i \subset v) \wedge v \text{ is transitive} \to [\varphi \leftrightarrow \varphi^v_*].$$

Given $x_0, \ldots, x_{n-1} \in M$ take $x \in M$ which is transitive and includes $x_0, \ldots, x_{n-1}$ as subsets. Then $M \models \varphi(x_0, \ldots, x_{n-1}) \leftrightarrow \varphi^x_*(x_0, \ldots, x_{n-1})$. By (ii), jx is transitive in N^* and includes $jx_0, \ldots, jx_{n-1}$; hence we get the analogous equivalence in N^*. It is, therefore, sufficient to show that $M \models \psi(z_1, \ldots, z_n) \Rightarrow N^* \models \psi(jz_1, \ldots, jz_n)$ where ψ is a Σ_0-formula in which all the quantifiers are restricted by the same free variable. We allow $\langle\, ,\, \rangle$ as a primitive function symbol.

Pull the quantifiers in front and proceed by induction on the number of $\exists$'s in the prefix. The claim, for universal formulas, has been proved and the passage from ψ to $\exists u \in v\psi$ is trivial. The essential step is to prove it for $\forall u_1 \in v \ldots \forall u_n \in v \, \exists u_{n+1} \in v \, \psi$, assuming its truth for formulas with fewer existential quantifiers. For simplicity, take $n = 1$. The general case is completely analogous. Assume $M \models \forall u_1 \in x \, \exists u_2 \in x \, \psi$ (ψ might include additional parameters from M, which are replaced by their images under j when passing to N^*). Use separation in M and get $z \subset x^2$ such that, in M, we have

$$\forall u_1 \in x \forall u_2 \in x[\langle u_1, u_2 \rangle \in z \rightarrow \psi],$$

$$\forall u_1 \in x \exists u_2 \in x[\langle u_1, u_2 \rangle \in z].$$

The first can be put in the form $\varphi(x, z, \ldots)$ where φ has one existential quantifier less; hence it holds in N^* (with x, z replaced by jx, jz, etc.). If we show that $N^* \models \forall u_1 \in jx \exists u_2 \in jx[\langle u_1, u_2 \rangle \in jz]$, we are finished, since this would imply the satisfaction in N^* of the original formula.

Let a_1 be a member of N such that $a_1 \in jx$. Get $b_1 \in M$ of cardinality $<\gamma$ in M such that $a_1 \in jb_1$. We can assume $b_1 \subset x$, for otherwise we can replace b_1 by $b_1 \cap x$. Use choice inside M and get a function h in M such that $\langle t, h(t) \rangle \in z$ for all $t \in b_1$. Put $b_2 = \{h(t) : t \in b_1\}$. Then b_2 is in M of cardinality $<\gamma$ and $M \models \forall u_1 \in b_1 \; \exists u_2 \in b_2[\langle u_1, u_2 \rangle \in z]$. Hence $N \models \forall u_1 \in jb_1 \; \exists u_2 \in jb_2[\langle u_1, u_2 \rangle \in jz]$, implying $N \models \exists u_2 \in jb_2[\langle a_1, u_2 \rangle \in jz]$. Let $a_2 \in jb_2$ be such that $\langle a_1, a_2 \rangle \in jz$. Since $M \models \kappa(b_2) < \gamma$, $a_2 \in N^*$. This proves the main claim.

The theorem follows now from

LEMMA. *If $\mathfrak{B}_1 \subset \mathfrak{B}_2$ and $\mathfrak{A}$ is a model of Z^+ which is a cofinal submodel of $\mathfrak{B}_1$ and an elementary submodel of both, then $\mathfrak{B}_1 \prec \mathfrak{B}_2$.*

This is proved by a very simple argument which has been given in the proof of Theorem II.4 (proof of the "claim"). Q.E.D.

In the following theorem M is transitive. The theorem and the proof can be obviously generalized to arbitrary M's.

THEOREM 2. *Let $j_i : M \rightarrow \mathfrak{A}^i$, $i = 1, 2$, be γ-cofinal elementary embeddings, where γ is an infinite cardinal in M. Assume that h_0 is a 1-1 mapping of $\{\xi : \mathfrak{A}^1 \models \xi \in j_1\gamma\}$ onto $\{\xi : \mathfrak{A}^2 \models \xi \in j_2\gamma\}$ such that* (i) $\mathfrak{A}^1 \models \langle \xi_1, \xi_2 \rangle \in j_1 z \Leftrightarrow \mathfrak{A}^2 \models \langle h_0\xi_1, h_0\xi_2 \rangle \in j_2 z$, *for every $z \subset \gamma^2$ in M and every ξ_1, ξ_2 which, in $\mathfrak{A}^1$, are $<j_1\gamma$.*

Then, there is a unique isomorphism $h : \mathfrak{A}^1 \simeq \mathfrak{A}^2$ which extends h_0 such that

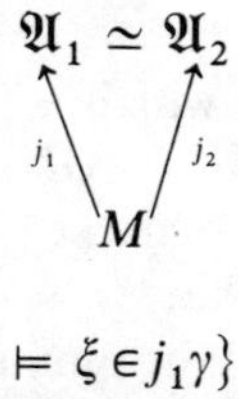

is commutative. If, moreover, $\{\xi : \mathfrak{A}^1 \models \xi \in j_1\gamma\}$ is well-ordered by $\in^{\mathfrak{A}^1}$, and h_0 maps it isomorphically onto the corresponding well-ordered set of $\mathfrak{A}^2$, then (i) *can be* *replaced by* (i') $\mathfrak{A}^1 \models \xi \in j_1 z \Leftrightarrow \mathfrak{A}^2 \models h_0\xi \in j_2 z$, *for all* $z \subset \gamma$ *in* M *and all* ξ *such that* $\mathfrak{A}^1 \models \xi \in j_1\gamma$.

A similar claim is true if the j_i *are* $<\gamma$*-cofinal*, h_0 *is defined on* $\{\xi : \mathfrak{A}^1 \models \xi \in j_1\delta$, *for some cardinal* $\delta < \gamma$ *in* $M\}$ *and satisfies the analogous property for all* $z \subset \delta^2$ *in* M, *and for all cardinals* $\delta < \gamma$ *in* M.

PROOF. We treat the case of γ-cofinal embeddings. The second case is proved in a completely analogous way.

Without loss of generality assume that h_0 is the identity.

If q is a mapping in M of y onto γ, let h_q be the mapping defined for all x which are, in $\mathfrak{A}^1$, members of $j_1 y$, such that

$$h_q(x) = (j_2 q)^{-1}(j_1 q)(x).$$

Here "$(j_1 q)(x)$" denotes the value of $j_1 q$ when applied in $\mathfrak{A}^1$ to x, and "$(j_2 q)^{-1}(z)$" denotes the value of the inverse in $\mathfrak{A}^2$ of $j_2 q$ when applied in $\mathfrak{A}^2$ to z. (Note that, in $\mathfrak{A}^1$, $(j_1 q)(x)$ is an ordinal $<j_1\gamma$; hence it is in $\mathfrak{A}^2$ in the range of $j_2 q$. Thus, h_q is well defined.)

A straightforward checking shows that any h which satisfies the requirement of the theorem must coincide with h_q for all $x \in D(h_q)$. Since j_1 is γ-cofinal, any $x \in |\mathfrak{A}^1|$ is in $\mathfrak{A}^1$ a member of some $j_1 y$, where $M \models \kappa(y) \leqq \gamma$. We can assume that $M \models \kappa(y) = \gamma$, for, otherwise, replace it by $y \cup \gamma$. Taking $q \in M$ which maps y onto γ we have $x \in D(h_q)$. Hence h is uniquely determined.

Let $Q = \{q \in M : q \text{ is a function onto } \gamma\}$. It remains to be shown that $\bigcup_{q\in Q} h_q$ has the desired properties.

Let y, y' be, respectively, the domains of q, q', where $q, q' \in Q$. Let $z = y \cup y'$ and let q^* be a mapping in M which maps z onto γ. Let $g = q^* \circ q^{-1}$, then $g \in M$ and $\forall x \in y\,[q^*x = g(qx)]$; hence $\mathfrak{A}^1 \models \forall x \in j_1 y[(j_1 q^*)(x) = (j_1 g)(j_1 q)(x)]$ and, similarly for $\mathfrak{A}^2$.

If $x \in D(h_{q^*})$, then $h_{q^*}(x) = (j_2 q^*)^{-1}(j_1 g)(j_1 q)(x)$. Now, $g \subset \gamma^2$. Hence (i) implies that $(j_1 g)(t) = (j_2 g)(t)$ for all t which in $\mathfrak{A}^1$ are $<j_1\gamma$. Consequently $h_{q^*}(x) = (j_2 q^*)^{-1}(j_2 g)(j_1 q)(x)$. But $\mathfrak{A}^2 \models (j_2 q^*)^{-1} = (j_2 q)^{-1} \circ (j_2 g)^{-1}$. Hence $h_{q^*}(x) = (j_2 q)^{-1}(j_1 q)(x) = h_q(x)$. This shows that $h_q \subset h_{q^*}$. By symmetry $h_{q'} \subset h_{q^*}$. Thus $\{h_q : q \in Q\}$ is directed by inclusion and this implies that $\bigcup_{q\in Q} h_q$ is a function. Call it h. Obviously,

$$h \text{ maps } |\mathfrak{A}^1| \text{ into } |\mathfrak{A}^2|.$$

Now exchange the roles of $\mathfrak{A}^1$ and $\mathfrak{A}^2$ and, for every $q \in Q$, put $\bar{h}_q(x) = (j_1 q)^{-1}(j_2 q)(x)$. A straightforward checking shows that $\bar{h}_q h_q(x) = x$ for all $x \in D(h_q)$ and, by symmetry, $h_q\bar{h}_q(x) = x$ for all $x \in D(\bar{h}_q)$. Hence $\bar{h}_q = h_q^{-1}$ and $\bar{h}$ (that is to say, $\bigcup_{q\in Q} \bar{h}_q$) is h^{-1}. Thus, h is 1-1 and onto.

It remains to show that $\mathfrak{A}^1 \models x_1 \in x_2 \Leftrightarrow \mathfrak{A}^2 \models hx_1 \in hx_2$, and similarly for other predicates in the language, if there are any. Take y of cardinality γ in M such that $\mathfrak{A}^1 \models x_i \in j_1 y, i = 1, 2$, and let q be a mapping in M of y onto γ. Let $z = \{\langle qz_1, qz_2\rangle : z_1, z_2 \in y \text{ and } z_1 \in z_2\}$. Then $z \in M$ and, by (i), for all ξ_1, ξ_2 which, in $\mathfrak{A}^1$, are $<j_1\gamma$, $\mathfrak{A}^1 \models \langle\xi_1, \xi_2\rangle \in j_1 z \Leftrightarrow \mathfrak{A}^2 \models \langle\xi_1, \xi_2\rangle \in j_2 z$. A routine checking shows that

$$\mathfrak{A}^1 \models \text{`}x_1 \in x_2\text{'} \Leftrightarrow \mathfrak{A}^1 \models \text{`}\langle(j_1 q)(x_1), (j_1 q)(x_2)\rangle \in j_1 z\text{'}$$
$$\Leftrightarrow \mathfrak{A}^2 \models \text{`}\langle(j_1 q)(x_1), (j_1 q)(x_2)\rangle \in j_2 z\text{'}$$

and that $(j_1 q)(x_i) = (j_2 q)(h_q(x_i))$, $i = 1, 2$. Since for every t_1, t_2 which, in $\mathfrak{A}^2$, are members of $j_2 y$, we have $\mathfrak{A}^2 \models \text{`}\langle(j_2 q)t_1, (j_2 q)t_2\rangle \in j_2 z \leftrightarrow t_1 \in t_2\text{'}$, the desired equivalence follows.

If the language has n-ary relations, where $n > 2$, observe that (i) implies the analogous condition for n-ary relations, since by standard coding techniques every n-ary relation can be encoded inside M into several binary relations; the same encoding works in $\mathfrak{A}^i$, $i = 1, 2$, and when applied to $j_i\gamma$ it is absolute (where we assume that h_0 is the identity).

Finally, we show that, under the additional assumptions, (i′) implies (i). Without loss of generality we can identify every ξ such that $\mathfrak{A}^i \models \xi \leqq j_i\gamma$ with an actual ordinal and let $\in^{\mathfrak{A}^i}$ on these ordinals be the real $\in$. Using a mapping in M of γ^2 onto γ one can encode, in M, binary relations over γ into subsets of γ. Moreover, one can use a mapping $\gamma^2 \to \gamma$, obtained by some diagonal procedure, which depends only on the ordinal γ, not on the model in which we are working. The same process will give the similar mapping, in $\mathfrak{A}^i$, of $(j_i\gamma)^2$ onto $j_i\gamma$. The encodings are absolute, from which (i) easily follows. Q.E.D.

THEOREM 3 (UNIQUE DECOMPOSITION THEOREM). *Let M, N be transitive models, and let $j: M \to N$ be a cofinal elementary embedding. For every infinite cardinal γ of M there is a unique transitive model $\tilde{N}$, and unique elementary embeddings $\tilde{j}: M \to \tilde{N}$, $k: \tilde{N} \to N$ satisfying*

(i) $M \xrightarrow{j} N = M \xrightarrow{\tilde{j}} \tilde{N} \xrightarrow{k} N$, *and*

(ii) $x \in \tilde{N} \Leftrightarrow \exists y[M \models \text{`}\kappa(y) \leqq \gamma\text{'}$ *and* $kx \in jy]$.

These embeddings (and the models) also have and are uniquely determined by (i) *and the following two properties*:

(iii) $\tilde{j}$ *is γ-cofinal, and*

(iv) $k(\beta) = \beta$ *for all* $\beta \leqq \tilde{j}\gamma$ *(unique determination continues to hold if* "$\beta \leqq \tilde{j}\gamma$" *is replaced by* "$\beta < \tilde{j}\gamma$").

$\tilde{j}: M \to \tilde{N}$ *is uniquely determined by* (i), (iii) *and*

(v) $\tilde{j}y = jy$ *for all* $y \subset \gamma$ *in* M.

An analogous claim is true for $<\gamma$-cofinal embeddings: Replace, in (ii), "$\kappa(y) \leqq \gamma$" *by* "$\kappa(y) < \gamma$," *in* (iii), "γ-*cofinal*" *by* "$<\gamma$-*cofinal*," *change* (iv) *to*

(iv′) $k\beta = \beta$ *for all* $\beta \leqq \tilde{j}\delta$ *whenever* $M \models \text{`}\delta = \kappa(\delta) < \gamma\text{'}$

and change (v) *in a similar way.*

DEFINITION 2. $\tilde{j}: M \to \tilde{N}$ is the *γ-approximation* (*$<\gamma$-approximation* in the case of a $<\gamma$-cofinal embedding) of $j: M \to N$.

PROOF OF THEOREM 3. It is easy to check that the only way to get embeddings satisfying (i) and (ii) is to let $\tilde{N}$ be the transitive model isomorphic to N' where

$N' = \{x \in N\colon$ for some y of cardinality $\leqq \gamma$ in $M, x \in jy\}$, to let $k(x) = h^{-1}(x)$ where h is the isomorphism $N' \simeq \tilde{N}$, and to let $\tilde{j}(x) = h(jx)$. Since, by Theorem 1, $N' \prec N$, these definitions yield elementary embeddings as required. Obviously, (iii) holds.

(iv) Since $M \models$ '$\kappa(\gamma) = \kappa(\gamma + 1)$,' every $\gamma' \in j(\gamma + 1)$ is in N'. Hence $h(\beta) = \beta$ for all $\beta \leqq j\gamma$ and consequently the same is true of k. Since $\tilde{j}\gamma \leqq j\gamma$ (obvious), we get $\beta \leqq \tilde{j}\gamma \Rightarrow k\beta = \beta$.

Now, (i) and (iv) imply, by themselves, (v) for we have (given $y \subset \gamma$ in M) $jy = k(\tilde{j}y), \tilde{j}y \subset \tilde{j}\gamma$ and, since k keeps fixed every $\beta \leqq \tilde{j}\gamma, k(\tilde{j}y) = \tilde{j}y$.

The uniqueness of $\tilde{j}\colon M \to \tilde{N}$ which satisfies (i), (iii) and (v) follows from Theorem 2 and the observation that every two isomorphic transitive models are identical. Finally, if $x \in \tilde{j}y$ and $M \models \kappa(y) \leqq \gamma$, let $q \in M$ be a mapping of γ onto y. Then, $\tilde{j}q$ is a mapping of $\tilde{j}\gamma$ onto $\tilde{j}y$; hence, for some $\beta < \tilde{j}\gamma, x = (\tilde{j}q)(\beta)$. Consequently $k(x) = (k(\tilde{j}q))(k\beta)$. Assuming (iv) (with an inequality, $<$) we have $k\beta = \beta$. Since $k(\tilde{j}q) = jq$ we get $kx = (jq)(\beta)$.

This shows that $k\colon \tilde{N} \to N$ is uniquely determined by (i), (iii) and (iv).

The treatment of $<\gamma$-approximations is completely analogous. Q.E.D.

REMARK 1. The γ-approximation of $j\colon M \to N$ is not the identity mapping $M \to M$ iff $\gamma \geqq$ first ordinal moved by j. This is obvious.

REMARK 2. By iterating the decomposition one can get a sequence of elementary embeddings $M_0 \to M_1 \to M_2 \to \cdots \to M_\alpha \to \cdots \to N$ such that $M_\xi \to M_\zeta = M_\xi \to M_\eta \to M_\zeta$, and similarly with M_ζ replaced by N; all the models are transitive; M_α, for limit ordinals, as well as N, is a direct limit of the preceding models with respect to these embeddings and, for every β, $M_\beta \to M_{\beta+1}$ is the μ_β-approximation of $M_\beta \to N$ where $\mu_\beta =$ first ordinal moved by $M_\beta \to N$ (or μ_β is chosen in some other way).

REMARK 3. The larger γ is the richer the model $\tilde{M}$ of the $<\gamma$-approximation. Many properties of $j\colon M \to N$ are shared by $<\gamma$-approximations if γ is sufficiently large. As we shall see, this and their characterizátion in terms of limit ultrapowers makes them useful.

Limit ultrapower extensions.

Terminology. From now on I is any set.

A partition of I is a set p of disjoint nonempty subsets of I such that $\bigcup p = I$.

p_1 is a refinement of p_2, in symbols "$p_1 \leqq p_2$", if every $x \in p_2$ is a union of members of p_1.

For two partitions p_1, p_2, put $p_1 \wedge p_2 = \{x_1 \cap x_2 \colon x_1 \in p_1, x_2 \in p_2, x_1 \cap x_2 \neq \varnothing\}$. Then $p_i \geqq p_1 \wedge p_2, i = 1, 2$, and every p which is a refinement of both p_1 and p_2 is a refinement of $p_1 \wedge p_2$.

A filter P of partitions is a nonempty family of partitions such that

$$p_1, p_2 \in P \;\&\; p \geqq p_1 \wedge p_2 \Rightarrow p \in P.$$

From now on we let "p", "p_1", ..., "p'", ... denote partitions of I and "P", "P'", ... etc. denote filters of partitions of I.

If $g \in A^I$ let $p_g = \{g^{-1}(a) : a \in A$ and $g^{-1}(a) \neq \emptyset\}$. Then p_g is a partition of I. We put $A^I_P = \{g \in A^I : g_p \in P\}$.

If $\mathfrak{F}$ is a filter over I we let $A^I_{P,\mathfrak{F}} = \{g/\mathfrak{F} : g \in A^I_P\}$, where $g/\mathfrak{F}$ represents the equivalence class of g modulo $\mathfrak{F}$. Similarly, $\mathfrak{A}^I_{P,\mathfrak{F}}$ is the substructure of the reduced power $\mathfrak{A}^I_{\mathfrak{F}}$ whose universe is $A^I_{P,\mathfrak{F}}$. The canonical embedding $\mathfrak{A} \to \mathfrak{A}^I_{P,\mathfrak{F}}$ is, as usual, the embedding which sends every $a \in |\mathfrak{A}|$ to $c_a/\mathfrak{F}$ where $c_a(i) = a$ for all $i \in I$. If $\mathfrak{F}$ is an ultrafilter the canonical embedding $\mathfrak{A} \to \mathfrak{A}^I_{P,\mathfrak{F}}$ is elementary. We call a l.u.p. extension (limit ultrapower extension) any embedding of the form $\mathfrak{A} \to \mathfrak{A}^I_{P,\mathfrak{F}} \simeq \mathfrak{B}$ where $\simeq$ is an isomorphism and $\mathfrak{A} \to \mathfrak{A}^I_{P,\mathfrak{F}}$ is the canonical embedding.

The well-known criterion for satisfaction of formulas in an ultrapower holds also for l.u.p.'s, from which we get

(1) $$P \subset P' \Rightarrow \mathfrak{A}^I_{P,\mathfrak{F}} \prec \mathfrak{A}^I_{P',\mathfrak{F}}.$$

Since $\kappa(p_g) \leqq \kappa(A)$ for all $g \in A^I$ we have

(2) $$\mathfrak{A}^I_{P,\mathfrak{F}} = \mathfrak{A}^I_{P^*,\mathfrak{F}} \quad \text{where } P^* = \{p \in P : \kappa(p) \leqq \kappa(\mathfrak{A})\}.$$

(Note that if $P_1 = \{p \in P : \kappa(p) < \gamma\}$ (where $\gamma \geqq \omega$) then P_1 is a filter of partitions.)

An elementary embedding $j : \mathfrak{A} \to \mathfrak{B}$ is *complete* if it can be extended to an elementary embedding $\bar{j} : (\mathfrak{A}, R_i)_{i \in \mathscr{I}} \to (\mathfrak{B}, \bar{j}R_i)_{i \in \mathscr{I}}$, where $\{R_i : i \in \mathscr{I}\} = \{R : R \subset |\mathfrak{A}|^n, n = 1, 2, \dots\}$ and $(\mathfrak{A}, R_i)_{i \in \mathscr{I}}$ is the expanded structure obtained by adding each R_i, $i \in \mathscr{I}$, as an additional relation (i.e., $\bar{j}(x) = j(x)$ for all $x \in |\mathfrak{A}|$ and $\bar{j}$ associates with every relation of $\mathfrak{A}$ the same relation associated by j).

Limit ultrapowers have been introduced by Keisler [**8**]. (He used, instead of filters of partitions, filters on $I \times I$; with filters of partitions one gets a 1-1 correspondence between the l.u.p.'s and the filters on which they are based.) He proved also the fundamental theorem for models which are sets:

An elementary extension is complete iff it is a l.u.p. extension.

THEOREM 4. *An elementary self-extension $j : V \to OV$ is a l.u.p. extension iff, for every x, $\{z : OV \models z \in x\}$ is a set and, for some infinite cardinal γ, j is $<\gamma$-cofinal. Furthermore a l.u.p. extension is $<\gamma$-cofinal ($\gamma \geqq \omega$) iff it is of the form $V \to V^I_{P,\mathfrak{F}} \simeq OV$ where $\kappa(p) < \gamma$ for all $p \in P$.*

The proof uses only ZC^+; hence the theorem can be generalized by replacing V by any model $\mathfrak{A}$ of ZC^+, where the conditions and the l.u.p. are to be interpreted within $\mathfrak{A}$.

PROOF. Assume $\kappa(p) < \gamma$ for all $p \in P$ (γ an infinite cardinal). Given $g \in V^I_P$ let $y = \{g(i) : i \in I\}$. Then, $\kappa(y) < \gamma$ and since $g(i) \in y = c_y(i)$ for all $i \in I$ we have $V^I_{P,\mathfrak{F}} \models g/\mathfrak{F} \in c_y/\mathfrak{F}$. Thus the l.u.p. extension is $<\gamma$-cofinal. Obviously $\{g/\mathfrak{F} : V_{P,\mathfrak{F}} \models g/\mathfrak{F} \in h/\mathfrak{F}\}$ is a set. This proves the easy direction.

Assume that $j : V \to OV$ is $<\gamma$-cofinal and satisfies the other condition. For every β let $\mathfrak{C}(\beta) = (\beta, R_i)_{i \in \mathscr{I}(\beta)}$ where $\{R_i : i \in \mathscr{I}(\beta)\} = \{R : R \subset \beta^n, n = 1, 2, \dots\}$. Put $\bar{\beta} = \{x : OV \models x \in j\beta\}$ and, for every $R \subset \beta^n$, put $\bar{R} = \{\langle x_1, \dots, x_u\rangle : OV \models \langle x_1, \dots, x_n\rangle \in jR\}$. Let $\bar{\mathfrak{C}}(\beta) = (\bar{\beta}, \bar{R}_i)_{i \in \mathscr{I}(\beta)}$, and let $j_\beta : \mathfrak{C}(\beta) \to \bar{\mathfrak{C}}(\beta)$ be the embedding which sends every $\alpha \in \beta$ to $j\alpha$. Obviously, j_β is an elementary embedding which is complete. Put $\beta = \gamma$. Since $\bar{\gamma}$ is a set, we get by Keisler's theorem I, P', $\mathfrak{F}$ and an isomorphism $h(\gamma)$ such that

$$j_\gamma = \mathfrak{C}(\gamma) \to \mathfrak{C}(\gamma)^I_{P',\mathfrak{F}} \overset{h(\gamma)}{\simeq} \overline{\mathfrak{C}}(\gamma).$$

If $\beta < \gamma$ then every $R \subset \beta^n$ is also an n-ary relation over γ and the corresponding relations in $\overline{\mathfrak{C}}(\beta)$ and $\overline{\mathfrak{C}}(\gamma)$ are the same, namely $\bar{R}$. Also, $j_\beta(\alpha) = j(\alpha) = j_\gamma(\alpha)$, for all $\alpha \in \beta$. Treating $\mathfrak{C}(\beta)$ and $\overline{\mathfrak{C}}(\beta)$ as substructures in the general sense (i.e., reducts of substructures) of $\mathfrak{C}(\gamma)$ and $\overline{\mathfrak{C}}(\gamma)$, respectively, we get the commutative diagram

$$\mathfrak{C}(\beta) \subset \mathfrak{C}(\gamma) \xrightarrow{j_\gamma} \overline{\mathfrak{C}}(\gamma) = \mathfrak{C}(\beta) \xrightarrow{j_\beta} \overline{\mathfrak{C}}(\beta) \subset \overline{\mathfrak{C}}(\gamma).$$

A similar diagram is obtained if we replace $\overline{\mathfrak{C}}(\beta)$ and $\overline{\mathfrak{C}}(\gamma)$ by the limit ultrapowers of $\mathfrak{C}(\beta)$ and $\mathfrak{C}(\gamma)$, respectively, and j_β, j_γ by the canonical embeddings.

Consequently, $h(\beta)$ which is defined by: $h(\beta)(g/\mathfrak{F}) = h(\gamma)(g/\mathfrak{F})$ (for all $g \in \beta^I_P$) is an isomorphism such that

$$\mathfrak{C}(\beta) \to \overline{\mathfrak{C}}(\beta) = \mathfrak{C}(\beta) \to \mathfrak{C}(\beta)^I_{P',\mathfrak{F}} \overset{h(\beta)}{\simeq} \overline{\mathfrak{C}}(\beta).$$

Put $P = \{p \in P' : \kappa(p) < \gamma\}$. Then, $\mathfrak{C}(\beta)^I_{P',\mathfrak{F}} = \mathfrak{C}(\beta)^I_{P,\mathfrak{F}}$ for all $\beta < \gamma$.

Now consider the canonical embedding $V \to V^I_{P,\mathfrak{F}}$ and $j: V \to OV$. If we put $h_0 = \bigcup \{h(\beta) : \beta = \kappa(\beta) < \gamma\}$, then the assumptions of Theorem 2 (for the case of $<\gamma$-cofinal embeddings) are satisfied and we get the desired isomorphism $V \simeq V^I_{P,\mathfrak{F}}$. Q.E.D.

COROLLARY. *Quantification of the form "$\exists$ elementary $<\gamma$-cofinal embedding $j: V \to M$ such that M is transitive and ..." is equivalent in ZC^+ to first-order quantification.*

(Here "..." stands for the formula which succeeds the quantifier; "γ", "j", "M" and other parameters can occur in it.)

This is so because one can quantify over I, P, $\mathfrak{F}$ which are first-order entities.

Definitions of large cardinals via elementary embeddings. Many properties, in fact, almost all natural properties of elementary embeddings $V \to M$, with M transitive, which one thinks of, are shared by their γ-approximations, if γ is large enough. Hence, using the last corollary, statements involving second-order quantification of the form "$\exists$ elementary embedding $j: V \to M$" can be translated, in these cases, to first-order statements. This makes possible many definitions of large cardinals, which use the concept of an elementary embedding of V. Here are some examples. In this section $M, M', \ldots$ range over transitive classes, and we identify them with the substructures of V which they determine.

Consider the following schema:

(I_X) $j: V \to M$ is an elementary embedding, μ is the first ordinal moved by j and $X \subset M$.

Here X is any *set*. X can be defined also in terms of μ and other first-order parameters; e.g., $X = R(\mu + \alpha)$.

It is easy to check that if $j: V \to M$ satisfies (I_X) so does its γ-approximation $\tilde{j}: V \to \tilde{M}$, provided that $\gamma \geqq \mu, \kappa(\bar{X})$, where $\bar{X}$ = transitive closure of X. Consequently quantifying over embeddings which satisfy (I_X) is first-order quantification.

A variation of (I_X) is obtained by letting X be defined in terms of j; for example, putting $X = R(j\mu + \alpha)$ we get

$j: V \to \tilde{M}$ is elementary, μ = first ordinal moved and $R(j\mu + \alpha) \subset M$.

Or we can put $X = R(\mu_n + \alpha)$ where $\mu_n = j^n\mu$, or $X = R(\mu_\omega)$, $\mu_\omega = \sup\{\mu_n : n < \omega\}$, or $X = R(j^n\alpha)$, or $X = \{\langle x, jx\rangle : x \in R(j\mu + \alpha)\}$, etc. (Recall the conditions in Theorems II.7, 8.) In all such cases the γ-approximation (for $\gamma \geqq \mu, \kappa(\bar{X})$) will share this property. Hence the resulting quantification is of the first order. This will be true as long as X is, so to speak, locally defined.

Conditions of the form (I_X), with $X = R(j(\mu + \alpha))$, are, of course, closely connected with the concept of an α-extendible cardinal. These connections will be clarified in the next section.

A local property of a different kind which is shared by suitable γ-approximations is

(II_α) $j: V \to M$ is elementary, μ = first ordinal moved by j and $M^\alpha \subset M$.

Here, $M^\alpha =_{\mathrm{Df}} \{\langle x_\xi\rangle_{\xi<\alpha} : x_\xi \in M\}$.

Claim. If $j: V \to M$ *satisfies* (II_α), *so does its* γ-*approximation*, $\tilde{j}: V \to \tilde{M}$, *provided that* $\kappa(\gamma^\alpha) = \gamma$.

PROOF. Let

$$V \xrightarrow{j} M = V \xrightarrow{\tilde{j}} \tilde{M} \xrightarrow{k} M.$$

Given $\langle x_\xi\rangle_{\xi<\alpha} \in \tilde{M}^\alpha$ choose for each ξ a set y_ξ of cardinality $\leqq \gamma$ such that $kx_\xi \in jy_\xi$ and let $y = \bigcup_{\xi<\alpha} y_\xi$. Put $y^* = \bigcup_{\delta \leqq \alpha} y^\delta$; then $M \models$ '$jy^* = \bigcup_{\delta \leqq j\alpha} (jy)^\delta$'.

Now $\langle kx_\xi\rangle_{\xi<\alpha} \in M^\alpha \subset M$. Since $M \models kx_\xi \in jy$ and $\alpha \leqq j\alpha$ we have $\langle kx_\xi\rangle_{\xi<\alpha} \in jy^*$. But y^* is, by our assumption, of cardinality $\leqq \gamma$. Hence, for some $z \in \tilde{M}$, $\langle kx_\xi\rangle_{\xi<\alpha} = kz$. Since k keeps fixed every ordinal $\leqq j\gamma$ (see Theorem 3) and $\alpha < \gamma \leqq j\gamma$ we have $z = \langle x_\xi\rangle_{\xi<\alpha}$. Q.E.D.

Consequently, one can quantify over embeddings satisfying (II_α). For $\alpha > \mu$, the existence of j satisfying (II_α) is well known to be equivalent to the requirement that μ is α-supercompact. (The concept has been introduced by Solovay. See **[16]**, **[17]**, **[13]** and **[14]**. For $\alpha \leqq \mu$ this means that μ is measurable). The original definition of α-supercompactness is a first-order one. Moreover, as proved in Reinhardt-Solovay **[17]**, every elementary embedding $j: V \to M$ satisfying (II_α) has a factorization into elementary embeddings $V \xrightarrow{j^*} M^* \xrightarrow{k} M$ such that j^* is an ultrapower extension satisfying (II_α) and μ is the first ordinal moved by it ($j^* = V \to V^I_{\mathfrak{F}} \simeq M^*$, where $I = \{x : x \subset \alpha \wedge \kappa(x) < \mu\}$, $\mathfrak{F} = \{x \in I : j''\alpha \in jx\}$ and, identifying M^* with the ultrapower, $k(g/\mathfrak{F}) = (jg)(j''\alpha)$).

The point of using γ-approximations is, however, that a translation into a first-order property is automatically provided whenever the condition on the embedding $j: V \to M$ is of a "local" character. (II_α) is just one case among many. We can consider a general schema of the form:

$j : V \to M$ is elementary and μ is the first moved by j and $\Phi(j, M, \mu, \ldots)$, where Φ is a condition involving $j : V \to M$ and, possibly, other parameters. Quantifying over j is equivalent to first-order quantification as long as Φ has a "local" character with respect to $j : V \to M$. For example, every formula of C is for our purposes "local", where C is the smallest class of formulas containing the following ones and closed under conjunctions, disjunctions and existential quantification:

$$M^\alpha \subset M, \quad \varphi(j, M, \ldots), \quad \psi(\mu, \ldots),$$

where φ is a Σ_1-formula involving the classes j (conceived as a binary relation) and M (and any additional first-order variables) and ψ is any formula not involving $j : V \to M$. The first-order parameters in the φ's and ψ's can, of course, include any parameters which occur in any of the other participating formulas. (A Σ_1-formula is, as defined in Lévy's hierarchy, a formula obtained by prefixing a block of existential unbounded quantifiers to a Σ_0-formula.) Note that '$x \subset M$', '$R(j(\mu + \alpha)) \subset M$', '$R\mu_\omega \subset M$', etc., as well as their negations, are all Σ_1-formulas; hence we did not mention them explicitly. '$M^\alpha \not\subset M$' is also Σ_1 (but not '$M^\alpha \subset M$'). For example the following formula is in C:

$$[R(j^2\mu) \subset M] \wedge [M^{j\mu+\alpha} \subset M] \wedge [M^{j\beta} \not\subset M] \wedge \exists n > m \exists \delta[\alpha = j^n\delta].$$

By now, it should not be difficult to verify that, for a suitable γ, the γ-approximation of $j : V \to M$ will share the property expressed by a formula of C.

Using only variations of (I_X) we get, among the rest, the following variants of the concept of extendibility:

(n, α)-c-extendibility: $R(\mu_n + \alpha) \subset M$.
$[n, \alpha]$-c-extendibility: $R(j^n(\mu + \alpha)) \subset M$.
(n^*)-c-extendibility: $\forall\alpha[\mu$ is (n, α)-c-extendible$]$.
$[n^*]$-c-extendibility: $\forall\alpha[\mu$ is $[n, \alpha]$-c-extendible$]$.

"c" is added here to distinguish these concepts from Reinhardt's concept which is based on elementary embeddings of the form $V_\alpha \to V_\beta$. The analogue of α-extendibility, in his sense, is $[1, \alpha]$-c-extendibility. The relationship between these is analyzed in the next section. $[1^*]$-c-extendibility and extendibility (i.e., α-extendibility, for all α) turn out to be equivalent. Measurability is equivalent to (0, 0)-extendibility. Using standard methods together with the lifting techniques of the next section it is not difficult to prove

THEOREM. *If μ is $(n + 1, 0)$-c-extendible then there are many $\mu' < \mu$ which in V_μ are $[n^*]$-c-extendible. They constitute a set in the normal ultrafilter associated with the embedding $V \to M$ which satisfies $(\mathrm{I}_{R(\mu_n+1)})$.*

One can get a few other variants along the same lines and determine similar relationships. Comparisons of the first cardinals which satisfy the given variants (assuming that there are any) are, in general, much more difficult and there are many open problems.

Variants of (II_α) can be similarly constructed, as well as mixed variants based on both (I_X) and (II_α).

An example of a "global" property which cannot be dealt with via γ-approximations is: μ is the first ordinal moved by $j : V \to M$ and $M = V$. This is, however, equivalent to a first-order property for other reasons—it is contradictory.

Axioms of infinity which cannot be rephrased as existence requirements of certain elementary embeddings having "local" properties of the kinds which we discussed are the following:

The class of all strongly inaccessible cardinals is unbounded.

The schema asserting that this class is stationary.

Vopenka's principle (see [**14**]), etc.

Note that the properties of cardinals which we get via γ-approximations of elementary embeddings relativize to V_α if α is a large enough strongly inaccessible cardinal.

Liftings of embeddings and extendible cardinals. From now on $M, M', \ldots, M_1, \ldots, N, N_1, \ldots$, etc. are transitive models. Again, we use the same names for structures and for their underlying universes.

DEFINITION 3. A *partial lifting* of an embedding $j : M \to N$ is an embedding $j^* : V \to N^*$ such that $j(x) = j^*(x)$ for all $x \in M$. A *lifting* is a partial lifting such that $N \subset N^*$. *j can be lifted* (*partially lifted*) if it has a lifting (partial lifting).

We shall always use these concepts for cases in which M and N are sets. By the method of the last section the existence of a lifting, or a partial lifting can be translated into a first-order condition. If a (partial) lifting exists then there is one which is a l.u.p. extension. Consequently, to find out whether there is a partial lifting one has to find whether some l.u.p. extension coincides with j on M and yields a well-founded structure when applied to V; in the case of a lifting the transitive isomorphic structure should include N as a subset.

We continue to use the terminology and notations of the last section concerning l.u.p.'s.

LEMMA 5. *If γ is a cardinal,* $\mathrm{cof}(\gamma) > \omega$ *and $\kappa(p) < \gamma$ for all $p \in P$ and if $(\lambda, \in|\lambda)^I_{P,\mathfrak{F}}$ is well-ordered* (*by its binary relation*) *for all $\lambda < \gamma$, then it is well-ordered for all λ and, consequently, $V^I_{P,\mathfrak{F}}$ is well-founded.*

Originally, the author proved this under the assumption that γ is regular, which is the only case where the lemma will be needed. Powell pointed out to him that a shorter and, indeed, almost trivial argument proves the more general claim. The argument runs as follows.

PROOF. If for some $\lambda \geqq \gamma$ the l.u.p. of $(\lambda, \in|\lambda)$ is not well-founded, let $g_1/\mathfrak{F} > g_2/\mathfrak{F} > \cdots > g_n/\mathfrak{F} > \cdots$ be a decreasing sequence (where $<$ is the relation of the l.u.p. which corresponds to $\in|\lambda$). Put $X_n = g_n''I$, $X = \bigcup_{n<\omega} X_n$. Since $\kappa(p_{g_n}) < \gamma$, $\kappa(X_n) < \gamma$; hence by our assumptions $\kappa(X) < \gamma$. Hence $(X, \in|X)$ is isomorphic to $(\lambda', \in|\lambda')$ for some $\lambda' < \gamma$. But $g_n \in X^I_P$; hence $(X, \in|X)^I_{P,\mathfrak{F}}$ is not well-ordered. Contradiction. Q.E.D.

DEFINITION 4. Let X be a subset of the model M, and let $\alpha \subset M$. We say that X^α *can be encoded in* M if there is a formula which defines in M a function $F^M(t, x, y)$ and there is $t \in M$ such that for every $\langle x_\xi \rangle_{\xi<\alpha} \in X^\alpha$ there is $x \in M$ satisfying $F^M(t, x, \xi) = x_\xi$ for all $\xi < \alpha$. F is called *the encoding* function. (Note that t is a fixed parameter in M.)

It is not difficult to show that

(3) $(R\alpha)^\omega$ *can be encoded in* V_α *iff* $\alpha > \omega$ *and* $\mathrm{cof}(\alpha) \neq \omega$ (including the case $\alpha = \alpha' + 1$).

($\mathrm{cof}(\alpha) = \omega \Rightarrow k(R\alpha)^\omega > k(R\alpha)$ hence the condition is necessary. $\mathrm{cof}(\alpha) > \omega \Rightarrow (R\alpha)^\omega \subset R\alpha$. For $\mathrm{cof}(\alpha) = 1$ put $\alpha = \beta + n$ where $1 \leqq n < \omega$ and $\beta = \bigcup \beta > 0$ and show by induction on n that an encoding exists; show that an encoding of $(R\delta)^\omega$ in V_δ can be used to get an encoding of $(R(\delta + 1))^\omega$ in $V_{\delta+1}$. If $\mathrm{cof}(\beta) = \omega$ take a sequence $\langle \beta_n \rangle_{n<\omega}$ whose limit is β and use $\langle R(\beta_n) \rangle_{n<\omega}$ as a parameter in $V_{\beta+1}$ to get an encoding of $(R(\beta + 1))^\omega$ in $V_{\beta+1}$.)

THEOREM 6. (I) *Let* $j: V_{\alpha+1} \to N$ *be an elementary embedding and let* $j\alpha = \beta$. *If* $(R^N\beta)^\omega$ *can be encoded in* N *then* j *has a partial lifting* $j^*: V \to N^*$ *such that* $N^*_\beta = N_\beta$.

(*Recall that* $R^N\beta = \{x : N \vDash \mathrm{rank}(x) < \beta\} = R\beta \cap N$ *and* $N_\beta =$ *substructure of* N *whose universe is* $R^N\beta$.)

(II) *If* $j: M \to N$ *is a complete elementary embedding where* M *and* N *are sets, and* N^ω *can be encoded in* N *then* j *has a lifting* j^* *such that* $j^*(M) = N$.

PROOF. (I) Put $\mathfrak{A} = (V_\alpha, X)_{X \subset R\alpha}$, $\mathfrak{B} = (N_\beta, jX)_{X \subset R\alpha}$. Let $\tilde{j}: \mathfrak{A} \to \mathfrak{B}$ be such that $\tilde{j}(x) = j(x)$. Obviously $\tilde{j}$ is elementary. Obviously $\beta \geqq \omega$; hence, $\alpha \geqq \omega$. Since $(R\alpha)^2$ can be encoded in V_α, one can encode in the standard way an n-ary relation over $R\alpha$ into subsets of $R\alpha$. It is not difficult to deduce from it that $\tilde{j}$ is complete. So let $\mathfrak{B} = \mathfrak{A}^I_{P,\mathfrak{F}}$ so that $\tilde{j}$ becomes the canonical embedding. By (2) we can also have $\kappa(p) \leqq \kappa(R\alpha)$, for all $p \in P$.

If $V^I_{P,\mathfrak{F}}$ is well-founded we are done because we can identify it with some transitive class N^* and a routine checking shows that $R^{N^*}\beta = R^N\beta$, and that the canonical embedding $V \to N^*$ coincides with j on $V_{\alpha+1}$.

(The identification of $V^I_{P,\mathfrak{F}}$ with N^* will extend the identification of $\mathfrak{A}^I_{P,\mathfrak{F}}$ with $(N_\beta, jX)_{X \subset R\alpha}$.)

Using Lemma 5 it suffices to show that $(\lambda, \in|\lambda)^I_{P,\mathfrak{F}}$ is well-ordered for all $\lambda < \kappa(R\alpha)^+$. Obviously, one can replace $(\lambda, \in|\lambda)$ by any isomorphic model. Since $\kappa(\lambda) \leqq \kappa(R\alpha)$, we can take $S \subset R\alpha \times R\alpha$ of order type λ. If α is a limit ordinal then $S \subset R\alpha$, (V_α, S) is a reduct of $\mathfrak{A}$ and we have $(V_\alpha, S)^I_{P,\mathfrak{F}} = (N_\beta, jS)$. We have to show that jS is well-ordered. The idea of the proof is to use the encoding of $(R^N\beta)^\omega$ in N in order to define the concept of an ω-sequence. Applying this concept in $V_{\alpha+1}$ we get: S is well-ordered. Hence, jS is well-ordered in N, and since all ω sequences are encoded, this is absolute. Carried out in detail, the argument is:

Let F be the encoding function of $(R^N\beta)^\omega$ in N. Then

$$V_{\alpha+1} \vDash \forall u \forall x \exists n < \omega [\langle F(u, x, n + 1), F(u, x, n) \rangle \notin S];$$

hence the same holds in N with S replaced by jS. Letting t be the parameter in N

which is used in the encoding we get

$$\forall x \in N\, \exists n < \omega[\langle F^N(t, x, n+1), F^N(t, x, n)\rangle \notin jS].$$

Since all ω sequences of members of $R^N\beta$ are encoded and $jS \subset R^N\beta \times R^N\beta, jS$ is well-ordered.

If α is not a limit ordinal we encode first the binary relation S into a subset S_1 of $R\alpha$. This is done by using a definable pairing function, say $[\,,]$, which encodes $(R\alpha)^2$ in V_α. The same function, applied in N_β, will encode there $(R^N\beta)^2$. (Actually, $[\ ,\ ]$ can be defined without using parameters of $R\alpha$; but even if parameters are used they can be replaced when passing to N_β by their images under j.) We put $S_1 = \{[x, y] : \langle x, y\rangle \in S]\}$; then $(\mathfrak{A}, S) \models \forall x, y(S(x, y) \leftrightarrow S_1([x, y]))$ and the same holds in $(\mathfrak{B}, \bar{S})$ with S replaced by $\bar{S}$, S_1 by jS_1 (where $\bar{S}$ is the relation corresponding to S under the l.u.p. extension). Now use the same argument with S_1 instead of S, and $[\ ,\ ]$ instead of $\langle\ ,\ \rangle$.

(II) Put $N = M^I_{P,\mathfrak{F}}$ so that j is the canonical embedding and $\kappa(p) \leqq \kappa(M)$ for all $p \in P$. If $V^I_{P,\mathfrak{F}}$ is well-founded we are done, for, then, identifying it with a transitive set N^*, a routine checking shows that the canonical embedding $j^* : V \to N^*$ coincides with j on M, and sends M to N. Using, as in (I), Lemma 5 (for $\gamma = \kappa(M)^+$), it suffices to prove that, for every well-ordering $S \subset M \times M$, if $(M, S)^I_{P,\mathfrak{F}} = (N, \bar{S})$ then $\bar{S}$ is a well-ordering. Since N^ω can be encoded in N we can use, as in (I), the encoding function to assert inside (M, S) that S is a well-ordering and to deduce that $\bar{S}$ is well-ordered. (The proof is actually shorter here than that of (I).) Q.E.D.

Our aim is to find which cofinal elementary embeddings $j : V_\alpha \to V_\beta$ can be lifted. The case of nonlimit ordinals is to be dealt with via Theorem 6(I). If α is a limit ordinal, Theorem 6(II) is to be used, but then one needs the additional assumption that j is complete. Hence we establish first necessary and sufficient conditions for that.

THEOREM 7. *Let $j : V_\alpha \to N$ be elementary and cofinal, where $\alpha = \bigcup \alpha > 0$. Let* $\alpha^* = \mathrm{cof}(\alpha)$.

(i) *If $\alpha^* = \alpha$ then j is complete.*

(ii) *If $\alpha^* < \alpha$ then j is complete iff $j''\alpha^*$ is cofinal in $j\alpha^*$ (where, as always, $j''\alpha^* = \{j\beta : \beta \in \alpha^*\}$).*

(iii) *If $\alpha^* < \alpha$ and $V_\alpha \not\models ZF$ (repeat—not a model of ZF) then $j''\alpha^*$ is cofinal in $j\alpha^*$, hence j is complete. In particular, j is complete if $\alpha^* < \alpha$ and $\alpha < \kappa(R\alpha)$.*

PROOF. For every $X \subset (R\alpha)^n$ put $\bar{X} = \bigcup_{\beta<\alpha} j(X \cap R\beta)$. It is easy to see that the embedding $(V_\alpha, R_i)_{i\in\mathscr{I}} \to (N, \bar{R}_i)_{i\in\mathscr{I}}$ which sends every x to jx is Σ_0-elementary. Indeed, this is the only way to correlate with relations over $R\alpha$ relations over N so as to get a Σ_0-elementary embedding which agrees with j.

Let $\{R_i\}_{i\in\mathscr{I}} = \{X : X \subset (R\alpha)^n, n = 1, 2, \ldots\}$. Put $\tilde{V}_\alpha = (V_\alpha, R_i)_{i\in\mathscr{I}}$, $\tilde{N} = (N, \bar{R}_i)_{i\in\mathscr{I}}$ and let $\tilde{j} : \tilde{V}_\alpha \to \tilde{N}$ be such that $\tilde{j}x = jx$. Then $\tilde{j}$ is Σ_0-elementary; j is complete iff $\tilde{j}$ is elementary.

Now apply the corollary of Theorem II.1 (i.e., we let the language have a name for each $R_i, i \in \mathscr{I}$). By (i) of this corollary, if $\tilde{V}_\alpha \models$ ZF then $\tilde{j}$ is elementary. Hence if α is strongly inaccessible j is complete. If α is regular but not strongly inaccessible then, for some $x \in R\alpha$, x is a well-ordering of order type α. The isomorphism of x onto α can be defined (by transfinite induction) inside V_α. Since j is elementary, jx is an ordering which is order-isomorphic to the ordinals of N. Since $j''\alpha$ is cofinal in the ordinals of N, it is not difficult to deduce that j maps x cofinally into jx. By (iii) of the corollary, $\tilde{j}$ is elementary.

It is obvious that if α is singular then $\tilde{V}_\alpha \not\models$ ZF (the cofinal embedding of α^* into α is a relation of $\tilde{V}_\alpha$). Hence, (ii) of our theorem follows from (ii) of the corollary to Theorem II.1.

Finally, if $\alpha^* < \alpha$ and $V_\alpha \not\models$ ZF then, for some well-ordering $x \in R\alpha$, one can define (possibly, from parameters) in V_α a cofinal embedding of x into α. Hence $\mathrm{cof}(x) = \alpha^*$, the cofinal embedding of α^* into x is in $R\alpha$ and, therefore, one can define in V_α a cofinal embedding of α^* into α. Consequently, in N, one can define a cofinal embedding of $j\alpha^*$ into the ordinals of N and this implies that j maps α^* cofinally into $j\alpha^*$. The last observation follows from $\alpha < \kappa(R\alpha) \Rightarrow V_\alpha \not\models$ ZF. Q.E.D.

A counterexample, pointed out to the author by Powell, shows that in (iii) one cannot omit "$V_\alpha \not\models$ ZF". Assume μ to be measurable and let $V_\alpha \models$ ZF with $\alpha > \mu$ and $\mathrm{cof}(\alpha) = \mu$. (Such V_α's exist if there is a strongly inaccessible cardinal $>\mu$.) Let N be the transitive model isomorphic to the "ultrapower of V_α" as defined inside V_α, using a $<\mu$-complete nonprincipal ultrafilter over μ. Then $V_\alpha \to N$ is elementary and cofinal but not complete (since μ is not mapped cofinally into its image). Note, however, that $V_\alpha \to N$ can be lifted.

We can now give an almost complete solution to the problem of lifting cofinal elementary embeddings $V_\alpha \to V_\beta$.

THEOREM 8. *Let $j: V_\alpha \to V_\beta$ be elementary and cofinal.*

(i) *If $\alpha = \alpha' + 1$ then j can be partially lifted to $j^*: V \to M$ such that $V_{\beta'} \subset M$, where $\beta' = j\alpha'$.*

(ii) *If α is a regular limit ordinal then j can be lifted.*

(iii) *If $\omega < \alpha^* = \mathrm{cof}(\alpha) < \alpha$ and if $j''\alpha^*$ is cofinal in $j\alpha^*$ then j can be lifted. In particular, j can be lifted if $j\alpha^* = \alpha^*$, or if $V_\alpha \not\models$ ZF.*

(i) and (ii), for strongly inaccessible α's, have been proved independently by Powell [**15**].

Note that, for $\alpha = \bigcup \alpha$, a partial lifting of a cofinal $j: V_\alpha \to N$ is also a lifting (because $N = \bigcup_{\beta<\alpha} j(R\beta)$).

Theorem 9 will show that one cannot, in general, improve upon (i) by making j^* a lifting (i.e., by stipulating that $V_\beta \subset M$) and that one cannot, in general, lift a nontrivial cofinal $V_\alpha \to V_\beta$, if $\mathrm{cof}(\alpha) = \omega$.

As (ii) and (iii) indicate, the case of a limit ordinal of cofinality $>\omega$ has almost always a positive solution. The only exceptions are those pathological cases where $\omega < \alpha^* = \mathrm{cof}(\alpha) < \alpha$ and j does not embed α^* cofinally into $j\alpha^*$. By Theorem 7(iii), V_α is then a model of ZF. Such examples can be constructed, assuming certain mildly strong axioms of infinity; but I do not know of an example which cannot be lifted, so this remains the only case where an improvement of the claim is not outruled. Such an improvement seems to me rather hard, if not unlikely.

[EXAMPLE. Let α_1, α_2 be strongly inaccessible, with $\alpha_1 < \alpha_2$, and assume that there is an elementary $k: V_{\alpha_2} \to V_{\beta_2}$ which moves α_1; this is possible if α_1 is α_2-extendible. Using reflection get α such that $\alpha_1 < \alpha < \alpha_2$, $\mathrm{cof}(\alpha) = \alpha_1$ and $V_\alpha \prec V_{\alpha_2}$. Let $\beta = \sup\{k\alpha' : \alpha' < \alpha\}$ and let $j : V_\alpha \to V_\beta$ be the embedding which sends each x to kx. By the corollary of Theorem II.1, j is elementary and it is easy to verify that $j''\alpha_1$ is not cofinal in $j\alpha_1$. Thus, j is what we called a pathological case. Still, j can be lifted. *Question*: Is there a transitive model N of ZF such that $j: V_\alpha \to V_\beta \in N$ and j cannot be lifted inside N?]

PROOF OF THEOREM 8. (i) follows from Theorem 6(I), the trivial observation that $\beta = \beta' + 1$, and the fact that, for infinite β', $(R\beta')^\omega$ can be encoded in $V_{\beta'+1}$.

(ii) follows from Theorem 7(i), Theorem 6(II) and the fact that $\mathrm{cof}(\beta) > \omega \Rightarrow (R\beta)^\omega \subset R\beta$. In a similar way, (iii) follows from Theorem 7(ii) and Theorem 6(II). That $V_\alpha \not\models$ ZF is sufficient follows from Theorem 7(iii). Q.E.D.

THEOREM 9. (I) *If an elementary nontrivial $j: V_{\alpha+1} \to V_{\beta+1}$ can be lifted then there is $\alpha' < \alpha$ and an elementary embedding $V_{\alpha'+1} \to V_{\alpha+1}$.*

(II) *If* $\mathrm{cof}(\alpha) = \omega < \alpha$ *and $j: V_\alpha \to V_\beta$ is cofinal and can be lifted, where $\alpha < \beta$, then there is $\alpha' < \alpha$ and an elementary cofinal embedding $V_{\alpha'} \to V_\alpha$.*

Solovay pointed out to the author that, using an iterated-extension process and an argument similar to that which proves (II), one gets an analogous version for the case $\alpha = \beta$:

(III) *If* $\mathrm{cof}(\alpha) = \omega$ *and $j: V_\alpha \to V_\alpha$ is nontrivial and can be lifted then there is $\alpha' <$ first ordinal moved by j and a nontrivial elementary embedding $V_{\alpha'} \to V_{\alpha'}$.*

It follows that $j: V_{\alpha+1} \to V_{\beta+1}$ cannot be lifted if α is the first ordinal for which an elementary embedding $V_{\alpha+1} \to V_{\beta+1}$ exists (otherwise get $\alpha' < \alpha$ and an elementary embedding $V_{\alpha'+1} \to V_{\alpha+1}$ and consider the composition $V_{\alpha'+1} \to V_{\alpha+1} \to V_{\beta+1}$). Similar claims follow from (II) and (III). Actually, one can get many ordinals $<\alpha$ with elementary embeddings $V_{\alpha'+1} \to V_{\alpha+1}$ (or $V_{\alpha'} \to V_\alpha$ if $\mathrm{cof}(\alpha) = \omega$ and $\alpha < \beta$, or $V_{\alpha'} \to V_{\alpha'}$). See the remarks to Theorems III.7, 8.

PROOF OF THEOREM 9. (I) Let $j^*: V \to M$ be a lifting of $j: V_{\alpha+1} \to V_{\beta+1}$. If $\alpha = \beta$ then $V_{\alpha+1} \subset M_{\alpha+2} \subset V_{\alpha+2}$ and the embedding $V_{\alpha+2} \to M_{\alpha+2}$ which sends every x to j^*x is elementary and nontrivial. By Kunen's proof [**12**] this is impossible. Hence $\alpha < \beta$. An elementary calculation shows that $\kappa(R(\beta+1)^{R(\alpha+1)}) = \kappa(R(\beta+1))$. This holds also inside M, and since $R^M(\alpha+1)$ and $R^M(\beta+1)$ are the real $R(\alpha+1)$ and $R(\beta+1)$, there is an encoding in M of $R(\beta+1)^{R(\alpha+1)}$ into $R(\beta+1)$. Consequently, every function from $R(\alpha+1)$ into $R(\beta+1)$ is in M and, in particular, $\{\langle x, jx\rangle : x \in R(\alpha+1)\}$. Therefore, the following is true in M:

There is an ordinal $\lambda < j^*\alpha$ and an elementary embedding $V_{\lambda+1} \to V_{j^*\alpha+1}$.

The same should be true in V, with $j^*\alpha$ replaced by α.

(II) Again, we show that, if $j^*: V \to M$ lifts $j: V_\alpha \to V_\beta$ then in M there is an elementary cofinal embedding of some V_λ into $V_{j^*\alpha}$, where $\lambda < j^*\alpha$, and then we argue as in the proof of (I).

Let $x = \langle \alpha_n \rangle_{n<\omega}$ where $\sup\{\alpha_n : n \in \omega\} = \alpha$. Since ω is not moved by j^* we have $j^*x = \langle j^*\alpha_n \rangle_{n<\omega} = \langle j\alpha_n \rangle_{n<\omega}$. Put $\beta_n = j\alpha_n$; since j is cofinal, we have $\sup\{\beta_n : n \in \omega\} = \beta$. Also $\alpha = \sup\{x(n) : n < \omega\} \Rightarrow M \models j^*\alpha = \sup\{(j^*x)(n) : n < \omega\}$, and this implies that $j^*\alpha = \beta$. Now define a tree $(H, <)$ as follows:

$h \in H$ iff for some α_n, h is a mapping of $R\alpha_n$ into $R\beta_n$ such that $h\alpha_i = \beta_i$ for all $i < n$ and $(V_\alpha, x)_{x \in R\alpha_n} \equiv (V_\beta, hx)_{x \in R\alpha_n}$ (where "$\equiv$" denotes elementary equivalence); $h_1 < h_2$ iff $h_1 \subset h_2$.

Every infinite branch in this tree determines an elementary cofinal embedding $V_\alpha \to V_\beta$. There is an infinite branch consisting of $\{\langle x, jx \rangle : x \in R\alpha_n\}$, $n = 0, 1, \ldots$. Since $\langle \beta_n \rangle_{n \in \omega} \in M$, and the other parameters used in the definition of $(H, <)$ are in M as well, the construction inside M yields the same tree. $(H, <)$ cannot be well-founded in M, for then one could assign inside M ordinals to its nodes. Hence there is in M a cofinal elementary embedding $V_\alpha \to V_\beta$.

To prove (III) one observes that $\alpha = \sup\{\mu_n : n \in \omega\}$ where $\mu_n = j^n\mu$ (otherwise one gets an elementary embedding $V_{\mu_\omega + 2} \to V_{\mu_\omega + 2}$ contradicting Kunen's result). After n iterations of the lifting $j^* : V \to M$, μ is pushed up to μ_n. After ω iterations μ is pushed up to α and, by Theorem III.7, if $V^{(\omega)}$ is the resulting universe, then $V_\alpha = V_{\mu_\omega} \subset V^{(\omega)}$. Now construct a non-well-founded tree, so that every infinite branch determines a nontrivial elementary embedding $V_{\alpha'} \to V_{\alpha'}$ for some $\alpha' \leqq \alpha$, and furthermore the definition is absolute for $V^{(\omega)}$. In $V^{(\omega)}$ the tree is not well-founded; hence in $V^{(\omega)}$ there exists $\lambda \leqq \mu_\omega$ with a nontrivial elementary embedding $V_\lambda \to V_\lambda$; the same holds true in V, with μ_ω replaced by μ. The tree is the following one: Its members are all elementary embeddings of the form $j_1 : V_{\lambda_1} \to V_{\lambda_2}$ such that $\lambda_1 < \lambda_2 < \alpha$ and, in addition, $V_{\lambda_1} \prec V_{\lambda_2}$. An immediate successor of j_1 is an embedding $j_2 : V_{\lambda_2} \to V_{\lambda_3}$ (which is in the tree) such that $j_2\lambda_1 = \lambda_2$. Q.E.D.

What can be said about lifting elementary embeddings $j : V_\alpha \to V_\beta$ without assuming cofinality? The question is interesting only for limit ordinals (because if $\alpha = \alpha' + 1$ then, in V_α, α' is the largest ordinal, implying that, in V_β, $j\alpha'$ is the largest ordinal, i.e., $\beta = j\alpha' + 1$ and j is cofinal). Much less than (ii) and (iii) of Theorem 8 can be stated in the general case, if cofinality is not assumed. Still, some information can be deduced. In the following claims we assume $j : V_\alpha \to V_\beta$ to be an elementary embedding where $\alpha = \bigcup \alpha > 0$.

(4) *If $\omega < \mathrm{cof}(\alpha)$ and j is complete, j can be lifted.*

(5) *If $\alpha < \kappa(R\alpha)$, j can be lifted only if it is complete.*

(6) *If $\omega < \mathrm{cof}(\alpha) = \alpha^* < \alpha$ and j embeds α^* cofinally into $j\alpha^*$ then j can be partially lifted.*

(7) *If α is strongly inaccessible, j can be partially lifted.*

(8) *Note that if $j^* : V \to M$ is any partial lifting of j, then $V_{\beta'} \subset M$ where $\beta' = \sup\{j\lambda : \lambda < \alpha\}$.*

PROOF. (4) follows from Theorem 6(II).

If $\alpha < \kappa(R\alpha)$ then in V_α we have a well-ordered set order-isomorphic to α, with the isomorphism defined inside V_α. j sends it to a set order-isomorphic to β. Consequently, any lifting of j sends α to β, and any subset of $R\alpha$ to a subset of $R\beta$; hence j is complete.

To prove (6) put $\beta' = \sup\{j\lambda : \lambda < \alpha\}$. Then, by the corollary to Theorem II.1, the induced embedding $V_\alpha \to V_{\beta'}$ is elementary and, by Theorem 8, can be lifted. Such a lifting is a partial lifting of j. (7) is proved by the same argument. (8) is obvious, because $j(R\lambda) \subset M$ for all $\lambda < \alpha$.

DEFINITION. μ is *λ-extendible* if there is an ordinal β and an elementary embedding $V_{\mu+\lambda} \to V_\beta$ such that μ = first ordinal moved.

(A different terminology is used in [**14**]. "μ is λ-extendible" becomes there "μ is α-extendible" where $\alpha = \mu + \lambda$. We follow [**17**]'s terminology.)

Extendible cardinals were defined by Reinhardt [**16**]. If we employ, instead, elementary embeddings of the whole universe we get the following correlated concept.

DEFINITION 6. λ is *completely λ-extendible* if there is an elementary embedding $j: V \to M$ such that μ is the first ordinal moved by j and $V_{j(\mu+\lambda)} \subset M$.

Obviously, complete λ-extendibility $\Rightarrow$ λ-extendibility.

(There is also an intermediate concept: $\exists$ elementary embedding $V_{\mu+\lambda} \to V_\beta$ which can be lifted, such that μ = first ordinal moved. This concept is equivalent to complete λ-extendibility, except, possibly, for the case where λ is a singular cardinal of cofinality $\geqq \mu$ such that $\kappa(R\lambda) = \lambda$, where the situation is unclear. The trouble is that under these conditions it is not clear that the lifting, j, of $V_{\mu+\lambda} \to V_\beta$, or some modification of it, does satisfy $V_{j(\mu+\lambda)} \subset M$.)

THEOREM 10. *Assume that μ is λ-extendible.*

(i) *If $\lambda = \lambda' + 1$ then μ is completely λ'-extendible.*

(ii) *If $\omega < \operatorname{cof}(\lambda) < \mu$ then μ is completely λ-extendible.*

(iii) *If λ is strongly inaccessible and $\lambda \neq \mu$ then μ is completely λ-extendible.*

(i) and (iii) for $\lambda > \mu$ have been also obtained by Powell [**15**].

PROOF. (i) follows immediately from Theorem 8(i).

To get (ii) notice that if $j: V_{\mu+\lambda} \to V_\gamma$ is elementary and μ = first ordinal moved, then, by (6), we have a partial lifting $j^*: V \to M$, with $V_\beta \subset M$, where $\beta = \sup\{j\alpha : \alpha < \mu + \lambda\}$. Letting $\mu + \lambda = \sup\{\alpha_\xi : \xi < \delta\}$ where $\delta = \operatorname{cof}(\lambda)$ and $\alpha_\xi < \mu + \lambda$, we have $j^*(\mu + \lambda) = \sup\{j\alpha_\xi : \xi < \delta\} = \beta$.

Finally, if λ is strongly inaccessible and $\lambda > \mu$ then, by (7), we have again a partial lifting $j^*: V \to M$. Letting β be as before it is obvious that $j^*(\mu + \lambda) = j^*(\lambda) \geqq \beta$. If $j^*(\lambda) > \beta$ consider $\tilde{j}: V \to \tilde{M}$ which is the $<\lambda$-approximation of j^*. Since $\lambda' < \lambda \Rightarrow \kappa(R\lambda') < \lambda$, every $R\lambda'$, with $\lambda' < \lambda$, is a subset of $\tilde{M}$; hence $V_\beta \subset \tilde{M}$. If $\eta \in \tilde{j}\lambda$ then, for some x of cardinality $<\lambda$, $\eta \in \tilde{j}x$, and therefore $\eta \in \tilde{j}(x \cap \lambda)$. By the regularity of λ, there is $\lambda' < \lambda$ such that $x \cap \lambda \subset \lambda'$; obviously, $\eta \in \tilde{j}\lambda'$. This shows that $\beta = \tilde{j}(\lambda)$. Q.E.D.

Using the proof of Theorem 9 one can show that if $\lambda = \lambda' + 1$ or if $\operatorname{cof}(\lambda) = \omega$, and if μ is the first ordinal moved by $j: V \to M$, where $V_{j(\mu+\lambda)} \subset M$, then, inside M, μ is λ-extendible. Hence, for the case $\operatorname{cof}(\lambda) \leqq \omega$, complete λ-extendibility is much stronger than λ-extendibility; the first completely λ-extendible cardinal is $>$ first λ-extendible cardinal, and much more can be said. (See in this connection the remarks to Theorems III.7, 8.)

If $\omega < \operatorname{cof}(\lambda) < \mu$, or if $\lambda \neq \mu$ and λ is strongly inaccessible, both concepts coincide. The other cases are open.

Problems. (i) By the theorem, $(\lambda + 1)$-extendibility $\Rightarrow$ complete λ-extendibility. Is the reverse implication true? (I think not.)

(ii) Let $j: V_\alpha \to V_\beta$ be elementary, α regular, $\alpha < \kappa(R\alpha)$. Is j complete? The point is that, since V_α has, as a member, a well-ordering of order type α, one can correlate with every relation over α a (unique) relation over β, so as to get an elementary embedding. Can one do it also for relations over $R\alpha$?

Postscript A

Since replacement is not included in ZC^+, one cannot, in general, identify inside models of ZC^+ cardinals with initial ordinals.

In Part IV the theorems of the first three sections involve a parameter γ which is supposed to be a cardinal in certain models of ZC^+. The formulation is such that γ is in those models an initial ordinal. Indeed, the reader is advised to read these theorems and proofs with this tacit assumption. Then he can easily verify that the assumption is not needed; all the claims and their proofs remain, modulo some obvious rephrasings, equally valid if γ is any cardinal in the model.

One way to handle the general situation is to let "OR" denote for the purpose of these theorems, not the class of all ordinals, but the well-ordered class of all isomorphism types of well-orderings. An isomorphism type of a well-ordering is to be defined, as usual, as the set of all isomorphic orderings of minimal rank. We get a natural ordering $<$ on OR which well-orders it, and letting $\alpha, \beta, \gamma, \ldots$ range over OR, we define α to be a cardinal if $\{\beta : \beta < \alpha\}$ is not equinumerous with $\{\beta : \beta < \alpha'\}$, for all $\alpha' < \alpha$. (Note that $\{\beta : \beta < \alpha\}$ is always a set.)

The proofs can now be left practically unchanged except that in the obvious places "$\in$" should be replaced by "$<$" and "γ" by "$\{\beta : \beta < \gamma\}$". In Theorem 3 an obvious modification should take care of the fact that now two "ordinals" of two transitive models can be, strictly speaking, different, even when representing the same well-ordering type. Note that γ can "exceed" all the ordinals of M, in which case the embedding $k : \tilde{N} \to N$, of Theorem 3, keeps fixed every "real" ordinal of $\tilde{N}$ (but one cannot, in general, deduce that it is the identity).

Postscript B

The first use of self-extensions in set theory was by Scott [**19**], the operation being an ultrapower formation, based on a ω-complete nonprincipal ultrafilter over some measurable cardinal. The process of iterations was discovered and applied to this operation by the author ([**2**] and [**4**]). In [**11**], Kunen extended the construction to certain cases where the ultrafilter need not be a member of the model in question. Assume $\mathfrak{A}$ to be a transitive model of ZFC (transitivity and even well-foundedness are, actually, not needed, but will be assumed for the sake of simplicity). Let α be a cardinal in $\mathfrak{A}$. We put, as before, $\kappa(x)$ = cardinality of x. An $\mathfrak{A}$-ultrafilter over α

is defined in [**11**] as a set, $\mathfrak{F}$, of subsets of α, such that $\varnothing \notin \mathfrak{F}$, and such that

(i) $X \in \mathfrak{F} \Rightarrow X \in |\mathfrak{A}|$,
(ii) $X \supset X_1 \cap X_2$ and $X_1, X_2 \in \mathfrak{F}$ and $X \in |\mathfrak{A}| \Rightarrow X \in \mathfrak{F}$,
(iii) $X \in |\mathfrak{A}| \Rightarrow X \in \mathfrak{F}$ or $\alpha - X \in \mathfrak{F}$,
(iv) $\mathfrak{A} \models$ '$\kappa(x) \leqq \alpha$' $\Rightarrow x \cap \mathfrak{F} \in |\mathfrak{A}|$,
(v) $\mathfrak{A} \models$ '$\kappa(Y) < \alpha$' and $Y \subset \mathfrak{F} \Rightarrow \cap Y \in \mathfrak{F}$.

Actually, to be able to iterate the operation, only (i)–(iv) are needed ((v) is connected with well-foundedness requirements). The "ultrapower of $\mathfrak{A}$ modulo $\mathfrak{F}$" is defined in the usual way except that $f/\mathfrak{F}$ is a member of the model (where $f \in |\mathfrak{A}|^\alpha$) iff $f \in |\mathfrak{A}|$. Letting $\mathfrak{A}^{(1)}$ be this "ultrapower," one can show (using (i)–(iv) and the fact that $\mathfrak{A} \models$ ZFC) that the canonical embedding $j : \mathfrak{A} \to \mathfrak{A}^{(1)}$ is elementary. Moreover, there is a $\mathfrak{A}^{(1)}$-ultrafilter over $j\alpha$, which is the natural correlate of $\mathfrak{F}$, call it $\mathfrak{F}^{(1)}$; it is determined as follows:

$$f/\mathfrak{F} \in \mathfrak{F}^{(1)} \Leftrightarrow \{\lambda : f(\lambda) \in \mathfrak{F}\} \in \mathfrak{F}.$$

(Note that (iv) implies that the set on the right-hand side is in $|\mathfrak{A}|$.)

Using $\mathfrak{F}^{(1)}$ one can construct the ultrapower of $\mathfrak{A}^{(1)}$, getting $\mathfrak{A}^{(2)}$ and $\mathfrak{F}^{(2)}$ and so on. Moreover, one can give a simple direct definition of $\mathfrak{A}^{(I)}$ and $\mathfrak{F}^{(I)}$, where I ranges over ordered classes and of the embedding $(\mathfrak{A}^{(I)}, \mathfrak{F}^{(I)}) \to (\mathfrak{A}^{(J)}, \mathfrak{F}^{(J)})$ which is induced by the inclusion $I \subset J$, so that the recursion conditions of Definitions I.1 and I.2 are satisfied. (See [**11**] for details.) Here $(\mathfrak{A}, \mathfrak{F})$ is the structure obtained by adding $\mathfrak{F}$ to $\mathfrak{A}$ as an additional 1-place relation.

What we actually get in this way is an operator $\boldsymbol{O}$, which associates with every pair $(\mathfrak{A}, \mathfrak{F})$, where $\mathfrak{A} \models$ ZFC and $\mathfrak{F}$ is a $\mathfrak{A}$-ultrafilter, another pair $(\mathfrak{A}^{(1)}, \mathfrak{F}^{(1)})$, and an embedding $(\mathfrak{A}, \mathfrak{F}) \to (\mathfrak{A}^{(1)}, \mathfrak{F}^{(1)})$ such that $\mathfrak{F}^{(1)}$ is a $\mathfrak{A}^{(1)}$-ultrafilter. Furthermore, certain embeddings $(\mathfrak{A}, \mathfrak{F}) \to (\mathfrak{B}, \mathscr{G})$ induce embeddings on the respective values, $(\mathfrak{A}^{(1)}, \mathfrak{F}^{(1)}) \to (\mathfrak{B}^{(1)}, \mathscr{G}^{(1)})$. In short, we get a category of structures of the form $(\mathfrak{A}, \mathfrak{F})$ and an operator $\boldsymbol{O}$ which operates on its structures and embeddings and satisfies all the requirements I.(1)–I.(4), as well as I.(5). $\boldsymbol{O}$ is, therefore, an example of what we called a natural extension operator [**6**]. (The satisfaction of I.(5) means that $\boldsymbol{O}$ is uniform.)

Note that although $\mathfrak{A} \models$ ZFC, $(\mathfrak{A}, \mathfrak{F})$ need not be a model of ZF; it need not even satisfy the separation axiom. (Of course, if $\mathfrak{F} \in |\mathfrak{A}|$ then $(\mathfrak{A}, \mathfrak{F}) \models$ ZFC, and we are within our old framework.)

Note also that the embedding $(\mathfrak{A}, \mathfrak{F}) \to (\mathfrak{A}^{(1)}, \mathfrak{F}^{(1)})$ need not be elementary, although the embedding $\mathfrak{A} \to \mathfrak{A}^{(1)}$ *is* elementary.

Can one construe $\boldsymbol{O}$ as a self-extension operator? The answer is positive. The only difficulty is that, although one can define inside $(\mathfrak{A}, \mathfrak{F})$ the equivalence relation $f \equiv g(\text{mod } \mathfrak{F})$, there is no general way of choosing inside this model, representatives of the equivalence classes, because the separation axiom need not hold in $(\mathfrak{A}, \mathfrak{F})$. The solution is, simply, to define $|\mathfrak{A}^{(1)}|$ as $\{f \in |\mathfrak{A}| : \mathfrak{A} \models$ 'f is a function $\wedge\ D(f) = \alpha$'$\}$ (where $D(f)$ = Domain of f) and to interpret in $\mathfrak{A}^{(1)}$ the equality symbol as the congruence relation: $f \equiv g(\text{mod } \mathfrak{F})$. Both the equality and the $\in$ of $\mathfrak{A}^{(1)}$ can be defined in $(\mathfrak{A}, \mathfrak{F})$ by the usual conditions: $\{\lambda : f(\lambda) = g(\lambda)\} \in \mathfrak{F}$, and

$\{\lambda : f(\lambda) \in g(\lambda)\} \in \mathfrak{F}$, and the same is true of any other predicates of the language of $\mathfrak{A}$. Finally, because of (iv), $\{\lambda : f(\lambda) \in \mathfrak{F}\} \in |\mathfrak{A}|$; hence, $\mathfrak{F}^{(1)}$ can be defined inside $(\mathfrak{A}, \mathfrak{F})$ as well. As we remarked, working with models in which equality is interpreted as a congruence relation causes no troubles, except for notational inconveniences. The whole mechanism of iteration can be applied in the same way.

Thus, Kunen's construction is another example of a self-extension operator. Now, however, the embeddings $f_{(\mathfrak{A},\mathfrak{F})} : (\mathfrak{A}, \mathfrak{F}) \to O(\mathfrak{A}, \mathfrak{F})$ are not, in general, elementary (they are elementary only on the reduct, $\mathfrak{A}$). It is, therefore, not obvious which models belong to the category $C_{e,*}$, i.e., on which models the operator can be iterated. Indeed, the definition of the iterated operator, as given by Kunen (and, independently, by Keisler), is not the inductive Definition I.1, but a special one which makes sense only for "ultrapower" extensions. (Then, one verifies that the inductive clauses are satisfied.)

I shall now show that the fact that we have an "ultrapower" extension here is not essential. The construction and its iteration can be viewed as a special case within a general framework which treats self-extensions which are not necessarily elementary, where the models in question are not necessarily models of ZF or even Z, but of the form $(\mathfrak{A}, R_1, \ldots, R_k)$, where $\mathfrak{A}$ satisfies a certain fraction of ZFC (i.e., ZC^+) and $R_1, \ldots, R_k$ are additional relations.

The obvious question is: Which general properties of the "ultrapower" extension $(\mathfrak{A}, \mathfrak{F}) \to (\mathfrak{A}^{(1)}, \mathfrak{F}^{(1)})$ make the iterations possible? In the following outline I isolate two such properties (given by the forthcoming Definition 1 and clause (4) of Definition 3). The outline should, however, be regarded as a possible rather than as a final answer.

The idea of the generalization can be visualized as follows:

Let $\mathfrak{F}$ be a $\mathfrak{A}$-ultrafilter. Put

$$R_0 = \{\langle f, g\rangle : \mathfrak{A} \models f \equiv g (\mathrm{mod}\ \mathfrak{F})\}, \qquad R_1 = \{\langle f, g\rangle : \mathfrak{A} \models f \in g (\mathrm{mod}\ \mathfrak{F})\}.$$

Assuming, for the sake of simplicity, that $\mathfrak{A}$ is a transitive model one can easily deduce from (iv):

$$\mathfrak{A} \models \kappa(x) \leqq \alpha \to x \cap R_i \in |\mathfrak{A}|, \qquad i = 0, 1.$$

(See the forthcoming Definition 1, for nontransitive $\mathfrak{A}$'s.) Now, the main point is to work not with the structure $(\mathfrak{A}, \mathfrak{F})$ but with $(\mathfrak{A}, R_0, R_1)$ (or $(\mathfrak{A}, R_0, \ldots, R_m)$, if there are additional predicates in the language of $\mathfrak{A}$, where R_i, $2 \leqq i \leqq m$, is defined in the analogous way). Note that $\mathfrak{F}$ can be recovered in $(\mathfrak{A}, R_0, R_1)$, but this is inessential to our construction.

As before, let $\text{ZC}^+ = \text{Z}^+ \cup \{\text{Axiom of Choice}\}$. Concerning the cardinals of models of ZC^+ see Postscript A. The remarks there apply equally well to all that will be done here. The structures to be treated are, to start with, sets.

DEFINITION 1. Let α be a cardinal in $\mathfrak{A}$. For a transitive $\mathfrak{A}$, say that a relation $R \subset |\mathfrak{A}|^n$ is a $(\alpha, \mathfrak{A})$-class ($(<\alpha, \mathfrak{A})$-class) if, for every $x \in |\mathfrak{A}|$ of cardinality $\leqq \alpha$ $(<\alpha)$ in $\mathfrak{A}$, we have $R \cap x \in |\mathfrak{A}|$.

For a nontransitive $\mathfrak{A}$ the requirement that $R \cap x \in |\mathfrak{A}|$ has to be replaced by its obvious generalization:

There is x^* which, inside $\mathfrak{A}$, is a set of n-tuples, such that

$$(\mathfrak{A}, R) \models \forall x_1, \dots, x_n[\langle x_1, \dots, x_n\rangle \in x^* \leftrightarrow R(x_1, \dots, x_n) \wedge \langle x_1, \dots, x_n\rangle \in x].$$

Notation of the form "$\langle, \dots, \rangle$" is to be read everywhere as "the ordered n-tuple of...". Thus, coming after "$\mathfrak{A} \models$" it denotes the tuple formation in the sense of $\mathfrak{A}$.

[REMARK. It is customary and convenient to identify 1-place relations over A with subsets of A, i.e., to put $A^1 = A$. This may lead to ambiguity if $A^n \subset A$, for then a subset of A^n can be regarded either as an n-ary relation or as a 1-ary relation. If $A = |\mathfrak{A}|$ and $\mathfrak{A}$ is a transitive model of Z, it makes no difference in Definition 1, whether we regard R as an n-ary relation or a 1-ary relation, the generalized condition will be satisfied in one case iff it is satisfied in the other case. For non-transitive models this is not necessarily so. One way of avoiding the ambiguity is to define $A^1 = \{\langle x\rangle : x \in A\}$, where $\langle x\rangle \neq x$. Another way is to let it stay, with the understanding that notation such as "$R \subset |\mathfrak{A}|^n$" and similar indications in the text imply that R is to be regarded as an n-ary relation. This last convention will be applied from now on.]

The structures to which we will apply the self-extension operator will be of the form $(\mathfrak{A}, R_0, \dots, R_m)$ where $\mathfrak{A} \models ZC^+$ and the R_i's are $(\alpha, \mathfrak{A})$-classes.

Consider a cofinal embedding $j: \mathfrak{A} \to \mathfrak{B}$ and let $R_0, \dots, R_m$ be relations over $|\mathfrak{A}|$. We want to associate with every R_i a relation jR_i over $\mathfrak{B}$ so as to get an embedding $(\mathfrak{A}, R_i)_{i \leqq m} \to (\mathfrak{B}, jR_i)_{i \leqq m}$ which will be nice, that is to say, will preserve as many first-order properties as possible.

DEFINITION 2. If $R \subset |\mathfrak{A}|^n$, then the j-correlate of R is the relation S which satisfies (for transitive $\mathfrak{A}$'s)

$$\langle x_1, \dots, x_n\rangle \in S \Leftrightarrow \exists x \subset R[x \in |\mathfrak{A}| \wedge \mathfrak{B} \models \langle x_1, \dots, x_n\rangle \in jx].$$

For nontransitive $\mathfrak{A}$'s the condition that $x \subset R$ is to be replaced by $\{\langle z_1, \dots, z_n\rangle : \mathfrak{A} \models \langle z_1, \dots, z_n\rangle \in x\} \subset R$.

We shall denote the j-correlate of R by "jR".

The embedding $(\mathfrak{A}, R_i)_{i \leqq m} \to (\mathfrak{B}, jR_i)_{i \leqq m}$ which sends every x to jx is called the natural expansion of j.

(Note that this expansion is indeed the natural one. For example, if $R_i \cap x \in |\mathfrak{A}|$, for all $x \in |\mathfrak{A}|$, then the natural expansion of j to $(\mathfrak{A}, R_0, \dots, R_m)$ is the only expansion which can be Σ_0-elementary.)

We are now in a position to define how the self-extending operator is to operate.

DEFINITION 3. $(\mathfrak{A}, R_i)_{i \leqq m}$ is an α-*-model if $\mathfrak{A} \models ZC^+$, α is an infinite cardinal in $\mathfrak{A}$ and the following is true:

(1) Each R_i is an $(\alpha, \mathfrak{A})$-class.

(2) Let $A^{(1)} = \{x : \langle x, x\rangle \in R_0\}$ and let $\mathfrak{A}^{(1)} = (A^{(1)}, R_0, \dots, R_{m-1})$, then $\mathfrak{A}^{(1)}$ is a model of the same similarity type as $\mathfrak{A}$, R_0 is the equality of $\mathfrak{A}^{(1)}$, R_1 is the $\in$ of $\mathfrak{A}^{(1)}$ and R_i, for $2 \leqq i < m$, are the other corresponding relations, if any.

(3) R_m is a binary relation which is an elementary embedding $j:\mathfrak{A} \to \mathfrak{A}^{(1)}$.

(4) The embedding $j:\mathfrak{A} \to \mathfrak{A}^{(1)}$ satisfies the condition:

For every x of cardinality $\leqq \alpha$ in $\mathfrak{A}$, there is y of cardinality $\leqq \alpha$ in $\mathfrak{A}$, such that

$$\{z \in A^{(1)}: \mathfrak{A} \models z \in x\} \subset \{z: \mathfrak{A}^{(1)} \models z \in jy\}.$$

(For a transitive $\mathfrak{A}$ this simply means that every member of $A^{(1)} \cap x$ is, in $\mathfrak{A}^{(1)}$, a member of jy.) Roughly speaking, (4) means that subsets of $|\mathfrak{A}^{(1)}|$ which are of cardinality $\leqq \alpha$ in $\mathfrak{A}$ can be covered by sets of the form jy, where y is of cardinality $\leqq \alpha$ in $\mathfrak{A}$.

It is easy to verify that $A^{(1)}$ is an $(\alpha, \mathfrak{A})$-class. (We can add $A^{(1)}$ as a 1-ary relation to the R_i's, but this is unnecessary, since it can be recovered from R_0.)

A $<\alpha$-$*$-model is defined in the same way, replacing in (1) "$(\alpha, \mathfrak{A})$" by "$(<\alpha, \mathfrak{A})$" and, everywhere in (4), "$\leqq \alpha$" by "$<\alpha$".

[Obviously if, in $\mathfrak{A}$, α^- is the successor cardinal of α, $(\mathfrak{A}, R)_{i \leqq m}$ is an α-*-model iff it is a $<\alpha^+$-*-model. However, because in $\mathfrak{A}$ not all "cardinals" are identified with initial ordinals, α^+ need not be an ordinal of $\mathfrak{A}$ (see Postscript A) and, to avoid complicating the formulation, we will continue to make explicit references to the α-concepts.]

A $*$-model is a model which for some α is an α-$*$-model or $<\alpha$-$*$-model.

Now define an operator $\boldsymbol{O}$, $\boldsymbol{O} = (O, f)$, to be called the *-extension operator, as follows:

Given a $*$-model $\overline{\mathfrak{A}} = (\mathfrak{A}, R_i)_{i \leqq m}$, let $\mathfrak{A}^{(1)} = (A^{(1)}, R_i)_{i<m}$ and $j:\mathfrak{A} \to \mathfrak{A}^{(1)}$ be as in Definition 3. Then, define $f_{\overline{\mathfrak{A}}}$ to be the embedding $(\mathfrak{A}, R_i)_{i \leqq m} \to (\mathfrak{A}^{(1)} jR_i)_{i \leqq m}$ which is the natural expansion of $j:\mathfrak{A} \to \mathfrak{A}^{(1)}$, and put $O\overline{\mathfrak{A}} = (\mathfrak{A}^{(1)}, jR_i)_{i \leqq m}$.

Obviously, $\boldsymbol{O}$ is a self-extension operator. It is easy to see what the self-extension schema e is going to be.

Now, define a category $\boldsymbol{C}$ as follows:

DEFINITION 4. (i) $\overline{\mathfrak{A}} \in \boldsymbol{C}$ iff $\overline{\mathfrak{A}}$ is a $*$-model.

(ii) Let $\overline{\mathfrak{A}} = (\mathfrak{A}, R_i)_{i \leqq m}$, $\overline{\mathfrak{B}} = (\mathfrak{B}, S_i)_{i \leqq m}$, then $h: \overline{\mathfrak{A}} \to \overline{\mathfrak{B}} \in \boldsymbol{C}$ iff for some α, $\overline{\mathfrak{A}}$ is an α-*-model ($<\alpha$-*-model), $\overline{\mathfrak{B}}$ is a $h\alpha$-*-model ($<h\alpha$-*-model) and h is the natural expansion of an elementary embedding $j:\mathfrak{A} \to \mathfrak{B}$ which is α-cofinal ($<\alpha$-cofinal), as defined in Part IV. (That is to say, every $x \in |\mathfrak{B}|$ is a member in $\mathfrak{B}$ of some jy where y is of cardinality $\leqq \alpha$ ($<\alpha$) in $\mathfrak{A}$.)

THEOREM $*$. *$\boldsymbol{C} \subset$ the domain of $\boldsymbol{O}$ and $\boldsymbol{O}$ acts within $\boldsymbol{C}$; that is to say,*

(1) $\overline{\mathfrak{A}} \in \boldsymbol{C} \Rightarrow O\overline{\mathfrak{A}} \in \boldsymbol{C}$,

(2) $\overline{\mathfrak{A}} \in \boldsymbol{C} \Rightarrow f_{\overline{\mathfrak{A}}} \in \boldsymbol{C}$,

(3) $h:\overline{\mathfrak{A}} \to \overline{\mathfrak{B}} \in \boldsymbol{C} \Rightarrow h$ *is* Γ_e*-elementary,*

where e is the self-extension schema (Γ_e *is the set of all subformulas which figure in e, as defined at the beginning of Part* I).

(4) $h:\overline{\mathfrak{A}} \to \overline{\mathfrak{B}} \in \boldsymbol{C} \Rightarrow Oh:O\overline{\mathfrak{A}} \to O\overline{\mathfrak{B}} \in \boldsymbol{C}$

(*where* $(Oh)(x) = h(x)$ *for all* $x \in |O\overline{\mathfrak{A}}|$).

We will not include the proof here. It is rather lengthy and involves the verifying of quite a few details some of which are far from obvious. (1) is the most difficult. First one shows that, if $\overline{\mathfrak{A}}$ is an α-*-model, $\overline{\mathfrak{A}} = (\mathfrak{A}, R_i)_{i \leqq m}$ and $O\overline{\mathfrak{A}} = (\mathfrak{A}^{(1)}$,

$jR_i)_{i \leqq m}$, then every jR_i is a $(j\alpha, \mathfrak{A}^{(1)})$-class. Then, one shows that $O\overline{\mathfrak{A}}$ satisfies (2) and (4) of Definition 3. This is the most confusing part, because one actually has to iterate the process and pass from $O\overline{\mathfrak{A}}$ to $O(O\overline{\mathfrak{A}})$; it is here that (4) of Definition 3 is needed. Then, using Theorem II.1 together with the techniques employed in the proof of Theorem IV.1 one shows that condition (3) of the definition holds, i.e., $j(R_m)$ is an elementary embedding of $\mathfrak{A}^{(1)}$ into $(\mathfrak{A}^{(1)})^{(1)}$. This step involves also, in the case where $\mathfrak{A} \not\models$ ZF, the checking of the satisfaction of $(*_{j,\varphi})$ of Theorem II.1.

(2) follows directly from (1) and the fact (which, by now, is proved) that if $\overline{\mathfrak{A}}$ is an α-*-model then $O\overline{\mathfrak{A}}$ is an $f_{\overline{\mathfrak{A}}}(\alpha)$-*-model.

(3) and (4) involve, again, the checking of several details, but they are easier than (1).

Theorem 6 enables one to iterate the operator O. Iterated "ultrapowers" (in which $\mathfrak{A}$-ultrafilters are used) are therefore special cases of this general process. Note that in the "ultrapower" extension only the equality and the $\in$ of the "ultrapower" are not definable in the original model $\mathfrak{A}$, and must be added as relations of an expanded model. The embedding itself, which sends every x to the constant function c_x, is definable in the original $\mathfrak{A}$. In the general process this need not be so, and we allow the embedding (which is the added relation R_m) not to be definable in $\mathfrak{A}$.

Note also that the embeddings $O^I\overline{\mathfrak{A}} \to O^J\overline{\mathfrak{A}}$ will be elementary on the reducts.

Can the iteration of O be defined inside $\overline{\mathfrak{A}}$? This is true for $\mathfrak{A}$-ultrapower extensions. I believe that this is true in general, provided that $\overline{\mathfrak{A}} \models \forall x[\{y : O\overline{\mathfrak{A}} \models y \in x\}$ is a set], but I have not checked the details. If true, it will provide a way of defining the iteration for proper classes.

The major problem which remains is to get a useful version of a well-foundedness theorem. The proof of Theorem II.5 will not work here unless some additional assumptions are made. What are these additional assumptions? Even for "ultra-powers" which are based on $\mathfrak{A}$-ultrafilters there is not, as far as I know, a proven analogue of Theorem II.5. The condition which is used in [**11**] is an exterior condition on the $\mathfrak{A}$-ultrafilter $\mathfrak{F}$, namely, $\bigcap X \neq \varnothing$ for every $X \subset \mathfrak{F}$ which is countable (in the actual world). By a satisfactory analogue of Theorem II.5, I mean a theorem of the form:

If $\overline{\mathfrak{A}}$ is transitive, if the ordinals of $O\overline{\mathfrak{A}}$ are order-isomorphic to those of $\mathfrak{A}$ and if ..., then $O^\lambda\mathfrak{A}$ is well-founded for every countable λ in $\mathfrak{A}$.

Here "..." stands for the missing additional condition.

(Of course, if $O^\lambda\mathfrak{A}$ is well-founded for all $\lambda < \omega_1$, then it is well-founded for all λ.)

References

1. P. J. Cohen, *Set theory and the continuum hypothesis*, Benjamin, New York, 1966. MR **38** #999.

2. H. Gaifman, *Measurable cardinals and constructible sets*, Notices Amer. Math. Soc. **11** (1964), 771. Abstract #64T-505. A paper under a similar title has been also contributed to the 1964 Congress of Logic, Methodology and the Philosophy of Science, held in Jerusalem and appeared in the volume of the contributed abstracts.

3. ———, *Uniform extension operators for models and their applications*, Sets, Models and Recursion Theory (Proc. Summer School Math. Logic and Tenth Logic Colloq., Leicester, 1965), North-Holland, Amsterdam, 1967, pp. 122–155. MR **36** #3640.

4. ———, *Pushing up the measurable cardinal*, Lecture notes of the AMS Summer Institute on Axiomatic Set Theory held at UCLA, 1967 (IVR1–IVR16).

5. ———, *Operations on relational structures, functors and classes*. I, Proc. Sympos. Pure Math., vol. 25, Amer. Math. Soc., Providence, R.I., 1974.

6. ———, *Iterated processes and their applications in model theory* (to appear).

7. W. Hanf and D. Scott, *Classifying inaccessible cardinals*, Notices Amer. Math. Soc. **8** (1961), 445. Abstract #61T-240.

8. H. J. Keisler, *Limit ultrapowers*, Trans. Amer. Math. Soc. **107** (1963), 382–408. MR **26** #6054.

9. ———, *The equivalence of certain problems in set theory with problems in the theory of models*, Notices Amer. Math. Soc., **9** (1962), 339. Abstract #62T-258.

10. H. J. Keisler and A. Tarski, *From accessible to inaccessible cardinals. Results holding for all accessible cardinal numbers and the problem of their extension to inaccessible ones*, Fund. Math. **53** (1963/64), 225–308. MR **29** #3385.

11. K. Kunen, *Some applications of iterated ultrapowers in set theory*, Ann. Math. Logic **1** (1970), 179–227. MR **43** #3080.

12. ———, *Elementary embeddings and cofinitary combinations*, preprint (to appear).

13. M. Magidor, *There are many normal ultrafilters corresponding to a singular cardinal*, Israel J. Math. **9** (1971), 186–192.

14. ———, *On the role of supercompact and extendible cardinals in logic*, Israel J. Math. **10** (1971), 147–157. MR **45** #4966.

15. W. C. Powell, *Elementary embeddings preserving well-ordering* (preprint).

16. W. N. Reinhardt, *Conditions on natural models of set theory*, Lecture notes of the AMS Summer Institute on Axiomatic Set Theory held at UCLA, 1967, (IVL1–IVL5).

17. W. N. Reinhardt and R. Solovay, *Strong axioms of infinity and elementary embedding* (to appear).

18. F. Rowbottom, *Some strong axioms of infinity incompatible with the axiom of constructibility*, Ann. Math. Logic **3** (1971), 1–44.

19. D. Scott, *Measurable cardinals and constructible sets*, Bull. Acad. Polon. Sci. Sér. Sci. Math. Astronom. Phys. **9** (1961), 521–524. MR **26** #1263.

20. J. Shoenfield, *The problem of predicativity*, Essays on the Foundations of Math., Magness Press, Hebrew Univ., Jerusalem, 1961, pp. 132–139. MR **29** #2177.

21. J. H. Silver, *Some applications of model theory in set theory*, Ann. Math. Logic **3** (1971), 45–110.

22. R. L. Vaught, *On a theorem of Cobham concerning undecidable theories*, Logic, Methodology and Philosophy of Science (Proc. 1960 Internat. Congr.), Stanford Univ. Press, Stanford, Calif., 1962, pp. 14–25. MR **28** #32.

HEBREW UNIVERSITY
JERUSALEM, ISRAEL

Proceedings of Symposia in Pure Mathematics
Volume 13, Part II, 1974

SET-THEORETIC FUNCTIONS FOR ELEMENTARY SYNTAX[1]

R. O. GANDY

INTRODUCTION

Over the last decade it has become increasingly clear that many of the constructions and methods used in set theory are analogous to well-understood techniques in recursion theory. A paradigm example is provided by Gödel's F-function which transfinitely enumerates the constructible sets. There is now a generally agreed set-theoretic analogue for "primitive recursive" (see the paper [**10**] by Jensen and Karp in these PROCEEDINGS), and a hierarchy of analogues for "partial recursive"—namely the functions whose graphs are Σ_1 over A, where A is an admissible class. These analogues emphasize the constructive nature of much work in set theory. In particular, they make it easy to see that certain theorems (for example, that L is absolute) follow simply from the fact that a given construction is of a certain *kind*.[2] More significantly, they have proved effective in guiding new and valuable research; a notable example is the compactness theorem of Barwise (see [**1**]). And just as the classification of number-theoretic predicates grew out of recursion theory, so are these analogues closely connected with Lévy's hierarchy (see [**12**]) of set-theoretic predicates.

In this paper we are concerned with analogues for the simplest kinds of recursive function; in particular, we are interested in constructing a class of set-theoretic functions which will suffice for describing the operations of elementary syntax. The goal is not fully achieved; the inadequacy of our proposals is discussed in § 4.

AMS (MOS) subject classifications (1970). Primary 02-02, 02K99, 02K15, 02K05, 02K30, 02F20, 02B10, 02B25.

[1] Some of the ideas in this paper were first adumbrated in a course of lectures given in 1967 while I was a visiting associate professor in the Philosophy Department at Stanford University; I record here my gratitude.

[2] An exposition of the properties of L based on the fact that it is V-recursively-enumerable (i.e., Σ_1 over V) will be found in Karp [**11**].

In number theory the functions and predicates necessary for the arithmetization of elementary syntax all lie in some class $\mathscr{C}$ which is a proper subclass of the primitive recursive functions. Various candidates for $\mathscr{C}$ have been considered: (i) Kalmár's elementary functions (this is the class $\mathscr{E}^3$ in Grzegorczyk's hierarchy [**8**]); (ii) Grzegorczyk's class $\mathscr{E}^0$; (iii) the rudimentary (or, equivalently, the constructive arithmetic) predicates introduced by Smullyan [**15**].

There are several ways of testing the adequacy of candidate $\mathscr{C}$:

(a) $\mathscr{C}$ must contain the proof-predicate (expressed in terms of a suitable Gödel-numbering) for any formal system;

(b) $\mathscr{C}$ must contain a predicate T such that $\{x:(\exists y)T(e, x, y)\}$ $(e = 0, 1, \ldots)$ is a complete enumeration of the recursively enumerable sets;

(c) all operations of "elementary" syntax shall be represented in $\mathscr{C}$.

As a starting point we consider the class Δ_0 of predicates defined by restricted quantification.[3] In § 1 we define and investigate the class $\mathscr{B}$ of "basic" set-theoretic functions which is associated with Δ_0 in a natural way; we also give an axiomatic characterization of sets which are closed under $\mathscr{B}$. In § 2 we construct a number of counterexamples which help to delimit the concepts introduced in § 1. In § 3 we show that if appropriate methods of coding are used, then $\mathscr{B}$ does indeed suffice for the description of the operations of elementary syntax of finitary languages. This necessitates the introduction of a new kind of set-theoretic numeral; for these numerals "rudimentary" and "Δ_0" coincide.[4] Finally, in § 4 we discuss the limitations of our concepts and a possible extension of them.

Why do all this? We can hardly hope that our investigations—unlike their number-theoretic counterparts—will cast light on problems of decidability. Further, for many purposes it is sufficient to know that syntactic operations can be represented in Δ_1.

Our long-term objective is to characterize the notion of elementary syntax, both for finite and infinitary languages. The present paper represents a first step in that direction. But there are also immediate applications of our results. Firstly, they facilitate the placing of a given predicate in the Lévy hierarchy. Secondly, our methods may be used where the notion of "primitive recursive" is too coarse. In his Habilitationschrift [**9**] Jensen proves and applies a number of rather delicate results about quantifier forms in ramified languages. For this purpose he devises a number of ingenious but *ad hoc* tricks. I believe that many of his results could be proved using the rather more uniform apparatus presented here.

ADDED IN PROOF. The theory developed in § 1 has been worked out independently by Jensen in [**16**]; see also the expositions of Jensen's work by Devlin [**17**], [**18**]. Jensen used "simple" for our "substitutable" and "rudimentary" for our "basic." His definition is like our 1.3.2 except that he uses $\lambda x . x_i$, π, $\lambda xy . x - y$ as initial functions. On p. 238 he gives a particularly transparent characterization of $\mathrm{Bc}(x)$.

[3] Lévy called this class "Σ_0," but, since it is symmetric between $\exists$ and $\forall$, it seems more logical to use "Δ_0."

[4] This statement assumes the truth of Conjecture 3.3.7.

Namely, consider x as a set of ur-elements: then $y \in \mathrm{Bc}(x)$ iff y is of finite rank over x and is hereditarily x-definable.

Finally, there is some spin-off. I think the "basic set theory" of § 1.4 can claim to be the weakest natural theory in which the cartesian product of two sets can be proved to be a set. In § 2.1 we show that quantifiers restricted to ω can be replaced by restricted quantifiers. Some of the counterexamples in § 2 may be of independent interest.

1. Basic Functions

1.1. Notations.

1.1.1. The formal language has primitive relation symbols "$\in$," "$=$." Its variables are denoted by lower-case roman letters: $\mathrm{a}, \mathrm{b}, \ldots, \mathrm{z}, \mathrm{a}_1, \ldots$. Its primitive logical symbols are "$\neg$," "$\vee$," "$(\exists \mathrm{x})$," "$(\exists \mathrm{u} \in \mathrm{x})$." Various standard notations are regarded as (metamathematical) abbreviations. We suppose that bound variables are always chosen so that collisions do not occur.

Formulae are denoted by $\phi, \psi, \theta, \chi, \phi_1, \ldots$. Attention is drawn to particular free variables of a formula by writing, for example, $\phi(\mathrm{x}, \mathrm{y})$ in place of "ϕ." The notation $\phi[\cdots]$ is used to indicate that *all* the free variables of ϕ come from the list....

1.1.2. A formula is said to be *restricted*, or *in* Δ_0, iff it contains no unrestricted quantifiers.

1.1.3. As far as possible, we use the material rather than the formal mode of speech. Set variables $a, b, \ldots, z, a_1, \ldots$ range over a transitive class V of well-founded sets. Some theory T for V is presupposed, and asserted statements are consequences of T. For definiteness T may be taken to be ZF. But in fact almost all assertions (except those in § 2) are consequences of the much weaker theory BST of §1.4.

By an abuse of language we also use $\phi, \psi, \theta, \chi, \phi_1, \ldots$ to denote statements which are obtained from formulae by substituting set variables for formal variables and interpreting in $\langle V, \in \rangle$.

1.1.4. Letters "P," "Q," "R," "S" denote set-theoretic predicates. We do not always identify a predicate with a relation (a class of ordered n-tuples), but we do write "$P \subset V^n$," etc., to indicate the number of arguments. Letters "F," "G," "H," "F_1,"... denote set-theoretic functions; we write "$F: V^n \to V$," etc. The characteristic function, taking values 0 for "true" and $\{0\}$ for "false," of a predicate P is denoted by cP.

1.1.5. A predicate $P \subseteq V^m$ is said to be *restricted* or to *belong to* Δ_0 iff there is a formula $\phi[\mathrm{x}_1, \ldots, \mathrm{x}_m]$ such that

$$(\forall x_1, \ldots, x_m)(Px_1, \ldots, x_m \leftrightarrow \phi[x_1, \ldots, x_m]);$$

that is, this statement is to be a consequence of T. The function F belongs to Δ_0 iff its graph does.

1.1.6. We use bold face letters to denote finite sequences of objects. In particular, we adopt the convention that $\boldsymbol{x}$, $\boldsymbol{y}$, $\boldsymbol{z}$ always stand for $x_1, \ldots, x_m$; $y_1, \ldots, y_n$;

$z_1, \ldots, z_p$, respectively. We extend this notation in various obvious ways: for example, "$\boldsymbol{u} \in \boldsymbol{x}$" stands for "$u_1 \in x_1 \& \ldots \& u_m \in x_m$."

1.1.7. Among the standard notations we use are "$\subseteq$" for inclusion, "$\subset$" for proper inclusion, and "$\mathfrak{P}$" for the power set operation. Where notations are not self-explanatory, reference may be made to the paper by Jensen and Karp. The set-theoretic functions $\mathfrak{C}$, ι, π, are defined as follows:

$$\mathfrak{C}xyuv = x \quad \text{if } u \in v, \qquad \iota x = \{x\},$$
$$= y \quad \text{otherwise}; \qquad \pi xy = \{x, y\}.$$

When describing a predicate or function we may omit the abstraction operator λ; if necessary, we then use quotes as demarcation symbols.

Let us illustrate some of the above usages and conventions:

"$(\exists \mathbf{x})\phi$" $\in \Delta_0$ (where $\phi = \phi[\mathbf{x}, \mathbf{y}]$), or equivalently $\lambda \boldsymbol{y}(\exists \boldsymbol{x})\phi[\boldsymbol{x}, \boldsymbol{y}] \in \Delta_0$ iff there is a formula $\psi(= \psi[\mathbf{y}])$ in Δ_0 such that

$$T \vdash (\forall \mathbf{y})[(\exists \mathrm{x}_1, \ldots, \mathrm{x}_m)\phi \leftrightarrow \psi[\mathrm{y}_1, \ldots, \mathrm{y}_n]].$$

(*N.B.* If the expression ϕ contains symbols for given functions, then we assume that the theory T contains the defining equations for those functions.)

1.2. Substitutable functions. It will be shown in § 2.2 that the collection of Δ_0 functions is not closed under substitution. We therefore define and investigate certain subcollections.

1.2.1. DEFINITION.[5] The function $F: V^m \to V$ is *substitutable* (and we write "$F \in \mathscr{S}$") iff for every $P\ (\subseteq V^{n+1}) \in \Delta_0$ there is a $Q\ (\subseteq V^{m+n}) \in \Delta_0$ such that

$$(\forall \boldsymbol{x}, \boldsymbol{y})[P(F\boldsymbol{x})\boldsymbol{y} \leftrightarrow Q\boldsymbol{x}\boldsymbol{y}].$$

If $m = 0$, then F is a *substitutable constant.*

1.2.2. LEMMA. *The function F is substitutable if and only if*

$$(*) \qquad \text{"}(\exists u \in F\boldsymbol{x})\phi\text{"} \in \Delta_0 \quad \textit{for every } \phi \textit{ in } \Delta_0.$$

PROOF. Necessity is obvious. Suppose then that (*) holds. If $\psi(\mathrm{y})$ is in Δ_0 and is atomic, then $\psi(F\boldsymbol{x}) \in \Delta_0$. For

(a) $v \in F\boldsymbol{x} \leftrightarrow (\exists u \in F\boldsymbol{x})(u = v)$;

(b) $v = F\boldsymbol{x} \leftrightarrow (\forall u \in v)(u \in F\boldsymbol{x}) \& (\forall u \in F\boldsymbol{x})(u \in v)$;

(c) $F\boldsymbol{x} \in v \leftrightarrow (\exists u \in v)(u = F\boldsymbol{x})$.

But now it follows by induction on the construction of ψ that $\psi(F\boldsymbol{x}) \in \Delta_0$ for any ψ in Δ_0. Q.E.D.

1.2.3. Let χ_r $(r = 0, 1, \ldots)$ enumerate all the formulae in Δ_0, and let n_r be the number of free variables in χ_r. We define $D_r: V^{n_r} \to V$ by

[5] In [**9**] Jensen introduced this concept under the title "predicative." By talking of a function F in formal contexts, we mean that there is some formula $\phi[\mathbf{x}, \mathrm{y}]$ which expresses the (often intuitively given) definition of the graph of F, and which satisfies $T \vdash (\forall \mathbf{x})(\exists \mathrm{y})\phi[\mathbf{x}, \mathrm{y}]$.

$$D_r x\boldsymbol{v} = \{u : u \in x \,\&\, \chi_r[u, \boldsymbol{v}]\};$$

thus $D_r x\boldsymbol{v}$ is the rth Δ_0 subset of x, using parameters $\boldsymbol{v}$.

1.2.4. THEOREM. *The following functions are substitutable:*

$$\lambda\boldsymbol{x}.x_i,\ 0,\ \pi,\ \bigcup,\ \mathfrak{C},\ D_r.$$

PROOF. By Lemma 1.2.2, it is sufficient to observe the following equivalences:

$$(\exists u \in 0)\phi \leftrightarrow 0 \in 0;$$

$$(\exists u \in \{x, y\})\phi(u) \leftrightarrow \phi(x) \vee \phi(y);$$

$$(\exists u \in \bigcup x)\phi \leftrightarrow (\exists v \in x)(\exists u \in v)\phi;$$

$$(\exists u \in \mathfrak{C}xyzw)\phi(u) \leftrightarrow (z \in w \,\&\, (\exists u \in x)\phi(u)) \vee (z \notin w \,\&\, (\exists u \in y)\phi(u));$$

$$(\exists u \in D_r x\boldsymbol{v})\phi \leftrightarrow (\exists u \in x)(\chi_r[u, \boldsymbol{v}] \,\&\, \phi). \quad \text{Q.E.D.}$$

1.2.5. THEOREM. *The collection of substitutable functions is closed under substitution.*

PROOF. Suppose, for example, that $G, H \in \mathscr{S}$, and that $Fx\boldsymbol{y} = G(Hx)\boldsymbol{y}$. Consider any ϕ in Δ_0; since $G \in \mathscr{S}$, there is a ψ in Δ_0 such that $(\exists u \in Gz\boldsymbol{y})\phi \leftrightarrow \psi(z, \boldsymbol{y})$. Hence

$$(\exists u \in G(H\boldsymbol{x})\boldsymbol{y})\phi \leftrightarrow \psi(H\boldsymbol{x}, \boldsymbol{y}).$$

But $H \in \mathscr{S}$, so the RHS $\in \Delta_0$. Therefore $F \in \mathscr{S}$ by Lemma 1.2.2. Q.E.D.

1.2.6. THEOREM. *The collection of substitutable functions is closed under the formation of images; i.e., if $G \in \mathscr{S}$, then so does F, where*

(I) $$Fx\boldsymbol{y} = \{Gu\boldsymbol{y} : u \in x\}.$$

For $(\exists v \in Fx\boldsymbol{y})\phi(v) \leftrightarrow (\exists u \in x)\phi(Gu\boldsymbol{y})$.

1.2.7. LEMMA. *If $F \in \mathscr{S}$, then "$u \in F\boldsymbol{x}$," "$y = F\boldsymbol{x}$," "$F\boldsymbol{x} = 0$" $\in \Delta_0$.*

1.3. Basic functions. The properties of $\mathscr{S}$ stated in Theorems 1.2.4–6 give an inductive definition of a rather natural class of functions, which we call *basic*. We give now two alternative definitions; other definitions are given in Theorems 1.4.5 and 1.5.5. A function F with arguments $\boldsymbol{x}$ is *explicitly defined* from the functions $\boldsymbol{G}$ iff there is a well-formed term τ constructed from the x_i and (symbols for) the functions $\boldsymbol{G}$ such that $F\boldsymbol{x} = \tau$.

1.3.1. DEFINITION. $\mathscr{B}_1$ is the least collection of functions which contains the initial functions π, $\bigcup$, D_r ($r \in \omega$), and is closed under explicit definition and the formation of images.

1.3.2. DEFINITION. $\mathscr{B}_2$ is the least collection of functions which contains the initial functions $\lambda\boldsymbol{x}.x_i$ ($1 \leqq i \leqq m$), $\lambda\boldsymbol{x}.0$, $\lambda xy.x \cup \{y\}$, $\mathfrak{C}$, and is closed under the substitution schemes

$$Fx\boldsymbol{y} = Gx(H\boldsymbol{x})\boldsymbol{y}, \qquad Fx\boldsymbol{y} = G(H\boldsymbol{x})\boldsymbol{y},$$

and the "union of image" scheme

(UI) $$Fx\boldsymbol{y} = \bigcup\{Gu\boldsymbol{y} : u \in x\}.$$

This second definition is obtained from the definition of "primitive recursive" given in Jensen and Karp's paper [**10**] by substituting the scheme (UI) for the recursion scheme.[6]

1.3.3. LEMMA. $\mathscr{B}_1 = \mathscr{B}_2$.

PROOF. It is obvious that $\mathscr{B}_2 \subseteq \mathscr{B}_1$. The proof of the converse proceeds by a number of steps; the details will mostly be found in Appendix I of Jensen and Karp's paper. (i) $\mathscr{B}_2$ is closed under explicit definition. (ii) By taking $G = \lambda u.u$ in (UI), we see that $\bigcup \in \mathscr{B}_2$. (iii) By taking $G = \lambda u.0 \cup \{Huy\}$ in (UI), we see that $\mathscr{B}_2$ is closed under scheme (I) of Theorem 1.2.6. (iv) By use of $\mathfrak{C}$ we show that if P is a boolean combination of the relations $\boldsymbol{R}$ whose characteristic functions $c\boldsymbol{R} \in \mathscr{B}_2$, then $cP \in \mathscr{B}_2$. Also c"$x \in y$" $= \mathfrak{C}01xy$ and c"$x = y$" $= \mathfrak{C}01x\{y\}$. (v) Using (UI) we now show that if $R \in \Delta_0$ then $cR \in \mathscr{B}_2$. (vi) Finally, by taking $G = \lambda uv.\mathfrak{C}\{u\}0(c\chi_r)\{0\}$ in (UI), we show that $D_r \in \mathscr{B}_2$. Q.E.D.

So we now set $\mathscr{B} = \mathscr{B}_1 = \mathscr{B}_2$. By Theorems 1.2.4–6 we see

1.3.4. THEOREM. *Every basic function is substitutable.*

Since the collection $\mathscr{P}$ of primitive recursive functions is closed under (UI), we also see

1.3.5. THEOREM. *Every basic function is primitive recursive.*

In § 2.1 we shall exhibit a function (actually the constant ω) which is substitutable but not primitive recursive, and *a fortiori* not basic.

1.3.6. THEOREM. *A relation* $R \in \Delta_0$ *iff* $cR \in \mathscr{B}$.

This follows from Lemma 1.2.7. and step (v) in the proof of Lemma 1.3.3.

1.3.7. THEOREM. *The following predicates and functions are basic*:
$\langle x, y\rangle$ ($=_{\mathrm{Df}}$ $\{\{x\}, \{x, y\}\}$), $\mathrm{Rel}_n\, x$ ($\leftrightarrow_{\mathrm{Df}}$ *x is an n-place relation*), $\mathrm{Fnc}_n\, x$ ($\leftrightarrow_{\mathrm{Df}}$ *x is an n-argument function*), $x\,\text{‘}\,y$, $x\,\text{“}\,y$, $x \restriction y$, $\mathrm{Dm}\,x$, $\mathrm{Rg}\,x$, $x \times y$, $x \cup y$, $x \cap y$, $x - y$, $\mathrm{Trans}\,x$ ($\leftrightarrow_{\mathrm{Df}}$ *x is transitive*), $\mathrm{Ord}\,x$ ($\leftrightarrow_{\mathrm{Df}}$ *x is an ordinal*), $x \in \omega$, $x = \omega$, $\omega \in x$, $x \subseteq \omega$, $\omega \subseteq x$, $\mathrm{Fsq}\,x$ ($\leftrightarrow_{\mathrm{Df}}$ $\mathrm{Fnc}\,x$ & $\mathrm{Dm}\,x \in \omega$), *the constant a for each* $a \in H_\omega$, *Gödel's functions* $\mathbf{F}_1$—$\mathbf{F}_8$.

The proofs are mostly straightforward; reference may be made to Levy [**12**] for the predicates. We give two examples.

(i) $x \times y = \{\langle u, v\rangle : u \in x \;\&\; v \in y\}$, so $\lambda xy.x \times y \in \mathscr{B}$ by two applications of scheme (I).

(ii) $x - y = \bigcup\{\mathfrak{C}0\{u\}uy : u \in x\}$, so $\lambda xy.x - y \in \mathscr{B}$ by scheme (UI).

[6] This very elegant definition of $\mathscr{B}$ is due to P. Aczel. My original characterization, after emendations suggested by J. B. Paris, was substantially the same as that given in Theorem 1.5.5.

1.4. Basic closure.

1.4.1. DEFINITION.[7] The set a is *basically closed* iff

(i) a is transitive,

(ii) $F\boldsymbol{x} \in a$ for all $F \in \mathscr{B}$, $\boldsymbol{x} \in a$.

1.4.2. DEFINITION. The *basic closure* $\mathrm{Bc}\,x$ of a set x is the smallest basically closed set a such that $x \subseteq a$.

Evidently $\mathrm{Bc}\,x = \{F\boldsymbol{y} : F \in \mathscr{B},\ \boldsymbol{y} \in \mathrm{Tc}\,x\}$, where $\mathrm{Tc}\,x$ denotes the transitive closure of x. Note that if $F \in \mathscr{B}$, then Fx belongs to $\mathrm{Bc}\{x\}$, but not necessarily to $\mathrm{Bc}\,x$.

If x is a set of primitive symbols, then all the finite formulae constructed from these symbols can be represented in $\mathrm{Bc}\,x$. This will be discussed further in § 3. Right now our aim is to give a simple axiomatization of basically closed sets (Theorem 1.4.7), a simple inductive definition of $\mathrm{Bc}\{x\}$ (1.5.6), and further characterizations of $\mathscr{B}$ (Theorems 1.4.5 and 1.5.5). For this purpose it is convenient to replace the scheme (I) in Theorem 1.2.6 by some new functions. We define

$$C_r = \lambda x\boldsymbol{y}.\{D_r x\boldsymbol{v} : \boldsymbol{v} \in \boldsymbol{y}\}.$$

Thus $C_r x\boldsymbol{y}$ is the set of those subsets of x which are determined by the rth Δ_0 formula when the parameters in it range respectively over the sets $\boldsymbol{y}$. Evidently $C_r \in \mathscr{B}$.

1.4.3. LEMMA. $D_r x\boldsymbol{y} = \bigcup C_r x\{y_1\} \ldots \{y_n\}$.

1.4.4. LEMMA. *If* (a) $C_r u\boldsymbol{v} \subseteq z$ *for all* $u \in x$, $\boldsymbol{v} \in \boldsymbol{y}$, *then, for an appropriate* r_1, (b) $C_r u\boldsymbol{v} = D_{r_1} zu\boldsymbol{v}$, *and* (c) $C_r u\boldsymbol{v} \in C_{r_1} zx\boldsymbol{y}$ *for all* $u \in x$, $\boldsymbol{v} \in \boldsymbol{y}$.

These follow straightforwardly from the definitions; an "appropriate" r_1 satisfies

$$\chi_{r_1}(t, u, \boldsymbol{v}) \leftrightarrow (\exists \boldsymbol{w} \in \boldsymbol{v})(t = D_r u\boldsymbol{w}).$$

1.4.5. THEOREM. $\mathscr{B}$ *is the least collection containing* π, $\bigcup$, C_r $(r \in \omega)$, *and closed under explicit definition.*

PROOF. Let $\mathscr{C}$ be the mentioned least collection. Evidently $\mathscr{C} \subseteq \mathscr{B}$. By Lemma 1.4.3, $D_r \in \mathscr{C}$, so to prove $\mathscr{B} \subseteq \mathscr{C}$ it is only necessary to prove that $\mathscr{C}$ is closed under the formation of images. We first prove that for every $F \in \mathscr{C}$ there is a bounding function $F^* \in \mathscr{C}$ such that

$$(*) \qquad \boldsymbol{u} \in \boldsymbol{x} \rightarrow F\boldsymbol{u} \in F^*\boldsymbol{x}.$$

(i) Set $\pi^* xy = C_p(x \cup y)xy$, where $\chi_p(t, u, v) \leftrightarrow .\, t = u \vee t = v$.

(ii) Set $\bigcup^* x = C_q(\bigcup\bigcup x)x$, where $\chi_p(t, u) \leftrightarrow (\exists w \in u)(t \in w)$.

(iii) To compute C_r^* it is only necessary, by Lemma 1.4.4, to find a set z such that

$$(1) \qquad u \in x \;\&\; \boldsymbol{v} \in \boldsymbol{y} .\rightarrow C_r u\boldsymbol{v} \subseteq z.$$

[7] In [9] Jensen called such sets "predicative."

Choose r_2 so that $\chi_{r_2}[s, u, \boldsymbol{w}] \leftrightarrow . s \in u \,\&\, \chi_r[s, \boldsymbol{w}]$; then if $u \in x$, $D_r u\boldsymbol{w} = D_{r_2}(\bigcup x)u\boldsymbol{w}$. But if $t \in C_r u\boldsymbol{v}$, then $t = D_r u\boldsymbol{w}$ for some $\boldsymbol{w} \in \boldsymbol{v}$. Thus if $u \in x$, $\boldsymbol{v} \in \boldsymbol{y}$ and $t \in C_r u\boldsymbol{v}$, then

$$t = D_{r_2}(\bigcup x)u\boldsymbol{w}, \quad \text{for some } w_1 \in \bigcup y_1, \ldots, w_n \in \bigcup y_n,$$

and so $t \in C_{r_2}(\bigcup x)x(\bigcup y_1)\ldots(\bigcup y_n)$ which determines a z satisfying (1).

(iv) If $F = \lambda\boldsymbol{u}.\tau$, where τ is a term built up from $\boldsymbol{u}$ and functions $\boldsymbol{G}$, take $F^* = \lambda\boldsymbol{x}.\tau^*$; here τ^* is obtained from τ by substituting the bounding functions $\boldsymbol{G}^*$ for $\boldsymbol{G}$.

Thus for every $F \in \mathscr{C}$ there is an $F^* \in \mathscr{C}$ satisfying (*). Now let $G = \lambda\boldsymbol{x}.\{F\boldsymbol{u} : \boldsymbol{u} \in \boldsymbol{x}\}$, where $F \in \mathscr{C}$. Then

$$Gx = \{v : v \in F^*\boldsymbol{x} \,\&\, (\exists \boldsymbol{u} \in \boldsymbol{x})(v = F\boldsymbol{u})\} = D_r(F^*\boldsymbol{x})\boldsymbol{x}$$

for some $r \in \omega$ (by Lemma 1.2.7, and Theorem 1.3.4). Hence, by Lemma 1.4.3, $G \in \mathscr{C}$, and we have shown that $\mathscr{C}$ is closed under (a generalization of) (I). Q.E.D.

1.4.6. COROLLARY. *Let b be a transitive set closed under π and $\bigcup$; then the following are equivalent*:

(i) *b is basically closed*;

(ii) *b is closed under all the C_r $(r \in \omega)$*;

(iii) *for each $r \in \omega$, $(\forall x \in b)(\exists w \in b)(\{D_r x\boldsymbol{v} : \boldsymbol{v} \in x\} \subseteq w)$.*

PROOF. The equivalence of (i) and (ii) follows immediately from the theorem. We prove (ii) → (iii) by taking $C_r xx \ldots x$ for "w." To prove (ii) from (iii), suppose the latter holds and that x, $\boldsymbol{v} \in b$ and $z = x \cup \{\boldsymbol{v}\}$. Then $z \in b$ (using π and $\bigcup$), so by (iii) there is a $w \in b$ such that $D_r x\boldsymbol{v}$ $(= D_{r_1} zx\boldsymbol{v}$, say$) \in w$. Hence, since b is transitive, it is closed under all the D_r. But (iii) also shows that if $x, y \in b$, then $C_r x\boldsymbol{y} \subseteq w$ for some $w \in b$. So b is closed under C_r, by Lemma 1.4.4. Q.E.D.

This corollary allows us to give axioms for basically closed sets. Consider the following axiom scheme:

(BR) $$(\forall x)(\exists w)(\forall \boldsymbol{v} \in x)(\exists t \in w)(\forall u)(u \in t \leftrightarrow . u \in x \,\&\, \phi[u, \boldsymbol{v}]),$$

where ϕ is any Δ_0 formula. It says that the collection $\{\{u : u \in x \,\&\, \phi[u, \boldsymbol{v}]\} : \boldsymbol{v} \in x\}$ of those Δ_0 subsets of a set x which are obtained by letting the parameters $\boldsymbol{v}$ in ϕ range over x is itself a set. It may be thought of as either a very weak form of the power set axiom, or as a special case of the Δ_0-replacement axiom, the function in question being $\lambda\boldsymbol{v}.\{u : u \in x \,\&\, \phi[u, \boldsymbol{v}]\}$. We call it the *basic replacement axiom*. *Basic set theory* has as its axioms extensionality, pair set, sum set and (BR). From Corollary 1.4.6 readily follows

1.4.7. THEOREM. *A transitive set b is basically closed iff $\langle b, \in\rangle \models$* BST.

The Δ_0-separation axiom is the scheme:

$$(\forall x)(\exists y)(\forall u)(u \in y \leftrightarrow u \in x \,\&\, \phi)$$

for any Δ_0 formula ϕ. By giving formal metamathematical proofs corresponding to those given in Theorem 1.4.5 and Corollary 1.4.6, we obtain a proof of

1.4.8. THEOREM. (i) *The Δ_0-separation axiom is a theorem of* BST.
(ii) *If $F \in \mathscr{B}$, then there is a Δ_0-formula $\phi[\boldsymbol{x}, y]$ defining the graph of F such that*

$$\text{BST} \vdash (\forall \boldsymbol{x})(\exists ! y)\phi[\boldsymbol{x}, y].$$

The construction of ϕ will of course mimic the definition of F according to Definitions 1.3.1 or 1.3.2. In particular, we see that we can prove in BST that the cartesian product of two sets is a set.

BST′ is obtained by adding an axiom of infinity to BST. In § 2.3 we shall compare it with other predicative set theories.

1.5. Constructible closure. In this section we consider another way of eliminating the formation of images from the definition of "basic"; we obtain a characterization of $\mathscr{B}$ (1.5.5) which is particularly easy to apply. Lemma 1.4.4 shows that the functions C_r can be replaced by the functions $\boldsymbol{D}_r$ provided that some bounding set (the "z" of the lemma) is available. Further, it is well known that the functions $\boldsymbol{D}_r$ can be constructed from Gödel's operations $\mathbf{F}_1$—$\mathbf{F}_8$. We first state this fact in a precise form. Let Tc_q, the qth "approximation" to the transitive closure, be defined by

$$\text{Tc}_q x = \bigcup_{0 \leqq n \leqq q} \left(\bigcup^n x\right).$$

1.5.1. LEMMA. *For each $r \in \omega$ there exist $p, q \in \omega$ and functions G, H explicitly defined from $\mathbf{F}_1$—$\mathbf{F}_8$ such that, if w satisfies*

(a) $\{z_1 \times \cdots \times z_p : \boldsymbol{z} \in \text{Tc}_q x\} \subseteq w$,

then

(b) $\{\langle u, \boldsymbol{v}\rangle : u, \boldsymbol{v} \in x \;\&\; \chi_r[u, \boldsymbol{v}]\} = Gwx$,

and, for all $u, \boldsymbol{v} \in x$,

(c) $D_r u\boldsymbol{v} = Hwu\boldsymbol{v}$.

We do not give the proof; it can be obtained by making appropriate modifications in the proof given by Gödel [**7**, pp. 8–11]. Now we make a construction which is similar to Gödel's; however, as we start with an arbitrary set, the members of a set need not be constructed before the set itself, nor need it be well-ordered. This entails certain differences. Let $\mathbf{F}_9 = \lambda xy.\bigcup x$.

1.5.2. DEFINITION. (i) $\text{Cc}_0\{\boldsymbol{x}\} = \{\boldsymbol{x}\}$; $\text{Cc}_{q+1}\{\boldsymbol{x}\} = \text{Cc}_q\{\boldsymbol{x}\} \cup \{\text{Cc}_q\{\boldsymbol{x}\}\} \cup \{\mathbf{F}_i uv : 1 \leqq i \leqq 9 \;\&\; u, v \in \text{Cc}_q\{\boldsymbol{x}\}\}$.

(ii) The *constructible closure* $\text{Cc}\{\boldsymbol{x}\}$ of the set $\{x_1, \ldots, x_m\}$ is defined by

$$\text{Cc}\{\boldsymbol{x}\} = \bigcup_{q \in \omega} \text{Cc}_q\{\boldsymbol{x}\}.$$

1.5.3. LEMMA. *For each $q \in \omega$,*

(i) $\lambda \boldsymbol{x}.\text{Cc}_q(\boldsymbol{x}) \in \mathscr{B}$,

(ii) $\text{Cc}_q\{\boldsymbol{x}\} \in \text{Cc}_{q+1}\{\boldsymbol{x}\} \;\&.\; y \in \text{Cc}_q\{\boldsymbol{x}\} \rightarrow y \subseteq \text{Cc}_{q+2}\{\boldsymbol{x}\}$,

(iii) $\lambda q\boldsymbol{x}.\text{Cc}_q\{\boldsymbol{x}\}$ *is primitive recursive.*

(i) follows from Theorem 1.3.7 and Definition 1.3.1. (ii) and (iii) are obvious.

1.5.4. LEMMA. *For each $F \in \mathscr{B}$ there is a $q \in \omega$ such that*

$$(\forall \boldsymbol{x})(F\boldsymbol{x} \subseteq \mathrm{Cc}_q\{\boldsymbol{x}\} \;\&.\; F\boldsymbol{x} \in \mathrm{Cc}_q\{\boldsymbol{x}\}).$$

PROOF. By Theorem 1.4.5, since $\pi = \mathbf{F}_1$ and $\bigcup y = \mathbf{F}_9 yy$, it is sufficient to prove (A) for each $p, r \in \omega$ there is a $q \in \omega$ such that

$$u, \boldsymbol{v} \in \mathrm{Cc}_p\{\boldsymbol{x}\} \rightarrow C_r u\boldsymbol{v} \in \mathrm{Cc}_q\{\boldsymbol{x}\}.$$

And to prove (A), by Lemma 1.4.4 it is sufficient to prove
(B) for each $p, r \in \omega$ there is a $q \in \omega$ such that

$$u, \boldsymbol{v} \in \mathrm{Cc}_p\{\boldsymbol{x}\} \rightarrow D_r u\boldsymbol{v} \in \mathrm{Cc}_q\{\boldsymbol{x}\}.$$

Finally, by using $w = \mathrm{Cc}_s\{x\}$ in Lemma 1.5.1 and applying Lemma 1.5.3(ii), we see that (B) will follow from
(C) for any $p, q, r \in \omega$ there is an $s \in \omega$ such that

$$y \in \mathrm{Cc}_r\{\boldsymbol{x}\} \rightarrow \{z_1 \times \cdots \times z_p : \boldsymbol{z} \in \mathrm{Tc}_q y\} \subseteq \mathrm{Cc}_s\{\boldsymbol{x}\}.$$

But (C) follows directly from the definition of Cc. Q.E.D.

1.5.5. THEOREM. *A function F is basic if and only if*
(i) "$v \in F\boldsymbol{x}$" $\in \Delta_0$, *and*
(ii) *there is a $q \in \omega$ such that* $(\forall \boldsymbol{x})(F\boldsymbol{x} \subseteq \mathrm{Cc}_q\{\boldsymbol{x}\})$.

PROOF. If the conditions are satisfied, then there is an r such that $F\boldsymbol{x} = D_r(\mathrm{Cc}_q\{\boldsymbol{x}\})\boldsymbol{x}$, so $F \in \mathscr{B}$ by Lemma 1.5.3. If $F \in \mathscr{B}$, then the conditions are satisfied by Lemmas 1.2.7 and 1.5.4. Q.E.D.

REMARKS. (a) The power set function $\mathfrak{P}$ satisfies (i) but not (ii); it is not basic.
(b) Condition (i) cannot be replaced by "$y = F\boldsymbol{x}$" $\in \Delta_0$; a counterexample is given in § 2.2.
(c) If (i) is satisfied, then (ii) is equivalent to the condition obtained from it by substituting "$\in$" for "$\subseteq$."
(d) Condition (ii) can be replaced by
(ii)′ there is a $G \in \mathscr{B}$ such that $(\forall \boldsymbol{x})(F\boldsymbol{x} \subseteq G\boldsymbol{x})$.
From remark (c) and Lemma 1.5.3 there follows

1.5.6. $\mathrm{Bc}\{\boldsymbol{x}\} = \mathrm{Cc}\{\boldsymbol{x}\}$.

1.5.7. THEOREM. *Let* $\mathbf{F}$ *be Gödel's function which enumerates L; then* $\mathbf{F}$ "α *is basically closed iff* $\alpha = \bigcup_{\gamma<\beta} \gamma^\omega$ *for some* $\beta \leqq \alpha$.

This follows from 1.5.6 and the fact that the stated condition on α holds just in case α is closed under Gödel's pairing function.

The next two propositions are useful for showing that a given function is not basic. Let $\rho(x)$ be the rank of x, and let $\eta(x)$ be the cardinal of $\mathrm{Tc}(x)$ (we are interested in the case where this is finite). Examination of the definition of Cc_{q+1} shows that

(1) $$\rho(\mathrm{Cc}_{q+1}\{\boldsymbol{x}\}) \leqq \rho(\mathrm{Cc}_q\{\boldsymbol{x}\}) + 1,$$

(2) $$\eta(\mathrm{Cc}_{q+1}\{\boldsymbol{x}\}) \leqq \eta(\mathrm{Cc}_q\{\boldsymbol{x}\})^2 + \eta(\mathrm{Cc}_q\{\boldsymbol{x}\}) + 1.$$

And from these the following propositions are readily inferred:

1.5.8. *If $F \in \mathscr{B}$, then there is a $k \in \omega$ such that*

$$\rho(F\boldsymbol{x}) < \mathrm{Max}\{\rho(x_i): 1 \leqq i \leqq m\} + k.$$

1.5.9. *If $F \in \mathscr{B}$, then there is a $k \in \omega$ such that*

$$\eta(F\boldsymbol{x}) < (\eta(\{\boldsymbol{x}\}) + 1)^k.$$

2. Counterexamples

2.1. A substitutable function which is not basic. The function in question is the constant (function of zero arguments) ω. By 1.5.9 a basic constant must be hereditarily finite. (And, by Theorem 1.3.5, conversely every $a \in H_\omega$ is a basic constant.) Indeed, it is not hard to show that H_ω is primitive-recursively closed, so that ω and the functions $\lambda\boldsymbol{x}.\omega$ do not belong to $\mathscr{P}$. Thus we shall prove

2.1.1. Theorem. *There are substitutable functions which are not primitive recursive.*

By Lemma 1.2.2 it will suffice to prove

2.1.2. Theorem. *For every ϕ in Δ_0 there is a ψ in Δ_0 such that* $(\exists \mathrm{y} \in \omega)\phi \leftrightarrow \psi$.

This is a kind of quantifier-elimination theorem, and our proof consists of a sequence of reductions. Perhaps a simpler proof could be constructed using techniques from model theory.

Proof. We assume ("IH") that the theorem is true for formulas θ having fewer quantifiers than ϕ. Throughout the proof y is the indicated variable and is assumed to be distinct from u, v, $\mathbf{x}$; similarly for the corresponding set variables.

First step. Starting from the inside and working outwards replace each part of ϕ (if any) having the form $(\exists \mathrm{u} \in \mathrm{y})\theta$ by $(\exists \mathrm{u} \in \omega)(\theta \;\&\; \mathrm{u} \in \mathrm{y})$. By IH the resulting formula belongs to Δ_0; so there is a ϕ_1 in Δ_0 which has no quantifiers restricted by y and which satisfies

$$(\exists \mathrm{y} \in \omega)\phi \leftrightarrow (\exists \mathrm{y} \in \omega)\phi_1.$$

Second step. Replace each atom of the form $\mathrm{u} \in \mathrm{y}$ in ϕ_1 by $(\mathrm{u} \in \omega \;\&\; \mathrm{u} \neq \mathrm{y} \;\&\; \mathrm{y} \notin \mathrm{u})$. By Theorem 1.5.5, the resulting formula belongs to Δ_0; so there is a ϕ_2 in Δ_0 in which y only occurs in atoms $\mathrm{u} = \mathrm{y}$, $\mathrm{y} \in \mathrm{u}$, and which satisfies

$$(\exists \mathrm{y} \in \omega)\phi \leftrightarrow (\exists \mathrm{y} \in \omega)\phi_2.$$

Third step. We suppose $\phi_2 = \phi_2[\mathrm{y}, \mathbf{x}]$. If a quantifier $(\exists \mathrm{u} \in \mathrm{v})$ appears in ϕ_2 within the scope of a quantifier $(\exists \mathrm{v} \in \mathrm{w})$ we say they are connected. Let n be the maximum length of any chain of connected quantifiers. By induction on n it is easy to prove

(1) the range of every quantifier in $\phi_2[\mathrm{y}, \mathbf{x}]$ is included in $\mathrm{Tc}_n(\{\mathbf{x}\})$ (cf. § 1.5).

Let $\phi_3[\mathbf{x}]$ be the formula obtained from $\phi_2[\mathrm{y}, \mathbf{x}]$ by replacing each atom of the form $\mathrm{u} = \mathrm{y}$ or $\mathrm{y} \in \mathrm{u}$ by $0 \in 0$. Then (1) implies

(2) $$y \notin \mathrm{Tc}_{n+1}(\{\boldsymbol{x}\}) \rightarrow (\phi_2[y, \boldsymbol{x}] \leftrightarrow \phi_3[\boldsymbol{x}]).$$

Let $\psi[\mathbf{x}]$ be a Δ_0 formula equivalent to

$$(\exists \mathrm{z} \in \mathrm{Tc}_{n+1}(\{\mathbf{x}\}))(\mathrm{z} \in \omega \ \& \ \phi_2[\mathrm{z}, \mathbf{x}]).$$

I claim

(3) $$(\exists y \in \omega)\phi_2[y, \boldsymbol{x}] \leftrightarrow \psi \vee (\omega \nsubseteq \mathrm{Tc}_{n+1}(\{\boldsymbol{x}\}) \ \& \ \phi_3).$$

For given $\boldsymbol{x}$, let $a = \mathrm{Tc}_{n+1}(\{\boldsymbol{x}\})$. We distinguish three cases. (A) $\psi[\boldsymbol{x}]$ holds; then both sides of (3) are true. (B) $\psi[\boldsymbol{x}]$ is false and $\omega \subseteq a$; then both sides of (3) are false. (C) $\psi[\boldsymbol{x}]$ is false, and so

(4) $$(\forall y \in \omega)(\phi_2[y, \boldsymbol{x}] \rightarrow y \notin a),$$

and

(5) $$(\exists y \in \omega)(y \notin a).$$

But (3) is a consequence of (2), (4), (5). So (3) holds in this case also. Finally, we observe that the RHS of (3) belongs to Δ_0; this completes the proof of the theorem. Q.E.D.

Although ω is not primitive recursive, it can be obtained by a primitive recursion from any set which is not hereditarily finite. For if $Fx = \{u : u \in \rho(x) \ \& \ u \in \omega\}$, then $x \notin H_\omega \rightarrow Fx = \omega$. This useful trick cannot be played with basic functions.

2.1.3. THEOREM. *There is a set* c, $c \notin H_\omega$, *such that if* $F \in \mathscr{B}$, *then* $Fc \neq \omega$.

PROOF. Let $c = \{\iota^n n : n \in \omega\}$. Let $P(x)$ stand for

$$(\exists k)(\forall m \geqq k)(m \notin \textstyle\bigcup^{m-k} x),$$

where k and m range over ω. It is easy to prove, by induction on q, that $x \in \mathrm{Cc}_q\{c\} \rightarrow P(x)$. So, by 1.5.6, $x \in \mathrm{Bc}\{c\} \rightarrow P(x)$. But evidently $P(\omega)$ is false. Q.E.D.

2.2. A Δ_0 function which is not substitutable.

2.2.1. THEOREM. *If* $P\ (\subseteq V)$ *belongs to* Δ_0, *then* $\omega \cap P$ *is finite or cofinite.*

PROOF. Let $P = \{u : \phi[u]\}$ where ϕ is in Δ_0. Let $\theta[m]$ be the formula in the language of $\langle \omega, < \rangle$ got from ϕ by replacing the variables by number variables and "$\in$" by "$<$". Since ω is transitive, θ contains only *bounded* quantifiers; and $\langle V, \in \rangle \models \phi(n) \leftrightarrow \langle \omega, < \rangle \models \theta[n]$ for any $n \in \omega$. But in $\langle \omega, <, 0, +1 \rangle$ quantifiers can be eliminated. Therefore

$$\langle \omega, <, 0, +1 \rangle \models \theta[n] \leftrightarrow \psi[n],$$

where $\psi[m]$ is a quantifier-free formula in the language of $\langle \omega, <, 0, +1 \rangle$. So $\{n : \psi[n]\}$ ($= P \cap \omega$) is finite or cofinite. Q.E.D.

2.2.2. COROLLARIES. (i) *The set of even numbers does not belong to* Δ_0.
(ii) *The graph of addition does not belong to* Δ_0.

For x is even iff $x \in \omega$ & $(\exists u \in x)(x = u + u)$. Already 1.5.8 suffices to show that addition is not basic; now we see that it is not substitutable, nor even Δ_0.

2.2.3. THEOREM. *There is a function* $F : V \to V$ *satisfying*
(i) $Fx \subseteq x$;
(ii) $F \in \Delta_0$;
(iii) $F \notin \mathscr{S}$ (*a fortiori* $F \notin \mathscr{B}$);
(iv) "$v \in Fx$" $\notin \Delta_0$.

PROOF. We define F by

$$Fx = \{v : x \in \omega \,\&\, v \in x \,\&\, v \text{ is even}\}.$$

(i) is immediate. To prove (ii) observe that

$$y = Fx \leftrightarrow [y = 0 \,\&\, x \notin \omega]$$
$$\vee\ [x \in \omega \,\&\, y \subseteq x \,\&\, 1 \notin y \,\&\, (\forall v \in x)(v \in y \leftrightarrow . v = 0 \vee (\exists u \in y)(v = u + 2)].$$

To prove (iv) observe that x is even $\leftrightarrow x \in \omega \,\&.\ x = 0 \vee (\exists u \in x)(\forall v \in x)(v \in Fx \to v \neq u \,\&\, (\forall w \in v)(w \neq u))$.

If (iv) were false, this equivalence would contradict 2.2.2 (i). Finally, (iii) follows from (iv) by Lemma 1.2.7. Q.E.D.

This theorem shows the significance of condition (i) in Theorem 1.5.5. The fact that the complexity of a function may be greater than the complexity of its graph is a familiar phenomenon in recursion theory; for example, any general recursive function can be majorized by a function whose graph is primitive recursive.

2.3. Predicative set theories. Each of the theories PZ, BST′, BRT, PZF contains the axioms of extensionality, pair set, sum set, and an axiom of infinity; for definiteness we take this to be

Ax.Inf $$(\exists y)(\forall x)(x \in \omega \to x \in y).$$

They differ in the axiom scheme of separation or replacement.

PZ (for "Predicative Zermelo") has the Δ_0-separation axiom. BST′ has the basic replacement axiom (cf. Corollary 1.4.6). BRT has the "bounded replacement axiom;" namely, for any $\phi \in \Delta_0$,

$$(\forall x \in u)(\exists ! y)(\phi(x, y) \,\&\, y \subseteq z) \to (\exists v)(\forall y)[y \in v \leftrightarrow (\exists x \in u)(\phi(x, y) \,\&\, y \subseteq z)].$$

(In words, the image of a set by a function whose values are all included in a set is itself a set). PZF (for "Predicative Zermelo-Fraenkel") has the Δ_0-replacement axiom.[8]

It is a trivial matter to prove

2.3.1. $\mathrm{PZ} \subseteq \mathrm{BST}' \subseteq \mathrm{BRT} \subseteq \mathrm{PZF}$.

We shall show that each of the inclusions is proper.

2.3.2. THEOREM. *There is a set A such that $\langle A, \in\rangle \models \mathrm{PZ}$, but $\langle A, \in\rangle \not\models \mathrm{BST}$.*

PROOF. (1) We define "$x \equiv \omega$" to mean that x differs from ω by a finite set. Thus $x \equiv \omega$ iff there is an $n \in \omega$ and a finite set y such that $x = y \cup (\omega - n)$. Let n be chosen as small as possible; then we call $\omega - n$ the *infinite part* of x (written $\mathbf{I}x$), and $x - \mathbf{I}x$ the *finite part* of x (written $\mathbf{F}x$). If x is a finite set we adopt the convention $\mathbf{I}x = 0$, $\mathbf{F}x = x$.

(2) We define the model A by

$$A = \{x : (\forall y \in \mathrm{Tc}\{x\})(\mathrm{card}\, y \in \omega \vee y \equiv \omega)\}.$$

Then the following are obvious:

(a) A is transitive, and therefore satisfies the axiom of extensionality;

(b) $x \in A \to \bigcup x \subseteq A$;

(c) $\omega \in A$, so A satisfies the axiom of infinity;

(d) $x \subseteq A \;\&\; \mathrm{card}\, x \in \omega . \to x \in A$, so A satisfies the pair axiom.

(3) For each $x \in A$ there is a basic function G_x such that $G_x \omega = x$. This is proved by induction on the rank. Let $\mathbf{F}x = \{y_1, \ldots, y_n\}$, and suppose the $G_{y_i} \in \mathscr{B}$ have been defined. Set

$$\begin{aligned} G_x u &= \{G_{y_1} u, \ldots, G_{y_n} u\} && \text{if } \mathbf{I}x = 0, \\ &= \{G_{y_1} u, \ldots, G_{y_n} u\} \cup (u - m) && \text{if } \mathbf{I}x = \omega - m. \end{aligned}$$

Then, in either case, $G_x \in \mathscr{B}$ by Theorem 1.3.5, and $G_x \omega = x$.

(4) But now it follows from Theorem 2.1.2 (which shows that ω is substitutable) that every $x \in A$ is a substitutable constant. Hence if $\phi[u, \mathbf{y}] \in \Delta_0$, and $\mathbf{y} \in A$, then there is a ψ in Δ_0 such that $(\forall u)(\phi[u, \mathbf{y}] \leftrightarrow \psi[u])$.

(5) Let $z = \{u : u \in x \;\&\; \phi[u, \mathbf{y}]\}$, where x, $\mathbf{y} \in A$. Then

$$z = \mathbf{F}x \cap \{u : \psi[u]\} \cup (\mathbf{I}x \cap \{u : \psi[u]\}).$$

By (2)(d), the first term belongs to A. And by Theorem 2.2.1, the second term is a finite or cofinite subset of ω. Therefore $z \in A$, and so A satisfies the Δ_0-separation axiom; by (2)(b), this implies that it satisfies the sum set axiom. So we have proved $\langle A, \in\rangle \models \mathrm{PZ}$.

(6) Finally, it is obvious that $\omega \times \omega \not\equiv \omega$, so $\omega \times \omega \notin A$. Hence, by Theorem 1.4.8, $\langle A, \in\rangle \not\models \mathrm{BST}$. Q.E.D.

2.3.3. THEOREM. *There is a set B such that $\langle B, \in\rangle \models \mathrm{BST}'$ but $\langle B, \in\rangle \not\models \mathrm{BRT}$.*

[8] Predicative versions of set theory have been discussed by Kripke (in unpublished lecture notes) and by Feferman in [**4**].

PROOF. We set $B = \mathrm{Bc}\{\omega\}$. By Theorem 1.4.8, $\langle B,\in\rangle \models \mathrm{BST}'$. Let $e = \{u : u \in \omega \,\&\, u \text{ is even}\}$. Then $e \notin B$. For if $e \in B$, then $e = Fx$ for some $F \in \mathscr{B}$ and $x \in \mathrm{Tc}(\{\omega\})$. But, since members of ω are basic constants, this means $e = G\omega$ for some $G \in \mathscr{B}$; which would imply, by Lemma 1.2.7 and Theorem 2.1.2, "$u \in e$" $\in \Delta_0$, contradicting Corollary 2.2.2(i).

On the other hand, suppose c is a transitive set such that $\langle c, \in\rangle \models \mathrm{BRT}$, and let F be the Δ_0 function of Theorem 2.2.3. Then $\omega \in c$ and $x \in \omega \to Fx \subseteq \omega$. Therefore $F``\,\omega$ and $\bigcup(F``\,\omega)$ belong to c; but $\bigcup(F``\,\omega) = e$. Thus $\langle B, \in\rangle \not\models \mathrm{BRT}$. Q.E.D.

REMARK. B is the smallest basically closed set which contains ω. The above argument shows that the only subsets of ω which are contained in B are the finite and cofinite ones. It would be nice to have a more transparent definition of B, or of the smallest basically closed set containing H_ω.

2.3.4. THEOREM. *There is a set C such that* $\langle C,\in\rangle \models \mathrm{BRT}$ *but* $\langle C,\in\rangle \not\models \mathrm{PZF}$.

PROOF. Take $C = R_{\omega.2} = \{x : \rho(x) < \omega.2\}$. Evidently all the axioms of BRT are satisfied in $\langle C, \in\rangle$. But $\omega.2$, which can be proved to be a set in PZF, does not belong to C. Q.E.D.

Thus we have shown that all the inclusions in 2.3.1 are proper. Observe that if we drop the restriction that ϕ be Δ_0 in the axiom schemata we obtain a sequence of four set theories stretching from Zermelo to Zermelo-Fraenkel; presumably these also are all distinct.

3. BASIC NUMERALS AND ELEMENTARY SYNTAX

3.1. A stumbling block. Suppose we code the expressions of a finitary language in the natural way—i.e., as finite sequences of elements of a set A. A is the alphabet of "primitive symbols"; naturally we suppose $A \in V$, but A need not be countable. At first sight it seems that $\mathscr{B}$ will be hopelessly inadequate for describing the operations of elementary syntax, as is shown by the following theorem.

3.1.1. THEOREM. *The graph of concatenation does not belong to* Δ_0.

For suppose $\phi[x, u, v] \in \Delta_0$ and satisfies

$$\mathrm{Fsq}\, u \,\&\, \mathrm{Fsq}\, v \,.\!\to.\, \phi[x, u, v] \leftrightarrow x = u \frown v.$$

Then, for any $m, n \in \omega$,

$$z = m + n \leftrightarrow \phi[z \times \{0\}, m \times \{0\}, n \times \{0\}];$$

so, by Theorems 1.3.7, 1.3.4, the RHS $\in \Delta_0$, contradicting Corollary 2.2.2(ii). Q.E.D.

The stumbling block here is the fact that, for $n \in \omega$, $\rho(n) = n$. But there is nothing sacred in von Neumann's representation of the natural numbers by sets. By choosing a different set $\bar{n}$ to represent the number n in such a way that $\rho(\bar{n})$ does not increase too rapidly, we can ensure that both addition and multiplication are basic.

3.2. The basic numerals. Our treatment is based on the dyadic representation of the natural numbers. This allows us to make use of the results obtained by Smullyan and his co-workers. Other systems of numeration might serve as well; in particular it would be interesting to explore a system based on Ackermann's mapping of the hereditarily finite sets on the natural numbers.

Throughout this section $i, j, k, m, n, p, q, r, s$ are metamathematical variables which range over the natural numbers.

3.2.1. DEFINITION. For each n the corresponding *number set* $\hat{n}$ and its *level* $\lambda(n)$ are defined as follows:

(i) $\hat{0} = 0, \quad \lambda(0) = 0$;

(ii) for $n > 0$ let $n - 1 = \sum_{i<k} a_i 2^i$ (where $a_i \in \{1, 2\}$) be the dyadic expansion of $n - 1$, then

$$\hat{n} = \{\{\hat{\imath} : i < k \,\&\, a_i = 2\}, \{\hat{\imath} : i < k\}\} \quad \text{and} \quad \lambda(n) = k.$$

Thus, for example,

$$\hat{1} = \{0\}, \lambda(1) = 0; \qquad \hat{2} = \{0, \{0\}\}, \qquad \hat{3} = \{\{0\}\}, \lambda(2) = \lambda(3) = 1;$$

$$\hat{4} = \{0, \{0, \{0\}\}\}, \qquad \hat{5} = \{\{0\}, \{0, \{0\}\}\}, \qquad \hat{6} = \{\{\{0\}\}, \{0, \{0\}\}\},$$

$$\hat{7} = \{\{0, \{0\}\}\}, \lambda(4) = \lambda(5) = \lambda(6) = \lambda(7) = 2.$$

Observe that if x is any subset of $\{i : i < k\}$, then $\{\hat{\imath} : i \in x\}$ is a member of a member of $\hat{n}$, provided $n > 2^{k-1}$. This allows some recursively defined predicates to be described by Δ_0 formulae.

We define $\hat{\omega} = \{\hat{n} : n \in \omega\}$. It is probably false that membership of $\hat{\omega}$ is a Δ_0 property. But, as we shall show, being an initial segment of $\hat{\omega}$ is Δ_0. It is these initial segments which we take for our numerals.

3.2.2. DEFINITION. The *basic numeral* $\bar{n}$ corresponding to the natural number n is given by $\bar{n} = \{\hat{m} : m < n\}$. Further, $\bar{\omega} =_{\mathrm{Df}} \{\bar{n} : n \in \omega\}$.

In order to express $x \in \bar{\omega}$ by a Δ_0 formula we need some preliminary definitions and lemmas. All the functions and predicates defined below are basic; the proofs which justify this statement are left to the reader.

3.2.3. DEFINITION.

$$y_0 = \{z : (\exists u, v \in y)(y = \{u, v\} \,\&\, u \subseteq v \,\&\, z \in u)\},$$

$$y_1 = \{z : (\exists u, v \in y)(y = \{u, v\} \,\&\, u \subseteq v \,\&\, z \in v)\},$$

$$y <_x z \leftrightarrow : y = 0 \,\&\, z \neq 0 . \vee . y_1 \subset z_1 . \vee . y_1 = z_1$$

$$\& \, (\exists w \in x)(w_1 \subset z_1 \,\&\, y_0 - z_0 \subseteq w_1 \,\&\, z_0 - y_0 \nsubseteq w_1).$$

Thus if $n > 0$ and $\lambda(n) = k$, then $\hat{n} = \{\hat{n}_0, \hat{n}_1\}$ and $\hat{n}_0 \subseteq \hat{n}_1 = \bar{k}$. Also it is readily checked that

$$m, n < p \rightarrow . \hat{m} <_{\bar{p}} \hat{n} \leftrightarrow m < n.$$

Indeed, rather weak conditions on x suffice to ensure that $\hat{m} <_x \hat{n}$ is equivalent to $m < n$, as the following lemma shows.

3.2.4. LEMMA. (i) *If* $\lambda(m) \neq \lambda(n)$, *then* $\hat{m} <_x \hat{n} \leftrightarrow m < n$.
(ii) *If* $\lambda(m) = \lambda(n) = k$, *then*

$$\cdot(\forall j < k)(\exists p)(\lambda(p) = j \,\&\, \hat{p} \in x) \rightarrow . \, m < n \rightarrow \hat{m} <_x \hat{n}.$$

(iii) $\mathrm{LO}(x) \,\&\, (\forall m < p)(\hat{m} \in x) . \rightarrow (\forall m, n < p)(m < n \leftrightarrow \hat{m} <_x \hat{n})$.

Here $\mathrm{LO}(x)$ is the Δ_0 formula which asserts that x is linearly ordered by $<_x$. The proof of the lemma is left to the reader.

3.2.5. DEFINITION.

$$y \leqq_x z \leftrightarrow . \, y <_x z \vee y = z.$$

$$\mathrm{Seg}_x\, w = \{z : z \in x \,\&\, z \leqq_x w\}.$$

$$\mathrm{Max}_x\, v = \{z : (\exists w \in x)(v = \mathrm{Seg}_x\, w \,\&\, z \in w)\}.$$

$$v/x = v - \{\mathrm{Max}_x\, v\}.$$

$$\phi_1(x) \leftrightarrow : x = 0 \vee . 0 \in x \,\&\mathrm{LO}(x) \,\&\, (\forall y \in x)(y = 0 \vee (\exists u, v \in y)$$
$$[y = \{u, v\} \,\&\, u \subseteq v \subseteq x \,\&. \, v = 0 \vee (\exists w \in v)(v = \mathrm{Seg}_x\, w)]).$$

$$\phi_2(x) \leftrightarrow (\forall y \in x)[\{y_1/x\} \in x \,\&\, (\forall u \in \textstyle\bigcup x)(\forall z \in x)(\forall t \in z_1)$$
$$(\{u - \{t\}, z_1/x\} \in x \,\&\, \{u, z_1\} <_x y . \rightarrow \{u, z_1\} \in x)].$$

Roughly speaking $\phi_1(x)$ says that members of x have the same form as number sets, and $\phi_2(x)$ ensures that if a number set belongs to x then so do all its predecessors. Together they suffice to give a Δ_0 characterization of the initial segments of $\bar{\omega}$:

3.2.6. THEOREM. $\phi_1(x) \,\&\, \phi_2(x) . \leftrightarrow . \, x \in \bar{\omega} \vee x = \bar{\omega}$.

It is straightforward to verify "$\leftarrow$"; to prove "$\rightarrow$" we need some lemmas.

3.2.7. LEMMA. $\phi_1(x) \,\&\, \phi_2(x) \,\&\, \hat{n} \in x . \rightarrow \bar{n} \subseteq x$.

PROOF. Assume the LHS. The RHS follows trivially if $n = 0, 1$. So we suppose $\hat{n} = \{u, \bar{k}\}$, where $k = j + 1$, and set $p = n - 1$. We also assume

(IH) $\qquad \hat{m} \in x \rightarrow \bar{m} \subseteq x \quad$ for all $m < n$.

We prove $\hat{p} \in x$; the lemma then follows by induction on n.

Since $\phi_1(x)$ holds, $\bar{k} \subseteq x$ and so, by Lemma 3.2.4, $<$ and $<_x$ coincide on $\bar{k}$. Therefore $\hat{\jmath} = \mathrm{Max}_x\, \bar{k}$ and $\hat{n}_1/x = \bar{\jmath}$. Hence, by the first clause of $\phi_2(x)$, $\{\bar{\jmath}\}$ $(= \hat{q}$, say$) \in x$. If $u = 0$, then $q = p$ and so $\hat{p} \in x$. Otherwise $\hat{p} = \{u', \bar{k}\}$, where $u' \subset \bar{k}$. Let $r = \mathrm{Max}\{i : i \in u'\}$ $(= 0$ if $u' = 0)$ and let s be defined by $\hat{s} = \{u' - \{\hat{r}\}, \bar{\jmath}\}$. Since $u' \subset \bar{k}$, $u' - \{\hat{r}\} \subset \bar{\jmath}$ and so $s < q$ and $\hat{s} \in \bar{q}$. But $\bar{q} \subseteq x$ by (IH); therefore $\hat{s} \in x$. Finally, $\hat{p} <_x \hat{n}$ by Lemma 3.2.4 and so $\hat{p} \in x$ by the second clause of $\phi_2(x)$. Q.E.D.

3.2.8. LEMMA. $\phi_1(x) \,\&\, \phi_2(x) \,\&\, y \in x . \rightarrow y \in \hat{\omega}$.

PROOF. Suppose $\phi_1(x)$ & $\phi_2(x)$; we prove $y \in x \to y \in \hat{\omega}$ by induction on the rank of y ($= \alpha$, say). If $y \neq 0$, and $y \in x$, then $y = \{u, v\}$, where $u \subseteq v \subseteq x$. If $v = 0$, then $y = \{0\} \in \hat{\omega}$. Otherwise, since $\phi_1(x)$, $v = \mathrm{Seg}_x\, w$ for some $w \in x$. But $w \in v \in y$ and so $\rho(w) < \alpha$, and therefore w is a number set—$\hat{k}$ say. By Lemmas 3.2.7 and 3.2.4 it follows that $v = \overline{k+1}$, so $y \in \hat{\omega}$. Q.E.D.

Now Theorem 3.2.6 follows immediately. And from it we get a Δ_0 expression for "$\in \bar{\omega}$"; namely

$$x \in \bar{\omega} \leftrightarrow . \phi_1(x) \;\&\; \phi_2(x) \;\&\; (\exists y \in x)(x = \mathrm{Seg}_x\, y).$$

3.3. Basic functions of basic numerals. The main aim of this section is to show that addition, dyadic concatenation and multiplication are all basic functions. We need a number of auxiliaries; all functions defined are basic.

3.3.1. DEFINITION.

$$(\mu\hat{t} \in x)\chi(t) = \{w : (\exists y \in x)\,[\chi(y) \;\&\; w \in y \;\&\; (\forall z <_x y) \neg\, \chi(y)]\}.$$

$$(\mu\bar{t} \leqq x)\chi(t) = \{y : y \in x \;\&\; y <_x (\mu\hat{z} \in x)\chi(\{u : u \in x \;\&\; u \leqq_x z\})\}.$$

$$\bar{\lambda}(x) = \bigcup\bigcup x.$$

$$L(x) = (\mu\hat{t} \in x)(\forall y \in x)(y \leqq_x t).$$

$$T_1(x) = (\mu\hat{t} \in x)(\forall y \in x)(y \leqq_x t \to \{\{z : z <_x y\}\} \in x).$$

$$T_2(x) = \{y : y \in x \;\&\; y \leqq_x T_1(x)\}.$$

$$T(x) = \bar{2} \cup x \cup \{\{z_0, T_2(x)\} : z \in x\} \cup \{\{z_0 \cup \{T_1(x)\}, T_2(x)\} : z \in x\}.$$

It is assumed that $\chi \in \Delta_0$. $(\mu\hat{t} \in \bar{n})$ selects the least number set in $\bar{n}$ having a given property; $(\mu\bar{t} \leqq \bar{n})$ selects the least numeral $\leqq \bar{n}$. In both cases the result is 0 if no selection is possible. If k is the level of the greatest number set in $\bar{n}$, then

$$\bar{\lambda}(\bar{n}) = \bar{k}, \qquad L(\overline{n+1}) = \hat{n}.$$

If $n > 3$, then $T_1(\bar{n})$ is the largest number set $\hat{k}$ such that $\{\bar{k}\} \in \bar{n}$, and $T_2(\bar{n}) = \overline{k+1}$. But $\{\bar{k}\}$ is the number set corresponding to $2^{k+1} - 1$, so $T_2(\bar{n})$ is the numeral for the exponent of the greatest power of 2 not greater than n. Now T adds to x all the number sets of level $k + 1$; so $T(\bar{n})$ is the numeral for the least power of 2 which is greater than n. If $n \leqq 3$, then $T(\bar{n}) = \bar{4}$. Thus we have proved

3.3.2. LEMMA. $T(\bar{n}) \in \bar{\omega}$ & $\bar{n} < T(\bar{n})$ & $\bar{2}\bar{n} < T(T(\bar{n}))$.

This enables us to show that successor is basic, for

$$\overline{n+1} = (\mu\bar{y} \leqq T(\bar{n}))(\bar{n} \subset y).$$

Now for addition. Suppose $m, n > 1$, and $\lambda(n) = k \geqq j = \lambda(m)$. We consider the process of dyadic addition of $m - 1$ and $n - 1$. Let $p - 1 = (m - 1) + (n - 1)$. At the ith place $(0 \leqq i \leqq k)$ there may be a carry of 1 or 2 to the next place. Let $y = \{\hat{\imath} : i\text{th carry is } 1\}$, $z = \{\hat{\imath} : i\text{th carry is } 2\}$. Then the truth values of "$\hat{\imath} \in \hat{p}_0$," "$\hat{\imath} \in y$," "$\hat{\imath} \in z$" for $0 \leqq i \leqq k$ can be determined in a Boolean way from the truth values of "$\hat{\imath} \in \hat{m}_0$," "$\hat{\imath} \in \hat{n}_0$," "$\hat{\imath} - 1 \in y$," "$\hat{\imath} - 1 \in z$," "$i < k$," "$i < j$"; and the level of p is $k + 1$ if $\hat{k} \in y \cup z$, k otherwise.[9] Hence there is a Δ_0 formula $\psi(m, n, x, y, z)$ which is uniquely satisfied by $x = \hat{p}_0$, y and z as above. But $\hat{p}_0, y, z$ are all members of members of $T^2(\overline{n + 1})(= g(\bar{n})$, say, where $g \in \mathscr{B}$). Thus, for $n \geqq m > 1$,

$$\overline{m + n} = \{w : w \in g(\bar{n}) \,\&\, (\exists x, y, z \in \bigcup g(n))(\psi(L(\overline{m + 1}), L(\overline{n + 1}), x, y, z) \\ \& \, w <_{g(\bar{n})} \{x, \{t : t <_{\bar{n}} L(y \cup z)\}\})\}.$$

The definition is readily extended to the other cases. Thus we have proved

3.3.3. THEOREM. *Addition is a basic function of basic numerals.*

We write "$\mp$" to denote this function, and "$\bar{2}x$" for "$x \mp x$." In order to define dyadic concatenation we need more auxiliaries.

3.3.4. DEFINITION.

$$y \hat{+}_x z = L(\{w : w \in x \,\&\, w \leqq_x y\} \mp z).$$

$$u \nearrow x = \{z : z \in \bar{2}x \,\&\, (\exists y \in u)(z = y \hat{+}_x x)\}.$$

$$x^\frown y = \{(x_0 \nearrow y_1) \cup y_0, x_1 \mp y_1\}.$$

$$S(x) = \bar{2}\{y^\frown z : y, z \in T(x)\}.$$

$$x \bar{*} y = (\mu \bar{t} \leqq S(x \mp y))(L(t \mp \bar{2}) = L(x \mp \bar{2})^\frown L(y \mp \bar{2})).$$

If $p = m + n$ and $m < q$, then $\hat{m} \hat{+}_{\bar{q}} \bar{n} = \hat{p}$. If $u \subseteq \bar{\omega}$, then $u \nearrow \bar{k}$ is the set obtained by translating u upwards by k steps. The dyadic concatenation $m * n$ is by definition the number p whose dyadic representation is obtained by writing down the dyadic representation of m followed by the dyadic representation of n. Hence, if m', n', p' are the successors of m, n, p, then $\hat{m}'^\frown \hat{n}' = \hat{p}'$. Also $\bar{m} \bar{*} \bar{n} = \bar{p}$. Finally, if $p = 2^{k+1}$ is the least power of 2 greater than m (and $m > 1$), then $T(\bar{m}) = \bar{p}$, $\{y^\frown z : y, z \in T(\bar{n})\}$ is $\overline{2^{2k+1}}$ and $S(\bar{m})$ is $\overline{p^2}$. Hence

3.3.5. LEMMA. $S(\bar{n}) \in \bar{\omega} \,\&\, \overline{n^2} < S(\bar{n})$.

A predicate of natural numbers is *rudimentary* if it can be defined using identity, the graph of $*$, propositional connectives and bounded quantification. Since

$$\{\bar{m} : \bar{m} \subseteq \bar{n}\} = \{\{y : y \in \bar{n} \,\&\, y <_{\bar{n}} z\} : z \in \bar{n}\},$$

the LHS is a basic function of $\bar{n}$. So we have proved

3.3.6. THEOREM. *Every rudimentary predicate of basic numerals is Δ_0.*

3.3.7. CONJECTURE. *Every Δ_0 predicate of basic numerals is rudimentary in the corresponding numbers.*

Bennett, in his dissertation [**3**], showed that the graphs of multiplication and exponentiation are rudimentary. Using Lemma 3.3.5 and Theorem 3.3.6, we get

[9] Further details will be found in Ritchie [**14**].

3.3.8. THEOREM. *Multiplication is a basic function of basic numerals.*

It is easy to prove that $m \leqq \eta(\bar{m}) < 2m$; hence, by 1.5.9,

3.3.9. THEOREM. *A basic function from $\bar{\omega}^m$ into $\bar{\omega}$ is bounded by a polynomial.*

In particular, exponentiation is not a basic function of basic numerals.

3.4. Applications. We first consider the application of $\mathscr{B}$ to the codification of elementary syntax. We do not attempt a general definition of this term; instead we bear in mind two paradigms—the substitution function and the proof predicate for a formal language. By a finitary language we mean one in which only finite strings of quantifiers appear, and in which all functions (including propositional functions) and relations have only a finite number of arguments. Theorem 3.3.6 and the results of Smullyan [**15**] suffice to prove

3.4.1. THEOREM. *If L is a countable finitary language and its expressions are coded dyadically and then represented by basic numerals, then all the predicates of the elementary syntax of L belong to Δ_0.*

Note that we cannot add "and functions" to "predicates"; for if an expression of dyadic length m is substituted at each of n occurrences of a variable, the resulting expression will have dyadic length greater than $m \cdot n$. This means that the substitution function will grow faster than any polynomial in its arguments, and so cannot be basic.

The dyadic coding is not of course suitable for uncountable languages, nor even for languages (e.g. for a countable transfinite hierarchy) which involve transfinite orderings. But we have now got round the difficulty discussed in § 3.1. A *basic finite sequence* is a function whose domain is a basic numeral. If we code expressions as basic finite sequences, it is easy to see that concatenation is a basic function. There is however still a difficulty about substitution. In expressing that x is the result of substituting y at all the occurrences of z in w, we need to know how many occurrences of z precede some given one. But it is far from clear that there are basic counting functions. We pose the question in a simple form.

3.4.2. PROBLEM. *Is there a predicate $P \in \Delta_0$ such that if $x \subseteq \bar{m}$ then $P\bar{m}x\bar{n}$ holds iff the cardinal of x is n?*

I believe that the answer is negative. To get round this difficulty we need a slightly more elaborate coding: Before forming the basic finite sequence corresponding to a given expression we tag the nth occurrence of a variable with the number n (for each variable and each relevant n). E.g., we could replace the variable v by $\langle \mathrm{v}, \bar{n} \rangle$. The number of occurrences of a variable in an expression is then a basic function of its code. It is easy to see that concatenation and the substitution function for this coding are both basic. Indeed, we can assert

3.4.3. THEOREM. *Let L_A be a finitary language whose expressions are words on the alphabet A. Suppose that "$x \in A$" and the classification of A into logical constants,* *various sort of variable, constant, relation symbol, etc. are all* Δ_0. *Then all the predicates and functions of the elementary syntax of* L_A *are basic.*

We conclude this section with an application to the indexing of basic functions. We allow for parameters by setting

$$\mathscr{B}^A = \{\lambda\mathbf{x} . F\mathbf{x}\mathbf{y} : F \in \mathscr{B} \mathbin{\&} \mathbf{y} \in A\},$$

$$\mathscr{B}_A = \{F \restriction A : F \in \mathscr{B}^A\}.$$

We describe the notion of indexing in some generality. Let $\mathscr{C}$ be a collection of functions which is given an inductive definition of the following form:

(i) a finite number of specified initial functions belong to $\mathscr{C}$;

(ii) $\mathscr{C}$ is closed under precisely defined schemata including:

(a) if $F\mathbf{x} = x_i$ then $F \in \mathscr{C}$, for each m and $1 \leq i \leq m$;

(b) the substitution schemata of Definition 1.3.2, for each m, n;

(c) if $F\mathbf{x} = c$ then $F \in \mathscr{C}$, for each m and each $c \in C$, where C is a given (possibly empty) class of "parameters."

Conditions (a)–(c) ensure that $\mathscr{C}$ is closed under explicit definitions using parameters from C. Now let D be a class and $\mathscr{D}$ a collection of D-theoretic functions and predicates.

3.4.4. DEFINITION. $\mathscr{C}$ has a $\mathscr{D}$-system of indexing iff to each definition of an $F \in \mathscr{C}$ there is assigned an index $f \in D$ (and we write $F = \{f\}$) in such a manner that the following operations all belong to $\mathscr{D}$:

(1) determining, for all $x \in D$, whether x is an index;

(2) determining the number of arguments of $\{f\}$;

(3) when F is defined by a schema, determining an index f for it from the precise specification of the schema (e.g., m and c in (c) above) and the indices of other functions occurring in it;

(4) conversely, determining from f whether $\{f\}$ is an initial function or is defined by a schema; if the latter, recovering from f the specification of the schema and the indices of the other functions;

(5) determining the effects of substituting parameters for arguments; that is, there shall be functions $S_n^m \in \mathscr{D}$ which satisfy

$$\{S_n^m f\mathbf{x}\} = \lambda\mathbf{y}.\{f\}(\mathbf{x}, \mathbf{y})$$

for all indices f for functions of $m + n$ arguments, all $\mathbf{x} \in C$ and for each m and n.

It is well known that if $\mathscr{C}$ has a $\mathscr{D}$-system of indexing and $\mathscr{D} \subseteq \mathscr{C}$, then the second recursion theorem holds uniformly for $\mathscr{C}$: for each m there is an $H^m \in \mathscr{D}$ such that

$$\{H^m f\} = \lambda\mathbf{x}.\{f\}(H^m f, \mathbf{x})$$

whenever f is an index for a function of $m + 1$ arguments.

REMARK. We have assumed that the functions of $\mathscr{C}$ are all total over some given domain; of course the definitions can be modified to allow for partial functions. Despite first appearances, the above conditions are very much weaker than those given by Wagner, Strong and others for abstract recursion theories.

3.4.5. THEOREM. *Let A be basically closed. Let $\mathscr{B}(A)$ denote the class of functions and predicates obtained by adding the characteristic function of "$x \in A$" to the initial functions of $\mathscr{B}$. Then*

(i) *there is a $\mathscr{B}(A)$-system of indexing for $\mathscr{B}^A$ and $\mathscr{B}_A$;*
(ii) *there is a $\mathscr{B}_A$-system of indexing for $\mathscr{B}^A$ and $\mathscr{B}_A$;*
(iii) *if $A \in V$, there is a $\mathscr{B}^{\{A\}}$-system of indexing for $\mathscr{B}^A$ and $\mathscr{B}_A$;*
(iv) *if "$x \in A$" is Δ_0, there is a $\mathscr{B}^A$-system of indexing for $\mathscr{B}^A$ and $\mathscr{B}_A$.*

3.4.6. COROLLARY. *If A is basically closed, the second recursion theorem holds for $\mathscr{B}_A$; it holds in addition for $\mathscr{B}^A$ provided "$x \in A$" is Δ_0.*

Notes. (1) A system of indexing for $\mathscr{B}^A$ also works for $\mathscr{B}_A$.

(2) Parts (ii)–(iv) of the theorem follow almost immediately from part (i). In particular, in $\mathscr{B}_A$ the variables are restricted to A and so "$x \in A$" is otiose.

(3) In proving the theorem, care must be exercised in the choice of indices. It will not do, for example, to choose $\langle l, m, n, g, h\rangle$ as an index for $\lambda xy.Gx(Hx)y$, since a Δ_0 predicate can only penetrate a fixed number of levels of the transitive closure of its arguments. Instead one must work with a "shallow" system of coding, so that, e.g., an index for F is a sequence of the indices for all the functions used in constructing F.

Here is an application of the theorem. In [**13**] Moschovakis gave the following inductive definition of a set H:

(a) $\langle 1, 0\rangle \in H$;
(b) $(\forall x \in A)(\{f\}_{\text{pr}}(x) \in H) \to \langle 2, f\rangle \in H$;
(c) $(\exists x \in A)(\{f\}_{\text{pr}}(x) \in H) \to \langle 3, f\rangle \in H$.

Here $\{f\}_{\text{pr}}$ is the "primitive recursive" function over A indexed by f ($\in A$). He showed that all subsets of A which are first-order monotonically inductively definable are reducible to H. In [**2**] Barwise, Gandy and Moschovakis showed that if $A = L_\alpha$ is admissible, then H provides notations for all the ordinals less than α^+, the next admissible ordinal. The proof that H has the required properties depends on the fact that the "primitive recursive" functions and their indices satisfy Theorems 3.4.5(ii) and 3.4.6. Hence "primitive recursive" can be replaced throughout by "basic." This is an extension and a refinement of the fact that in the definition of Kleene's $\mathcal{O}$ it is sufficient to use indices for functions which are elementary in the sense of Kalmár.

4. DISCUSSION

In this paper we started out from the notion of Δ_0, extended it to embrace functions, and then exploited it as far as we could. But, however convenient, Δ_0 is a somewhat arbitrary notion. The essential feature of a Δ_0 predicate is that it can be decided without looking outside the transitive closure of its argument—the universe, so to speak, to which one is committed by considering that argument. The additional restriction to a finite number of levels of the transitive closure arises from the particular formalism used. It would seem therefore a sensible move to replace "Δ_0" by "Δ_*": a formula is in Δ_* iff the only quantifiers it contains are of the form $(Qx \in^* y)$, where $\in^*$ is the ancestral of $\in$; $(Qx \in^* y)$ is equivalent to $(Qx \in \mathrm{Tc}(y))$. A model-theoretic justification for this move is available. Using the definitions and methods introduced by Feferman and Kreisel in [**6**] (see also Feferman [**5**]) it is easy to prove

4.1. THEOREM. *If ϕ is $\bigcap$-invariant relative to a theory T then there is $\psi \in \Delta_*$ such that $T \vdash \phi \leftrightarrow \psi$.*

In [**6**] a similar theorem is stated concerning Δ_0; but it involves placing a strong condition on T. Under this condition Δ_0 coincides with Δ_*.

Evidently $\Delta_* = \Delta_0(\mathrm{Tc})$; this suggests considering $\mathscr{B}_* = \mathscr{B}(\mathrm{Tc})$ (alternatively $\mathscr{B}_*$ can be defined by replacing "$\bigcup$" by "Tc" in Definition 1.3.1 or 1.3.2). One finds that many of our results (e.g., 1.3.3–1.3.6) can be carried over *mutatis mutandis.*[10] In particular, 1.5.8 and 1.5.9 hold for $\mathscr{B}_*$, and so $\mathscr{B}_*$ is also inadequate for handling functions on ω. On the other hand, simplifications can be made in § 3, since, presumably, "$x \in \hat{\omega}$" $\in \Delta_*$.

I had hoped that $\mathscr{B}_*$ would suffice for the elementary syntax of the infinitary languages $L_{A,\omega}$, where A is an admissible set; but there are technical difficulties about the substitution predicate which I have not been able to overcome. There is an argument, however, for the feasibility of this programme, or some modification of it; the argument involves a rather more precise characterization of "elementary syntax" and runs as follows. The expressions from which a language is constructed constitute an inductively generated set of objects; syntactic concepts are therefore naturally defined by induction (or recursion) with respect to the generative processes. Now it seems that the distinguishing feature of *elementary* syntactic concepts (e.g., well-formed formula, proof) is that, when an expression falls under the concept, it contains within itself all the essential ingredients for a proof that it satisfies the relevant inductive definition. Thus a well-formed formula contains as parts all the well-formed subformulae from which it is constructed; while, for example, the value of a primitive recursive function for a particular argument need not "contain within itself" the relevant values of all the auxiliary functions. To some extent the proposed characterization is normative. For example, it may be necessary to insist that an elementary proof contain not only a sequence of formulae, but also references to the rules and previous steps by which the formulae are derived; such requirements are a familiar feature of finitist accounts of metamathematics. The proposed characterization reduces the problem to determining what is meant by "contains within itself"; this need not be taken absolutely literally. One might allow certain ways of extracting the necessary information. But it does seem plausible that if the objects concerned are well-founded sets then these extraction processes should belong to $\mathscr{B}_*$ and hence that the elementary syntactic concepts *should* all belong to Δ_*.

---

[10] The details have been worked out by P. W. Grant.

Note added in 2nd proof. The author is not responsible for the frequent idiotic transposition of punctuation and quotation marks.

REFERENCES

1. J. Barwise, *Infinitary logic and admissible sets*, J. Symbolic Logic **34** (1969), 226–252.

2. J. Barwise, R. O. Gandy and Y. N. Moschovakis, *The next admissible set*, J. Symbolic Logic **36** (1971), 108–120.

3. J. H. Bennett, *On spectra*, Doctorial dissertation, Princeton Univ., Princeton, N.J., 1962.

4. S. Feferman, *Predicative provability in set theory*, Bull. Amer. Math. Soc. **72** (1966), 486–489. MR **33** #1230.

5. ———, *Persistent and invariant formulas for outer extensions*, Compositio Math. **20** (1968), 29–52. MR **37** #6170.

6. S. Feferman and G. Kreisel, *Persistent and invariant formulas relative to theories of higher order*, Bull. Amer. Math. Soc. **72** (1966), 480–485. MR **33** #1229.

7. K. Gödel, *The consistency of the continuum hypothesis*, Ann. of Math. Studies, no. 3, Princeton Univ. Press, Princeton, N.J., 1940. MR **2**, 66.

8. A. Grzegorczyk, *Some classes of recursive functions*, Rozprawy Mat. **4** (1953), 46 pp. MR **15**, 667.

9. R. B. Jensen, *Stufen den konstruktiblen Hierarchie*, Habilitationschrift, mimeographed, Bonn, 1967.

10. R. B. Jensen and C. Karp, *Primitive recursive set functions*, Proc. Sympos. Pure Math., vol. 13, Part I, 1971, pp. 143–176.

11. C. Karp, *A proof of the relative consistency of the continuum hypothesis*, Sets, Models and Recursion Theory (Proc. Summer School Math. Logic and Tenth Logic Colloq., Leicester, 1965), North-Holland, Amsterdam, 1967, pp. 1–32. MR **36** #1320.

12. A. Lévy, *A hierarchy of formulas in set theory*, Mem. Amer. Math. Soc. No. 57 (1965), 76 pp. MR **32** #7399.

13. Y. N. Moschovakis, *Abstract first order computability*. II, Trans. Amer. Math. Soc. **138** (1969), 465 504. MR **39** #5362.

14. R. W. Ritchie, *A rudimentary definition of addition*, J. Symbolic Logic **30** (1965), 350–354. MR **33** #2541.

15. R. M. Smullyan, *Theory of formal systems*, Ann. of Math. Studies, no. 47, Princeton Univ. Press, Princeton, N.J., 1961. MR **22** #12042.

16. R. B. Jensen, *The fine structure of the constructible hierarchy*, Ann. Math. Logic **4** (1972), 229–308.

17. K. J. Devlin, *Aspects of constructibility*, Lecture Notes in Mathematics, no 354, Springer-Verlag Berlin, 1973.

18. ———, *An introduction to the fine structure of the constructible hierarchy*, Generalized recursion theory (Proc. 1972 Oslo Symposium), North-Holland, Amsterdam, 1974, pp. 123–163.

THE UNIVERSITY
MANCHESTER, ENGLAND

MATHEMATICAL INSTITUTE
OXFORD, ENGLAND

Proceedings of Symposia in Pure Mathematics
Volume 13, Part II, 1974

SECOND-ORDER CARDINAL CHARACTERIZABILITY

STEPHEN J. GARLAND*

Scott [**13**] has pointed out that a close relationship exists between the axiomatizations of set theory and of the theory of types. In this paper we examine the relationship between the model theory of set theory and the model theory of a particular piece of the theory of types, namely pure second-order logic. It will be seen that set-theoretical techniques play a key role in the study of second-order definability, while various considerations in this study raise problems concerning the foundations of set theory.

The axiom of choice is assumed throughout the following discussion, which can be formalized in Gödel-Bernays set theory. Greek letters $\alpha, \beta, \gamma, \delta$ range over ordinals; κ, λ, μ range over cardinals, which are initial ordinals; κ^+ is the cardinal successor of a cardinal κ, while $\kappa + 1$ is its ordinal successor. Since an ordinal α equals the set of its predecessors, the structure $\langle \alpha, \in \cap\, \alpha \times \alpha \rangle$ (hereafter denoted by "$\langle \alpha, \in \rangle$") is a well-order structure of type α. For any sets A and B, ${}^AB$ is the set of all functions from A into B, and card(A) is the cardinal equinumerous with A.

1. Second-order languages. We shall consider primarily pure second-order relational languages with identity. It is well known that every second-order sentence is logically equivalent to a *full prenex sentence*, i.e., to a prenex sentence in which all relation quantifiers precede any individual quantifier. Among the quantifier interchange rules which allow economical translations of sentences into full prenex sentences, the following consequence of the axiom of choice is particularly useful: for any formula φ, any individual variable $\boldsymbol{v}$, any n-ary

AMS (MOS) subject classifications (1970). Primary 02B15, 02F35; Secondary 02K35, 02K05, 02K30.

* Many of the results of this paper are contained in the author's doctoral dissertation, written in 1967 under the supervision of Professor J. W. Addison at the University of California at Berkeley with the support of a National Science Foundation Graduate Fellowship and NSF grant GP-5632. Preparation of this manuscript was supported by a Faculty Fellowship from Dartmouth College while the author was on leave at UCLA during 1970–1971.

relation variable $\boldsymbol{R}$, and any $(n + 1)$-ary relation variable $\boldsymbol{S}$, $\bigwedge \boldsymbol{v} \bigvee \boldsymbol{R} \varphi$ is logically equivalent to $\bigvee \boldsymbol{S} \bigwedge \boldsymbol{v} \psi$, where ψ is obtained from φ by replacing each subformula of the form $\boldsymbol{R v}_1 \ldots \boldsymbol{v}_n$ by $\boldsymbol{S v v}_1 \ldots \boldsymbol{v}_n$.

Prenex sentences fall into a natural prefix classification: a full prenex sentence is $\bigvee_n^1$ ("exists-one-n") or $\bigwedge_n^1$ ("all-one-n") iff it begins respectively with an existential or a universal relation quantifier and has n homogeneous blocks of relation quantifiers. By associating with each sentence its class of models we induce a classification of the second-order definable classes of structures. A class of structures is $\Diamond_n^1$ ("diamond-one-n") iff it is both $\bigvee_n^1$ and $\bigwedge_n^1$. Thus we have the following hierarchy of definable classes of structures:

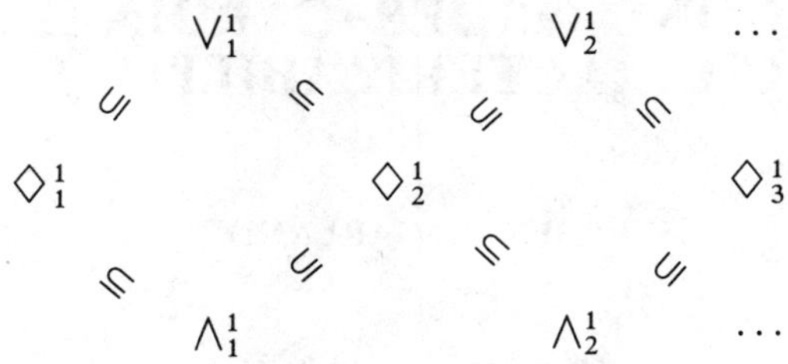

We use the $\bigvee$-$\bigwedge$-$\Diamond$ notation to distinguish this hierarchy in pure logic from the analogous *analytical hierarchy*

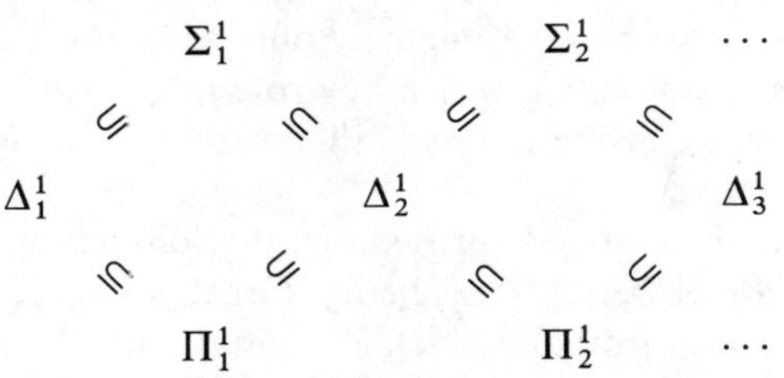

of sets definable over the structure $\langle \omega, \in \rangle$ in an applied second-order language for number theory, since we feel that the Σ-Π-Δ notation is overworked and since we shall have occasion to refer to both hierarchies.

The influence of the axioms for set theory on the properties of the analytical hierarchy have been studied by several authors (for example, see Addison [**1**] or Moschovakis [**11**]). Their results carry over in a sense to the pure second-order hierarchy for languages which contain a binary relation parameter, since the class of isomorphs of $\langle \omega, \in \rangle$ is $\bigwedge_1^1$ definable. However, these results do not apply to weaker second-order languages, and in particular to the parameter-free language. It is this language which will be our primary object of study.

2. Characterizable cardinals. Since any model of a parameter-free sentence φ is determined up to isomorphism by the cardinality of its universe, instead of discussing the class of models of φ it suffices to discuss the *spectrum* of φ, i.e., the class of all cardinals κ such that the structure $\langle \kappa \rangle$ is a model of φ. In the case that the spectrum of φ contains a single cardinal κ, we say that φ *characterizes* κ.

Which cardinals are characterizable by sentences in the various second-order prefix classes? Using the compactness and Löwenheim-Skolem theorems, it is easy to see that the spectrum of a propositional combination of $\bigvee^1_1$ sentences contains an infinite cardinal iff it contains all infinite cardinals; hence the only cardinals characterizable by propositional combinations of $\bigvee^1_1$ sentences are the finite cardinals. On the other hand, we shall see that many infinite cardinals are $\Diamond^1_2$ characterizable. We note first that the following notions are definable using only one relation quantifier:

(1) for any set A, $\text{card}(A) \geqq \aleph_0$ iff there is an unbounded ordering of A [an $\bigvee^1_1$ condition];

(2) for any sets A, B, $\text{card}(A) = \text{card}(B)$ iff there is a one-one function on A onto B [an $\bigvee^1_1$ condition]; and

(3) for any binary relation R, R is a well-ordering iff R is a linear ordering and

$$\forall X[X \neq \varnothing \Rightarrow \exists x \in X \forall y \in X \langle y, x\rangle \notin R]$$

[an $\bigwedge^1_1$ condition].

LEMMA 2.1. *If* $\{\lambda : \lambda < \kappa\}$ *is a* $\Diamond^1_2$ *spectrum, then so are* $\{\lambda : \lambda \leqq \kappa\}$ *and* $\{\kappa\}$.

PROOF. Note that, for any set A, $\text{card}(A) \leqq \kappa$ iff

$$\exists R[\underbrace{R \text{ well orders } A}_{\bigwedge^1_1 \text{ by (3)}} \;\&\; \forall x(\underbrace{\text{card}(R^{-1}\{x\}) = \text{card}(A)}_{\bigvee^1_1 \text{ by (2)}} \text{ or } \underbrace{\text{card}(R^{-1}\{x\}) < \kappa}_{\Diamond^1_2 \text{ by hypothesis}})]$$

$$\text{iff} \quad \forall B[B \subseteq A \quad \Rightarrow \quad \overbrace{\text{card}(B) = \text{card}(A)} \quad \text{or} \quad \overbrace{\text{card}(B) < \kappa}].$$

Hence $\{\lambda : \lambda \leqq \kappa\}$ is a $\Diamond^1_2$ spectrum, as is $\{\kappa\}$ since it equals $\{\lambda : \lambda \leqq \kappa\} \sim \{\lambda : \lambda < \kappa\}$.

By (1), the set of finite cardinals is an $\bigwedge^1_1$ spectrum. Hence one can show by induction using Lemma 2.1 that, for any $n < \omega$, $\{\lambda : \lambda < \aleph_n\}$ is a $\Diamond^1_2$ spectrum and $\aleph_n$ is $\Diamond^1_2$ characterizable. What about $\aleph_\omega$? It too is $\Diamond^1_2$ characterizable. To see this, let $\Psi(A, R, S)$ be the following $\Diamond^1_2$ condition on a set A, a binary relation R, and a set S:

$$\begin{aligned} &R \text{ well-orders } A \;\&\; \forall x(\text{card}(A) \neq \text{card}(R^{-1}\{x\})) \\ &\qquad\qquad \& \;\forall x[x \in S \Leftrightarrow \text{card}(R^{-1}\{x\}) \geqq \aleph_0 \\ &\qquad\qquad\qquad \& \;\forall y(\langle y, x\rangle \in R \Rightarrow \text{card}(R^{-1}\{x\}) \neq \text{card}(R^{-1}\{y\}))] \end{aligned}$$

(i.e., R and S "code" the set of infinite cardinals less than $\text{card}(A)$). Then, for any set A, $\text{card}(A) < \aleph_\omega$ iff

(4a) $\exists R \exists S[\Psi(A, R, S) \;\&\; \text{card}(S) < \aleph_0]$ iff

(4b) $\forall R \forall S[\Psi(A, R, S) \Rightarrow \text{card}(S) < \aleph_0]$.

Since (4a) and (4b) are $\bigvee^1_2$ and $\bigwedge^1_2$ conditions respectively, $\aleph_\omega$ is $\Diamond^1_2$ characterizable by Lemma 2.1. Continuing in this manner, one may show that a large number of cardinals are $\Diamond^1_2$ characterizable.

The extent of the class of $\Diamond^1_2$ characterizable cardinals is even more remarkable in view of the following theorem and its corollary.

THEOREM 2.2. *For any second-order spectrum $\mathscr{S}$ which contains no finite cardinal,*

(a) $\{2^\kappa : \kappa \in \mathscr{S}\}$ *is an* $\bigvee_2^1$ *spectrum, and*

(b) $\{(2^\kappa)^+ : \kappa \in \mathscr{S}\}$ *is a* $\Diamond_2^1$ *spectrum.*

COROLLARY 2.3. *For any second-order characterizable cardinal κ, 2^κ is $\bigvee_2^1$ characterizable and $(2^\kappa)^+$ is $\Diamond_2^1$ characterizable.*

Part (a) of the theorem was shown by Zykov [**17**]. We sketch here a proof of both (a) and (b). Let n be a positive integer. For any $(n + 1)$-ary relation C and any x, let $C_x = \{\langle x_1, \ldots, x_n\rangle : \langle x, x_1, \ldots, x_n\rangle \in C\}$. We say that C codes a set of n-ary relations over a set B iff

$$\forall x[C_x \subseteq B \,\&\, \forall y(y \neq x \,\&\, C_x \neq \varnothing \Rightarrow C_x \neq C_y)].$$

This is clearly a first-order condition, and the condition that C codes all n-ary relations over B is an $\bigwedge_1^1$ condition. Note that if C codes all n-ary relations over a set of cardinality κ, then the domain of C has cardinality $2^{\kappa^n} - 1$, which equals 2^κ if κ is infinite.

For part (a), suppose that φ is a second-order sentence whose spectrum $\mathscr{S}$ contains only infinite cardinals. Let l be the maximum of the ranks of relation variables occurring in φ. Then a set A has cardinality 2^κ for some κ in $\mathscr{S}$ iff there is a subset B of A and, for each $n \leqq l$, an $(n + 1)$-ary relation C_n such that

(5) the domain of C_l has cardinality card(A),

(6) for all $n \leqq l$, C_n codes all n-ary relations over B, and

(7) φ is true when its individual variables are interpreted as ranging over B and its n-ary relation variables are interpreted as ranging over the relations coded by C_n.

Since (5) is an $\bigvee_1^1$ condition, (6) is $\bigwedge_1^1$, and (7) is first-order (being just a relativization of φ), $\{2^\kappa : \kappa \in \mathscr{S}\}$ is an $\bigvee_2^1$ spectrum.

Note that one cannot show in a similar fashion that $\{2^\kappa : \kappa \in \mathscr{S}\}$ is the $\bigwedge_2^1$ spectrum $\mathscr{T}$ which contains the cardinal of a set A iff for any relations $B, C_1, \ldots, C_l$ over A, conditions (5) and (6) imply condition (7). Two things go wrong with such an attempt. In the first place, $\mathscr{T}$ contains all cardinals not of the form 2^κ for some cardinal κ. This defect could be remedied if $\{2^\kappa : \kappa \geqq 0\}$ were an $\bigwedge_2^1$ spectrum, but in general it is not, as will be seen below. Under certain strong assumptions, such as the generalized continuum hypothesis, this class is an $\bigwedge_2^1$ spectrum (e.g., it is the class of successor cardinals) and so is $\{2^\kappa : \kappa \in \mathscr{S}\}$. In the second place, $\mathscr{T}$ will omit a cardinal 2^κ for some κ in $\mathscr{S}$ if $2^\kappa = 2^\lambda$ for some other cardinal λ not in $\mathscr{S}$. This defect can be remedied, however, as we shall do below.

The difficulties encountered in attempting to show that $\mathscr{T} = \{2^\kappa : \kappa \in \mathscr{S}\}$ suggest a proof of part (b). Since, for any sets A and B,

(8) card$(A) < (2^{\mathrm{card}(B)})^+$ iff card$(A) \leqq 2^{\mathrm{card}(B)}$ iff there is a binary relation C with domain a subset of A which codes a set of unary relations over B [an $\bigvee_1^1$ condition],

the condition that card$(A) = (2^{\mathrm{card}(B)})^+$ is a $\Diamond_2^1$ condition by the methods in the proof of Lemma 2.1. Now, for any φ, $\mathscr{S}$, and l as above, a set A has cardinality

$(2^\kappa)^+$ for some κ in $\mathscr{S}$ iff there are relations $B, C_1, \ldots, C_l$ over A such that card(A) $= (2^{\mathrm{card}(B)})^+$ and conditions (6) and (7) hold [an $\bigvee_2^1$ condition] iff for any well-ordering R of A there is an x in A such that card$(A) = (2^{\mathrm{card}(R^{-1}\{x\})})^+$ and for any relations $B, C_1, \ldots, C_l$ over A, if $B = R^{-1}\{x\}$ and condition (6) holds, then (7) holds also [an $\bigwedge_2^1$ condition]. Hence $\{(2^\kappa)^+ : \kappa \in \mathscr{S}\}$ is a $\diamondsuit_2^1$ spectrum.

The hypothesis that $\mathscr{S}$ contain only infinite cardinals can be eliminated at the expense of replacing "2^κ" by "2^{κ^l}" in the conclusion, where l is any sufficiently large integer. We have not done so here in order to keep the notation simple and since this generalization is not needed for the corollary.

As noted above, under the assumption of the generalized continuum hypothesis, Corollary 2.3 can be strengthened to conclude that 2^κ is $\diamondsuit_2^1$ characterizable. This strengthening is not possible in general, since Kunen [**8**] has shown that it is consistent relative to Zermelo-Fraenkel set theory that $2^{\aleph_0}$ is not $\bigwedge_2^1$ characterizable. Hence, by our remarks above, it is also consistent relative to Zermelo-Fraenkel set theory that $\{2^\kappa : \kappa \geqq 0\}$ is not an $\bigwedge_2^1$ spectrum. Kunen's result is not surprising if one considers the fact that the "natural" characterizations of $2^{\aleph_0}$ as, for example, the cardinality of the power set of ω or the cardinality of a least-upper-bound complete linear ordering with a countable dense subset are all $\bigvee_2^1$ characterizations.

Several questions arise as a result of these considerations. We have noticed that it is consistent that there is a cardinal (e.g., $2^{\aleph_0}$) which is $\bigvee_2^1$ but not $\bigwedge_2^1$ characterizable. In § 3 we show that such a cardinal always exists, and that in fact there is a genuine prefix hierarchy of second-order characterizable cardinals. On the other hand, the general problem of which cardinals are second-order characterizable is in a sense reducible to the problem of which cardinals are $\diamondsuit_2^1$ characterizable by Corollary 2.3. Furthermore, if one makes the assumption that $2^{\aleph_0}$ possesses no "unnatural" characterizations, i.e., that $2^{\aleph_0}$ is not $\bigwedge_2^1$ characterizable, then our preliminary results about $\diamondsuit_2^1$ characterizability indicate that $2^{\aleph_0}$ is greater than $\aleph_\omega$. Hence we are led to a study of the $\diamondsuit_2^1$ characterizable cardinals in an attempt to gain further insight into both the class of second-order characterizable cardinals and the size of the continuum.

What is the extent of the class of $\diamondsuit_2^1$ characterizable cardinals? In the first place, there are only countably many such cardinals, so that the class has an upper bound. Also, if we replace "card$(S) < \aleph_0$" by "card$(S) < \aleph_1$" in (4a) and (4b), then we see that $\aleph_\Omega$ is $\diamondsuit_2^1$ characterizable, so that the class is not an initial segment of the sequence of cardinals. This situation leads us to ask the following two questions:

(I) What is the smallest cardinal which is not $\diamondsuit_2^1$ characterizable?

(II) What is the supremum of the class of $\diamondsuit_2^1$ characterizable cardinals?

In order to investigate question (I), we digress in § 4 to discuss $\diamondsuit_2^1$ *characterizable ordinals*, i.e., ordinals α such that the class of isomorphs of $\langle \alpha, \in \rangle$ is a $\diamondsuit_2^1$ class. The importance of these ordinals can be seen as follows: if in (4a) and (4b) we replace "card$(S) < \aleph_0$" by

(9) $R \cap S \times S$ has order type α

then we can show that $\aleph_\alpha$ is $\diamondsuit_2^1$ characterizable for any $\diamondsuit_2^1$ characterizable ordinal α.

In § 5 we apply the results of § 4 to study the characterizability of cardinals in various $\diamondsuit_2^1$ spectra. Finally, in § 6, we examine the relationship between the extent of the class of $\diamondsuit_2^1$ characterizable cardinals and the model theory of set theory.

3. A hierarchy of characterizable cardinals. Standard universal set arguments show that the analytical hierarchy is indeed a true hierarchy, i.e., that for all $n > 0$ there are Σ_n^1 sets which are not Π_n^1 and Δ_n^1 sets which are neither Σ_n^1 nor Π_n^1 (cf. [**14**, § 7.8]). These results also establish a hierarchy theorem for pure languages which contain, say, one individual and one binary relation parameter. In this section we strengthen this result to establish a hierarchy of second-order characterizable cardinals, employing in our proof a method which is similar to the use of universal set arguments in number theory.

We show first, by way of illustration, that for any $n > 0$ there is a $\diamondsuit_{n+1}^1$ characterizable cardinal which is not $\bigvee_n^1$ characterizable. Call a cardinal κ *weakly* $\bigvee_n^1$ *characterizable* iff κ is the smallest cardinal in some $\bigvee_n^1$ spectrum, and let κ_n be the smallest infinite cardinal which is not weakly $\bigvee_n^1$ characterizable. Then κ_n is obviously not $\bigvee_n^1$ characterizable, so it remains to show that κ_n is $\diamondsuit_{n+1}^1$ characterizable.

For any cardinal κ and any sentence φ, define the κ*-spectrum* of φ to be the set of all cardinals less than κ in the spectrum of φ.

LEMMA 3.1. *For any $n > 1$ there is an $\bigvee_n^1$ sentence ψ_n involving a binary relation parameter $<$ and individual parameters $\boldsymbol{\lambda}$, $\boldsymbol{m}$ such that, for any infinite cardinal κ and any set S of cardinals, S is the κ^+-spectrum of an $\bigvee_n^1$ sentence iff there is an m in ω such that*

$$S = \{\lambda : \langle \kappa + 1, \in, \lambda, m \rangle \models \psi_n\}.$$

PROOF. The proof proceeds by an evaluation of the complexity of the truth definition for $\bigvee_n^1$ sentences. Assume that these sentences are in a language which has the connectives $\sim$ and $\wedge$, the quantifier $\bigvee$, the identity symbol $\approx$, the individual variables $\boldsymbol{v}_i$ for i in ω, and the j-ary relation variables $\boldsymbol{R}_{i,j}$ for i, j in ω. Let $\langle \varphi_m^0 : m \in \omega \rangle$ be a recursive enumeration of all first-order formulas in this language and, for each $n > 0$, let $\langle \varphi_m^n : m \in \omega \rangle$ be a recursive enumeration of all prenex $\bigvee_n^1$ formulas which contain no free individual variables.

We define first an $\bigwedge_1^1$ class $\mathscr{C}_0$ of structures such that for any infinite cardinal κ and any cardinal $\lambda \leqq \kappa$, $\langle \kappa + 1, \in, \lambda, w, +, \cdot, \gamma, B, R, T \rangle \in \mathscr{C}_0$ if and only if

(1) $w = \omega$, i.e., w is the $\in$-least element of $\kappa + 1$ with an $\in$-predecessor but no immediate $\in$-predecessor [a first-order condition],

(2) $+$ and $\cdot$ are the relations of addition and multiplication on ω [a first-order condition since $+, \cdot$ are implicitly definable over $\langle \omega, \in \rangle$],

(3) γ is the maximum of λ and w [a first-order condition],

(4) B is a subset of $\gamma \times \omega \times \lambda$ which codes all finite sequences of elements of λ,

i.e., if B_α is $\{\langle m, \beta\rangle : \langle \alpha, m, \beta\rangle \in B\}$ for any α in γ, then $\bigcup\{{}^n\lambda : n \in \omega\} = \{B_\alpha : \alpha \in \gamma\}$ [an $\bigwedge_1^1$ condition],

(5) R is a subset of $\omega \times \omega \times \gamma \times \omega \times \lambda$ which codes fixed relations $R_{i,j} \subseteq {}^j\lambda$ for any i, j in ω, where $R_{i,j} = \{\langle \alpha_0, \ldots, \alpha_{j-1}\rangle : \exists\beta < \gamma \forall k < j \langle i, j, \beta, k, \alpha_k\rangle \in R\}$ [a first-order condition], and

(6) T is a subset of $\omega \times \gamma$ such that, for any $m < \omega$ and any $\alpha < \gamma$, $\langle m, \alpha\rangle \in T$ iff φ_m^0 is satisfied in $\langle\lambda\rangle$ when $\boldsymbol{v}_i$ is interpreted as $B_\alpha(i)$ and $\boldsymbol{R}_{i,j}$ is interpreted as $R_{i,j}$, i.e., iff the domain of B_α includes every integer i such that $\boldsymbol{v}_i$ has a free occurrence in φ_m^0 and

(a) φ_m^0 is $\boldsymbol{v}_i \approx \boldsymbol{v}_j$ and $B_\alpha(i) = B_\alpha(j)$, or

(b) φ_m^0 is $\boldsymbol{R}_{i,j}\boldsymbol{s}$ for some sequence $\boldsymbol{s}$ of j variables $\boldsymbol{v}_{s(0)}, \ldots, \boldsymbol{v}_{s(j-1)}$ and $\langle B_\alpha(s(0)), \ldots, B_\alpha(s(j-1))\rangle \in R_{i,j}$, or

(c) φ_m^0 is $\sim\varphi_{m'}^0$ and $\langle m', \alpha\rangle \notin T$, or

(d) φ_m^0 is $\varphi_{m'}^0 \wedge \varphi_{m''}^0$ and $\langle m', \alpha\rangle, \langle m'', \alpha\rangle \in T$, or

(e) φ_m^0 is $\bigvee \boldsymbol{v}_i \varphi_{m'}^0$ and there is a β in γ such that $B_\alpha(j) = B_\beta(j)$ for all $j \neq i$, i is in the domain of B_β, and $\langle m', \beta\rangle \in T$ [a first-order condition since the recursive enumeration of the first-order formulas is definable over $\langle \omega, +, \cdot\rangle$].

Now for each $n > 0$ we define a $\Diamond_{n+1}^1$ class $\mathscr{C}_n$ such that for any λ, κ as above, $\langle \kappa + 1, \in, \lambda, w, +, \cdot, \gamma, B, R, T\rangle$ is in $\mathscr{C}_n$ if and only if conditions (1) through (5) hold and

(7) T is a subset of $\omega \times \gamma$ such that, for any $m < \omega$ and any $\alpha < \gamma$, $\langle m, \alpha\rangle \in T$ iff φ_m^n is satisfied in $\langle\lambda\rangle$ when $\boldsymbol{R}_{i,j}$ is interpreted as $R_{i,j}$, i.e., iff φ_m^n is $\bigvee \boldsymbol{R}_{i_1,j_1} \cdots \bigvee \boldsymbol{R}_{i_k,j_k} \sim \varphi_{m'}^{n-1}$ and there are R', T' such that $\langle \kappa + 1, \in, \lambda, w, +, \cdot, \gamma, B, R', T'\rangle \in \mathscr{C}_{n-1}$, $\langle m', \alpha\rangle \notin T'$, and $R_{i,j} = R'_{i,j}$ whenever $\langle i, j\rangle \notin \{\langle i_1, j_1\rangle, \ldots, \langle i_k, j_k\rangle\}$.

Finally, for any $n > 1$, let ψ_n be an $\bigvee_n^1$ sentence such that, for any κ, λ as above and any m in ω, $\langle \kappa + 1, \in, \lambda, m\rangle$ is a model of ψ_n iff there are α, w, γ in $\kappa + 1$ and relations $+, \cdot, B, R, T$ such that

$\langle \kappa + 1, \in, \lambda, w, +, \cdot, \gamma, B, R, T\rangle \in \mathscr{C}_{n-1}$,

φ_m^n is a sentence $\bigvee \boldsymbol{R}_{i_1,j_1} \ldots \bigvee \boldsymbol{R}_{i_k,j_k} \sim \varphi_{m'}^{n-1}$, and

$\langle m', \alpha\rangle \notin T$.

Using Lemma 3.1 we can construct a $\Diamond_{n+1}^1$ class $\mathscr{C}$ such that, for any cardinal κ, $\langle \kappa + 1, \in\rangle \in \mathscr{C}$ iff $\kappa = \kappa_n$, i.e., iff $\kappa \geqq \aleph_0$,

$$\forall m < \omega[\underbrace{\langle \kappa + 1, \in, \kappa, m\rangle \models \sim\psi_n}_{\bigwedge_n^1} \text{ or } \underbrace{\exists\lambda < \kappa\langle \kappa + 1, \in, \lambda, m\rangle \models \psi_n}_{\bigvee_n^1}] \tag{8}$$

and

$$\forall\lambda < \kappa\exists m < \omega[\underbrace{\langle \kappa + 1, \in, \lambda, m\rangle \models \psi_n}_{\bigvee_n^1} \,\&\, \underbrace{\neg\exists\mu < \lambda\langle \kappa + 1, \in, \mu, m\rangle \models \psi_n}_{\bigwedge_n^1}]. \tag{9}$$

Furthermore, the class $\mathscr{D}$ of all structures $\langle A, R\rangle$ isomorphic to $\langle \text{card}(A) + 1, \in\rangle$ is a $\Diamond_2^1$ class since $\langle A, R\rangle \in \mathscr{D}$ iff

(10) A is infinite, R well-orders A with a last element y, and, for any $x \neq y$ in A, $\text{card}(A) \neq \text{card}(R^{-1}\{x\})$.

Thus κ_n is $\Diamond^1_{n+1}$ characterizable as the cardinality of any set A such that

(11) for some R, $\langle A, R\rangle \in \mathscr{D} \cap \mathscr{C}$ [an $\bigvee^1_n$ condition] or

(12) A is infinite and, for any R, if $\langle A, R\rangle \in \mathscr{D}$ then $\langle A, R\rangle \in \mathscr{C}$ [an $\bigwedge^1_n$ condition].

Hence we have shown that there is a $\Diamond^1_{n+1}$ characterizable cardinal which is not $\bigvee^1_n$ characterizable.

The above construction can be modified to show that there is an $\bigwedge^1_n$ characterizable cardinal which is not $\bigvee^1_n$ characterizable. In order to do this, we first strengthen Lemma 3.1 by using the techniques in the proof of Theorem 2.2.

LEMMA 3.2. *For any $n > 1$ there is an $\bigvee^1_2$ sentence ψ'_n and an $\bigwedge^1_2$ sentence ψ''_n such that, for any infinite cardinal κ and any λ, if $2^\lambda \leqq \kappa$ then, for any m in ω, $\langle \kappa + 1, \in, \lambda, m\rangle$ is a model of ψ_n iff it is a model of ψ'_n iff it is a model of ψ''_n.*

PROOF. In a manner similar to that used in the proof of Lemma 3.1, we define an $\bigwedge^1_1$ class $\mathscr{C}_0$ of structures such that, for any infinite cardinal κ and any λ such that $2^\lambda \leqq \kappa$, $\langle \kappa + 1, \in, \lambda, w, +, \cdot, \gamma, B, R, T\rangle$ is in $\mathscr{C}_0$ iff conditions (1) through (4) hold and

(13) R is a subset of $\kappa \times \omega \times \omega \times \gamma \times \omega \times \lambda$ which codes all ω-sequences of finitary relations over λ (such an R exists since $2^\lambda \leqq \kappa$); i.e., if

$$R_{\beta,i,j} = \{\langle \alpha_0, \ldots, \alpha_{j-1}\rangle : \exists \delta < \gamma \forall k < j \langle \beta, i, j, \delta, k, \alpha_k\rangle \in R\},$$

then for any R' as in (5) there is a β in κ such that, for all i, j in ω, $R'_{i,j} = R_{\beta,i,j}$ [an $\bigwedge^1_1$ condition]; and

(14) T is a subset of $\omega \times \kappa \times \omega \times \gamma$ such that, for any m, n in ω, any $\beta < \kappa$, and any $\alpha < \gamma$, $\langle n, \beta, m, \alpha\rangle \in T$ iff

(a) $n = 0$ and φ^0_m is satisfied in $\langle \lambda\rangle$ when $\boldsymbol{v}_i$ is interpreted as $B_\alpha(i)$ and $\boldsymbol{R}_{i,j}$ is interpreted as $R_{\beta,i,j}$ [a first-order condition as in (6)], or

(b) $n > 0$ and φ^n_m is satisfied in $\langle \lambda\rangle$ when $\boldsymbol{R}_{i,j}$ is interpreted as $R_{\beta,i,j}$, i.e., φ^n_m is $\bigvee \boldsymbol{R}_{i_1,j_1} \ldots \bigvee \boldsymbol{R}_{i_k,j_k} \sim \varphi^{n-1}_{m'}$ and there is a β' in κ such that $\langle n - 1, \beta', m', \alpha\rangle \notin T$ and $R_{\beta',i,j} = R_{\beta,i,j}$ whenever $\langle i, j\rangle \notin \{\langle i_1, j_1\rangle, \ldots, \langle i_k, j_k\rangle\}$ [a first-order condition].

Now for any $n > 1$, let ψ'_n be an $\bigvee^1_2$ sentence and ψ''_n be an $\bigwedge^1_2$ sentence such that, for any κ, λ as above and any m in ω, $\langle \kappa + 1, \in, \lambda, m\rangle \models \psi'_n$ iff there are $\alpha, \beta, w, \gamma, +, \cdot, B, R, T$ such that $\langle \kappa + 1, \in, \lambda, w, +, \cdot, \gamma, B, R, T\rangle$ is in $\mathscr{C}_0$ and $\langle n, \beta, m, \alpha\rangle \in T$, and $\langle \kappa + 1, \in, \lambda, m\rangle \models \psi''_n$ iff, for any $\alpha, \beta, w, \gamma, +, \cdot, B, R, T$, if $\langle \kappa + 1, \in, \lambda, w, +, \cdot, \gamma, B, R, T\rangle \in \mathscr{C}_0$, then $\langle n, \beta, m, \alpha\rangle \in T$

In order to apply Lemma 3.2, let

$$\mathscr{S} = \{\kappa : \forall \lambda < \kappa\, (2^\lambda < \kappa)\}$$

and define κ'_n to be the smallest member of $\mathscr{S}$ which is not weakly $\bigvee^1_n$ characterizable. By § 2(8), $\mathscr{S}$ is a $\Diamond^1_2$ spectrum since for any set A, card$(A) \in \mathscr{S}$ iff

(15) for any $B \subseteq A$, if card$(B) \neq$ card(A), then $2^{\text{card}(B)} <$ card(A) iff

(16) there is a well-ordering R of A such that for any x in A, if card$(R^{-1}\{x\})$ $\neq$ card(A), then $2^{\text{card}(R^{-1}\{x\})} <$ card(A).

Using Lemma 3.2 and the fact that $\mathscr{S}$ is a $\Diamond^1_2$ spectrum, we can replace the $\bigvee^1_n$ clauses in (8) and (9) by $\Diamond^1_2$ clauses to obtain an $\bigwedge^1_n$ class $\mathscr{C}'$ such that, for any cardinal κ, $\langle \kappa + 1, \in \rangle \in \mathscr{C}'$ iff $\kappa = \kappa'_n$. Hence, by (12), κ'_n is $\bigwedge^1_n$ characterizable.

By similar modifications of (8) and (9) using "universal" $\bigwedge^1_n$ sentences in addition to the "universal" $\bigvee^1_n$ sentences ψ_n, we obtain the following hierarchy theorem.

THEOREM 3.3. *For any $n > 1$ there is a $\diamondsuit^1_n$ characterizable cardinal which is neither $\bigvee^1_{n-1}$ nor $\bigwedge^1_{n-1}$ characterizable, an $\bigvee^1_n$ characterizable cardinal which is not $\bigwedge^1_n$ characterizable, and an $\bigwedge^1_n$ characterizable cardinal which is not $\bigvee^1_n$ characterizable.*

In conclusion, we mention an open problem concerning the notion of a κ-spectrum. It follows immediately from Theorem 3.3 that for any $n > 1$ there is a cardinal κ such that there is an $\bigvee^1_n$ κ-spectrum which is not an $\bigwedge^1_n$ κ-spectrum. What is the least such cardinal κ? In particular, is the least such κ always ω? Asser [**2**] raised this "spectrum problem" for $n = 1$ and $\kappa = \omega$; Bennett [**3**] considered it for arbitrary n, but could not show that there was any second-order ω-spectrum which was not already an $\bigvee^1_1$ ω-spectrum. Both authors were investigating ω-spectra in response to the problem raised by Scholz [**12**] of characterizing the class of $\bigvee^1_1$ ω-spectra.

4. Characterizable ordinals. As in our study of characterizable cardinals, we shall be interested in determining both the smallest ordinal not characterizable in a given manner and also the supremum of the set of ordinals characterizable in that manner. The first problem leads to a study of the characterizability of countable ordinals, while the second turns out to be equivalent to its version for cardinal characterizability (cf. Theorem 4.1).

For any linear ordering R, let $|R|$ be the order type of R, and for any x in the field of R, let $R \upharpoonright x$ be the initial segment of R determined by x, i.e., $\{\langle y, z\rangle: \langle y, z\rangle \in R \;\&\; \langle z, x\rangle \in R\}$. Note that the following conditions are $\bigvee^1_1$ definable:

(1) for any linear orderings R, S, $|R| = |S|$ iff there is an order-preserving map of the field of R onto the field of S, and

(2) for any linear ordering R and any well-ordering S, $|R| < |S|$ iff $|R| = |S \upharpoonright x|$ for some x in the field of S.

THEOREM 4.1. *For any $n > 1$ and any cardinal κ, κ is $\bigvee^1_n$ (or $\bigwedge^1_n$) characterizable iff the order structure $\langle \kappa, \in\rangle$ is $\bigvee^1_n$ (or $\bigwedge^1_n$) characterizable.*

PROOF. If κ is characterizable, then $\langle \kappa, \in\rangle$ is characterizable as that well-order structure $\langle A, R\rangle$ such that $\mathrm{card}(A) = \kappa$ and for any x, $\mathrm{card}(A) \neq \mathrm{card}(R^{-1}\{x\})$. Conversely, if $\langle \kappa, \in\rangle$ is $\bigvee^1_n$ characterizable, then κ is obviously $\bigvee^1_n$ characterizable; if $\langle \kappa, \in\rangle$ is $\bigwedge^1_n$ characterizable, then κ is $\bigwedge^1_n$ characterizable as the cardinality of a set A such that, for any well-ordering R of A, if $\mathrm{card}(A) \neq \mathrm{card}(R^{-1}\{x\})$ for any x in A, then $|R| = \kappa$.

By the compactness theorem, the only $\bigvee^1_1$ characterizable ordinals are the finite ordinals. Hence at least an $\bigwedge^1_1$ sentence is needed to characterize an infinite ordinal, and such a sentence characterizes $\langle \omega, \in\rangle$ (cf. § 2(3) and § 3(1)). Though we shall be interested primarily in the $\diamondsuit^1_2$ characterizability of countable ordinals, we shall return to the $\bigwedge^1_1$ characterizable ordinals later.

The characterizability of countable ordinals is closely related to the definability of sets of well-orderings of ω in second-order number theory, as the following theorem shows. For any countably infinite ordinal α, let WO_α be the set of all well-orderings of ω with order type α.

THEOREM 4.2. *For any $n > 1$ and any countably infinite ordinal α, α is $\bigvee_n^1$ (or $\bigwedge_n^1$) characterizable iff WO_α is a Σ_n^1 (or a Π_n^1) set.*

PROOF. If α is $\bigvee_n^1$ or $\bigwedge_n^1$ characterizable by a sentence φ, then φ also defines WO_α as a Σ_n^1 or a Π_n^1 set. Conversely, if WO_α is Σ_n^1, then $\langle\alpha, \in\rangle$ is $\bigvee_n^1$ characterizable as that well-order structure $\langle A, R\rangle$ such that there exist an x in A and relations $+, \cdot, S$ on A such that

(3) $|R \upharpoonright x| = \omega$ [a first-order condition by § 3(1)],

(4) $+$ and $\cdot$ are the relations of addition and multiplication on $R^{-1}\{x\}$ [a first-order condition by § 3(2)],

(5) S is an ordering of $R^{-1}\{x\}$ with $|S| = |R|$ [an $\bigvee_1^1$ condition], and

(6) S is "in" WO_α [an $\bigvee_n^1$ condition since WO_α is Σ_n^1].

Also, if WO_α is Π_n^1, then $\langle\alpha, \in\rangle$ is $\bigwedge_n^1$ characterizable as that well-order structure $\langle A, R\rangle$ such that card$(A) = \aleph_0$ [an $\bigwedge_2^1$ condition] and, for any $x, +, \cdot, S$ satisfying conditions (3) through (5), condition (6) holds.

Later we shall extend Theorem 4.2 to $\bigwedge_1^1$ characterizability, but first we derive some of its consequences. For any $n > 0$ let δ_n be the least ordinal which is not the order type of any Δ_n^1 well-ordering of ω, and let $\delta_{(\omega)} = \sup\{\delta_n : n \in \omega\}$.

THEOREM 4.3. *For any $n > 0$ and any infinite ordinal α, if $\alpha < \delta_n$ then WO_α is a Δ_n^1 set.*

PROOF. Let S be a Δ_n^1 well-ordering of ω with order type α. Then

$$\begin{aligned} WO_\alpha &= \{R : \exists T(T = S \,\&\, |T| = |R|)\} \\ &= \{R : \forall T(T = S \Rightarrow |T| = |R|)\}, \end{aligned}$$

so that WO_α is a Δ_n^1 set.

The converse of Theorem 4.3 is true for $n = 1$ since WO_α is not a Σ_1^1 set for any $\alpha \geqq \delta_1$ (otherwise, by standard arguments, every Π_1^1 set would also be Σ_1^1; cf. [**14**, p. 184]); it is also true for $n \geqq 2$ whenever the set of Δ_n^1 relations is a basis for the set of all Σ_n^1 sets of relations.

COROLLARY 4.4. *For any countable ordinal α, α is $\Diamond_2^1$ characterizable iff α is $\bigvee_2^1$ characterizable iff $\alpha < \delta_2$.*

PROOF. If WO_α is a Σ_2^1 set, then it contains a Δ_2^1 well-ordering by the basis theorem for Σ_2^1 (cf. [**14**, p. 190]). Hence the corollary follows from Theorems 4.2 and 4.3.

Having determined which countable ordinals are $\bigvee_1^1$ and $\bigvee_2^1$ characterizable, we now turn to $\bigwedge_1^1$ and $\bigwedge_2^1$ characterizability. We shall show that the $\bigwedge_1^1$ characterizable ordinals are cofinal with the $\bigvee_2^1$ characterizable ordinals and that δ_2 is $\bigwedge_2^1$ characterizable. In order to do this, we shall use several facts relating Δ_2^1

definability to implicit Π_1^1 definability. For any $n > 0$, call a subset A of ω a Π_n^1 (or a Σ_n^1) *singleton* iff $\{A\}$ is a Π_n^1 (or a Σ_n^1) set. It is easy to see that A is a Σ_n^1 singleton iff A is a Δ_n^1 set (cf. [**14**, p. 188] where the proof of a weaker result actually establishes this fact). Suzuki [**16**] showed that a subset of ω is Δ_2^1 iff it is hyperarithmetical (i.e., Δ_1^1-definable) in a Π_1^1 singleton.

Indices can be assigned to Π_1^1 singletons by applying the Novikov-Kondô-Addison uniformization theorem [**14**, p. 188] to a universal Π_1^1 set P_1^1 which enumerates all Π_1^1 sets of subsets of ω [**14**, p. 175] to obtain a Π_1^1 set P such that a subset A of ω is a Π_1^1 singleton iff, for some e in ω, $\langle e, A\rangle \in P$. We note in passing that the set of Π_1^1 singletons is thereby a Π_1^1 set and that, by the basis theorem for Σ_2^1, there is a Δ_2^1 set which is not a Π_1^1 singleton; hence we obtain an alternate proof of Suzuki's [**16**], Corollary 2. More to the point, the set $\{e : \exists A\langle e, A\rangle \in P\}$ of indices of Π_1^1 singletons is a Σ_2^1 set. To each index e in this set I we assign an ordinal $|e|$ as follows. Since P is Π_1^1, there is a recursive set Q such that for any e, A the set

$$Q_{e,A} = \{\langle m, n\rangle : \langle e, A, m, n\rangle \in Q\}$$

is a linear ordering and is furthermore a well-ordering iff $\langle e, A\rangle \in P$ (cf. Kleene [**6**, XXII]). For any e in I, let $|e|$ equal $|Q_{e,A}|$ for the unique A such that $\langle e, A\rangle \in P$.

LEMMA 4.5. *$\{|e| : e \in I\}$ is cofinal with δ_2.*

PROOF. For any e in I, the unique well-ordering R such that $R = Q_{e,A}$ for some A has order type $|e|$ and, being a Σ_2^1 singleton, is a Δ_2^1 well-ordering; hence $|e| < \delta_2$. Conversely, suppose that $\{|e| : e \in I\}$ is bounded by $|R|$ for some Δ_2^1 well-ordering R. Then, for any Π_1^1 singleton A and any index e of A,

$$\{A\} = \{B : |Q_{e,B}| < |R|\},$$

so that A is a Σ_1^1 singleton relative to R and hence hyperarithmetical in R. Now the hyperjump of R is also a Δ_2^1 set, and so by Suzuki's theorem it must be hyperarithmetical in some Π_1^1 singleton and thereby hyperarithmetical in R, which is impossible. Hence no such R exists.

We now extend Theorem 4.2 to $\bigwedge_1^1$ characterizability.

THEOREM 4.6. *For any countably infinite ordinal α, α is $\bigwedge_1^1$ characterizable iff WO_α is a Π_1^1 set.*

PROOF. Necessity is proved as in Theorem 4.2. Sufficiency is also proved in the same manner once we replace the $\bigwedge_2^1$ condition on $\langle A, R\rangle$ that $\text{card}(A) = \aleph_0$ by membership of $\langle A, R\rangle$ in an $\bigwedge_1^1$ class of countable well-order structures containing $\langle \alpha, \in\rangle$. Since WO_α is Π_1^1, α is less than δ_2 by the basis theorem for Σ_2^1. By the lemma, there is an e in I such that $\alpha < |e|$. Then an infinite well-order structure $\langle A, R\rangle$ has order type less than $|e|$ iff there do not exist any $x, +, \cdot, S, B$ satisfying (3) and (4) such that $|Q_{e,B}| \leqq |S| \leqq |R|$, so that the class of all such structures is an $\bigwedge_1^1$ class as desired.

In order to apply Theorem 4.6 to describe the class of $\bigwedge_1^1$ characterizable ordinals, we first prove several lemmas.

LEMMA 4.7. *For any* e *in* I, $WO_{|e|}$ *is a* Π_1^1 *set.*

PROOF. For any e in I and any R, R is in $WO_{|e|}$ iff R is a well-ordering and there exists an A such that $|R| = |Q_{e,A}|$. But there exists an A such that $|R| = |Q_{e,A}|$ iff there exists a unique such A iff there exists such an A with $\{A\}$ Σ_1^1 in R iff there exists an A hyperarithmetical in R [a Π_1^1 condition by the methods of Kleene [7]] such that $Q_{e,A}$ is a well-ordering and neither $|R| < |Q_{e,A}|$ nor $|Q_{e,A}| < |R|$ [a Π_1^1 condition by (2)]. Hence $WO_{|e|}$ is a Π_1^1 set.

For any set W of countably infinite ordinals and any $n > 0$, call W a Π_n^1 (or a Σ_n^1) set of ordinals iff $\{R : |R| \in W\}$ is a Π_n^1 (or a Σ_n^1) set.

LEMMA 4.8. $\{|e| : e \in I\}$ *is a* Σ_2^1 *set of ordinals.*

PROOF. Let $W = \{|e : e \in I\}$. Then $\{R : |R| \in W\}$ equals $\{R : \exists e \exists A\,(R$ is a well-ordering and $|R| = |Q_{e,A}|)\}$.

LEMMA 4.9. $\{\alpha : \omega \leqq \alpha < \delta_2$ & WO_α *is not* $\Pi_1^1\}$ *is a* Σ_2^1 *set of ordinals cofinal with* δ_2.

PROOF. Let R_0 be any Δ_2^1 well-ordering of ω. The following three conditions are all Σ_2^1 conditions on any well-ordering R:

(7) $WO_{|R|}$ is not Π_1^1 iff $\neg \exists e \forall S(|R| = |S| \Leftrightarrow \langle e, S\rangle \in P_1^1)$;

(8) $|R_0| < |R|$ iff $\exists S(S = R_0$ & $|S| < |R|)$;

(9) $|R| < \delta_2$ iff $\exists e \exists A(\langle e, A\rangle \in P$ & $|R| < |Q_{e,A}|)$.

By (7), (8), and the basis theorem for Σ_2^1, for any $\alpha < \delta_2$ there is a β such that $\alpha < \beta < \delta_2$ and WO_β is not Π_1^1, so the set of ordinals in question is cofinal with δ_2. It is a Σ_2^1 set by (7) through (9) and § 2(3).

Noting that condition (7) is still Σ_2^1 if P is replaced by any Δ_2^1 enumerating set, we see that the lemma can be generalized to show, for example, that the set of infinite $\alpha < \delta_2$ such that WO_α is not a Boolean combination of Π_1^1 sets is a Σ_2^1 set of ordinals cofinal with δ_2. Theorem 4.6 together with Lemmas 4.5, 4.7, and 4.9 shows that the set of Δ_2^1 ordinals is split into two cofinal subsets, one containing precisely the ordinals α such that α is Λ_1^1 characterizable. The following theorem establishes the order types of these cofinal subsets. Call a countably infinite ordinal α Σ_2^1 *regular* iff any cofinal Σ_2^1 subset of α has order type α.

THEOREM 4.10. δ_2 *is* Σ_2^1 *regular.*

PROOF. Let W be a cofinal Σ_2^1 subset of δ_2 and let $W' = \{R : |R| \in W\}$. Suppose that the order type of W is $|R_0|$ for some Δ_2^1 well-ordering R_0. For any ternary relation S and any n, let $S_n = \{\langle i, j\rangle : \langle n, i, j\rangle \in S\}$ and let Z be

$$\{S : \forall n S_n \in W' \;\&\; \forall m \forall n(\langle m, n\rangle \in R_0 \Rightarrow |S_m| < |S_n|)\}.$$

Then Z is a Σ_2^1 set and hence contains a Δ_2^1 member S. By construction, $\{|S_n| : n \in \omega\}$ is cofinal in W and hence also in δ_2. But then the ordering

$$\langle a, b\rangle < \langle c, d\rangle \Leftrightarrow a < c \quad \text{or} \quad (a = c \;\&\; \langle b, d\rangle \in S_a)$$

of pairs of integers is a Δ_2^1 well-ordering with order type at least δ_2, which is impossible.

By the remarks preceding Theorem 4.10, and by Lemmas 4.8, 4.9, and Theorem 4.10, we obtain the desired description of the class of $\bigwedge_1^1$ characterizable ordinals.

THEOREM 4.11. *The countable $\bigwedge_1^1$ characterizable ordinals form a cofinal subset of δ_2 with order type δ_2, and the complement of this set relative to δ_2 is also a cofinal subset of δ_2 with order type δ_2.*

Finally, we show that there is a countable $\bigwedge_2^1$ characterizable ordinal which is not $\bigvee_2^1$ characterizable.

THEOREM 4.12. *δ_2 is $\bigwedge_2^1$ characterizable.*

PROOF. By (9), $\{\alpha:\alpha \geqq \delta_2\}$ is a Π_2^1 set of ordinals. Since the set of countable $\bigwedge_1^1$ characterizable ordinals is cofinal in δ_2, the set of limits of countable sequences of these ordinals contains δ_2. But this set is a Π_2^1 set, since it equals $\{R:\forall n WO_{|R\restriction n|}$ is $\Pi_1^1\}$, which is Π_2^1 by (7). Hence WO_{δ_2} is a Π_2^1 set, and δ_2 is $\bigwedge_2^1$ characterizable by Theorem 4.2.

5. Characterizability of cardinals in $\Diamond_2^1$ spectra. In the manner indicated at the end of § 2, our results about ordinal characterizability enable us to show that $\aleph_\alpha$ is $\Diamond_2^1$ characterizable for any $\alpha < \delta_2$. The same techniques also enable us to establish a more general result: by replacing the $\aleph$ function, which enumerates the class of all infinite cardinals, by the enumerating function F of an arbitrary $\Diamond_2^1$ spectrum, we shall show that F_α is a $\Diamond_2^1$ characterizable cardinal for any $\Diamond_2^1$ characterizable ordinal α. Consequently, by showing that several common classes of cardinals are $\Diamond_2^1$ spectra, we shall obtain $\Diamond_2^1$ characterizations for many cardinals which are not "too far out" in these spectra.

While ordinal characterizability will play a key role in what follows, in some cases it can be replaced by a weaker notion, for if F enumerates a class of cardinals, then in general F_α will be much larger than card(α) and the "extra room" in a structure of cardinality F_α may allow us to define a $\Diamond_2^1$ well-ordering with order type α even though α is not a $\Diamond_2^1$ characterizable ordinal. Accordingly, for any ordinal α and any cardinal κ, call α *$\Diamond_2^1$-definable over universes of power at least κ* iff there is an $\bigvee_2^1$ sentence ψ and an $\bigwedge_2^1$ sentence ψ' such that for any structure $\langle A, R\rangle$ with card(A) $\geqq \kappa$, $\langle A, R\rangle \models \psi$ iff $\langle A, R\rangle \models \psi'$ iff $|R| = \alpha$. It is easy to see that an ordinal α is $\Diamond_2^1$ characterizable iff it is $\Diamond_2^1$ definable over all universes of power at least card (α). Using the techniques of § 2, one can extend this result as follows.

THEOREM 5.1. *For any second-order characterizable ordinal α, α is $\Diamond_2^1$ definable over universes of power at least $2^{\mathrm{card}(\alpha)}$.*

PROOF. Suppose that φ is a second-order sentence which is true in a structure $\langle B, R\rangle$ iff $\langle B, R\rangle$ is isomorphic to $\langle \alpha, \in\rangle$. As in the proof of Theorem 2.2, let l be the maximum of the ranks of relation variables occurring in φ, and let ψ be an $\bigvee_2^1$ sentence such that $\langle A, R\rangle \models \psi$ iff there are relations $B, C_1, \ldots, C_l$ over A

such that B is the field of R and conditions (6) and (7) of § 2 hold; likewise, let ψ' be an $\bigwedge_2^1$ sentence such that $\langle A, R\rangle \models \psi'$ iff, for all relations $B, C_1, \ldots, C_l$ over A such that B is the field of R and condition (6) of § 2 holds, condition (7) also holds. Then ψ and ψ' define α over universes of power at least $2^{\mathrm{card}(\alpha)}$.

COROLLARY 5.2. *For any $\alpha < \delta_{(\omega)}$, α is $\Diamond_2^1$ definable over universes of power at least $2^{\aleph_0}$.*

PROOF. The corollary follows immediately from Theorems 4.2, 4.3, and 5.1.

Theorem 5.1 together with our results on ordinal characterizability will be the principal tools for obtaining $\Diamond_2^1$ characterizations of cardinals. Before proceeding in this direction, we obtain a useful "converse" to Lemma 2.1.

LEMMA 5.3. *For any $\Diamond_2^1$ characterizable cardinal κ, $\{\lambda:\lambda < \kappa\}$ and $\{\lambda:\lambda \leqq \kappa\}$ are $\Diamond_2^1$ spectra.*

PROOF. For any set A, card$(A) < \kappa$ iff there does not exist a subset B of A with cardinality κ [an $\bigwedge_2^1$ condition since κ is $\bigvee_2^1$ characterizable] iff there is a well-ordering of A in which no proper initial segment has cardinality κ [an $\bigvee_2^1$ condition since κ is $\bigwedge_2^1$ characterizable]. The rest of the lemma follows from Lemma 2.1.

THEOREM 5.4. *Let F enumerate a $\Diamond_2^1$ spectrum in increasing order. For any ordinal α in the domain of F, if there is a $\Diamond_2^1$ characterizable cardinal $\kappa \leqq F_\alpha$ such that α is $\Diamond_2^1$ definable over universes of power at least κ, then F_α is a $\Diamond_2^1$ characterizable cardinal.*

PROOF. Let F enumerate a $\Diamond_2^1$ spectrum $\mathscr{S}$. In the definition of the $\Diamond_2^1$ condition $\Psi(A, R, S)$ in § 2, replace "card$(R^{-1}\{x\}) \geqq \aleph_0$" by "card$(R^{-1}\{x\}) \in \mathscr{S}$"; also, in (4a) and (4b) of § 2, replace "card$(S) < \aleph_0$" by condition (9) of § 2. By Lemma 5.3, $\{\lambda:\lambda \geqq \kappa\}$ is a $\Diamond_2^1$ spectrum, and hence F_α is $\Diamond_2^1$ characterizable as that cardinal in $\{\lambda:\lambda \geqq \kappa\}$ satisfying the modified conditions (4a) or (4b).

COROLLARY 5.5. *Let F enumerate a $\Diamond_2^1$ spectrum in increasing order. Then, for any ordinal α in the domain of F,*

(a) *if α is $\Diamond_2^1$ characterizable, then so is F_α;*
(b) *if $\alpha < \delta_2$, then F_α is $\Diamond_2^1$ characterizable; and*
(c) *if $2^{\aleph_0} < F_\alpha$ and $\alpha < \delta_{(\omega)}$, then F_α is $\Diamond_2^1$ characterizable.*

PROOF. Part (a) is immediate, (b) follows from (a) and Corollary 4.4, and (c) follows from the theorem and Corollaries 2.3 and 5.2.

In order to apply Corollary 5.5, we now show that several classes of cardinals are $\Diamond_2^1$ spectra.

LEMMA 5.6. *For any class $\mathscr{S}$ of infinite cardinals, $\mathscr{S}$ is a $\Diamond_2^1$ spectrum iff there is an $\bigvee_2^1$ class $\mathscr{C}$ of structures $\langle A, R, S\rangle$ such that, for any ordinal α and any $S \subseteq \alpha$, $\langle \alpha, \in, S\rangle \in \mathscr{C}$ iff $S = \mathscr{S} \cap \alpha$.*

PROOF. If $\mathscr{S}$ is a $\Diamond_2^1$ spectrum, let $\mathscr{C}$ be the $\Diamond_2^1$ class of structures $\langle A, R, S\rangle$ such that, for any x, $x \in S$ iff card$(R^{-1}\{x\}) \in \mathscr{S}$ and, for any y, if $\langle y, x\rangle \in R$, then

$\mathrm{card}(R^{-1}\{y\}) < \mathrm{card}(R^{-1}\{x\})$. Conversely, if $\mathscr{C}$ is an $\bigvee_2^1$ class, then $\mathscr{S}$ is $\Diamond_2^1$ since for any set A, $\mathrm{card}(A) \in \mathscr{S}$ iff there are relations R, S and an $x \in A$ such that R well-orders A, $\mathrm{card}(A) = \mathrm{card}(R^{-1}\{x\})$, $\langle A, R, S\rangle \in \mathscr{C}$, and $x \in S$ [an $\bigvee_2^1$ condition] iff A is infinite and for any R, S, x, if R well-orders A, x is the R-least element of A such that $\mathrm{card}(A) = \mathrm{card}(R^{-1}\{x\})$, and $\langle A, R, S\rangle \in \mathscr{C}$, then $x \in S$ [an $\bigwedge_2^1$ condition].

THEOREM 5.7. *The classes of infinite cardinals, limit cardinals, regular cardinals, successors of beths, weakly inaccessible cardinals, and inaccessible cardinals are all* $\Diamond_2^1$ *spectra.*

PROOF. The infinite cardinals form an $\bigvee_1^1$ spectrum by § 2(1). Let $\mathscr{C}$ be the class of structures $\langle A, R, S\rangle$ such that, for any x, $x \in S$ iff $R^{-1}\{x\}$ is infinite and

$$\forall y[\langle y, x\rangle \in R \Rightarrow \exists z(\langle y, z\rangle, \langle z, x\rangle \in R \ \&\ \mathrm{card}(R^{-1}\{z\}) < \mathrm{card}(R^{-1}\{x\}))];$$

$\mathscr{C}$ is a $\Diamond_2^1$ class and hence the class of limit cardinals is a $\Diamond_2^1$ spectrum by the lemma. Next let $\mathscr{C}$ be the class of structures $\langle A, R, S\rangle$ such that, for any x, $x \in S$ iff $R^{-1}\{x\}$ is infinite and for every subset C of $R^{-1}\{x\}$, if

$$\forall y[\langle y, x\rangle \in R \Rightarrow \exists z(\langle y, z\rangle, \langle z, x\rangle \in R \ \&\ z \in C)],$$

then $\mathrm{card}(C) < \mathrm{card}(R^{-1}\{x\})$; $\mathscr{C}$ is a $\Diamond_2^1$ class and hence the class of regular cardinals is a $\Diamond_2^1$ spectrum by the lemma. Using § 2(8) and the definition

$$\beth_\alpha = \sup\{2^{\beth_\beta} : \beta < \alpha\} \cup \aleph_0$$

of the beth function, one can show that there is an $\bigvee_2^1$ class $\mathscr{C}$ of structures $\langle A, R, S\rangle$ such that $\langle \alpha, \in, S\rangle \in \mathscr{C}$ iff there is a binary relation T such that S is the range of T and $T = \{\langle \beta, \beth_\beta^+\rangle : \beth_\beta^+ \leqq \alpha\}$; hence the successors of the beths form a $\Diamond_2^1$ spectrum by the lemma. The rest of the theorem follows now, since a cardinal κ is weakly inaccessible iff it is an uncountable regular limit cardinal, and is inaccessible iff it is weakly inaccessible and $\forall \lambda < \kappa\ (2^\lambda < \kappa)$ [a $\Diamond_2^1$ condition by § 3(15), (16)].

COROLLARY 5.8. (a) *For any* $\alpha < \delta_2$, $\aleph_\alpha$ *and the* α*th weakly inaccessible cardinal (if it exists) are* $\Diamond_2^1$ *characterizable.*

(b) *For any* $\Diamond_2^1$ *characterizable cardinal* κ, $\aleph_\kappa$ *is* $\Diamond_2^1$ *characterizable.*

(c) *For any* $\alpha < \delta_{(\omega)}$, *if* $2^{\aleph_0} < \aleph_\alpha$ *then* $\aleph_\alpha$ *is* $\Diamond_2^1$ *characterizable.*

(d) *For any* $\alpha < \delta_{(\omega)}$, $\beth_\alpha^+$ *and the* α*th inaccessible cardinal (if it exists) are* $\Diamond_2^1$ *characterizable.*

PROOF. The proof is immediate from Theorem 4.1, Corollary 5.5, and Theorem 5.7.

As noted before, Corollary 5.8 shows that cardinals which are not "too far out" in several $\Diamond_2^1$ spectra are $\Diamond_2^1$ characterizable. Characterizations of cardinals "farther out" in these spectra can be obtained also by utilizing various operators to "throw away" large numbers of cardinals in these spectra in such a manner that the remaining cardinals still form $\Diamond_2^1$ spectra to which Corollary 5.5 can be applied. We shall examine two such operators: the fixed point operator F and the

Mahlo operator M (cf. Lévy [**9**]), where, for any class $\mathscr{F}$ of ordinals,

$$F\mathscr{F} = \{\alpha \in \mathscr{F} : \mathscr{F} \cap \alpha \text{ has order type } \alpha\},$$

and

$$M\mathscr{F} = \{\alpha \in \mathscr{F} : \text{every closed, unbounded subset of } \alpha \text{ intersects } \mathscr{F}\}.$$

Call an operator T on the collection of all classes of ordinals into itself a *thinning* iff, for any class $\mathscr{F}$ of ordinals and any ordinal α,

$$T\mathscr{F} \cap \alpha = T(\mathscr{F} \cap \alpha) \subseteq \mathscr{F}.$$

Obviously both F and M are thinnings. For any thinning T, define the *graph* of T to be the class of all structures $\langle A, R, C, D\rangle$ such that

(1) R well-orders A, and

(2) "$D = TC$", i.e., $\{|R \upharpoonright x| : x \in D\} = T\{|R \upharpoonright x| : x \in C\}$.

A thinning is said to be an $\bigvee_2^1$ thinning if its graph is an $\bigvee_2^1$ class. We shall show that F and M are $\bigvee_2^1$ thinnings, and that $\bigvee_2^1$ thinnings preserve $\Diamond_2^1$ spectra.

THEOREM 5.9. *The fixed point operator and the Mahlo operator are* $\bigvee_2^1$ *thinnings.*

PROOF. It suffices in each case to show that condition (2) is an $\bigvee_2^1$ condition. First, "$D = FC$" holds iff

$$\forall x(x \in D \Leftrightarrow x \in C \;\&\; |R \upharpoonright x| = |(R \cap C \times C) \upharpoonright x|),$$

which is a $\Diamond_2^1$ condition by § 4(1). Second, a subset X of an ordinal α is closed iff $\forall \beta < \alpha \sup (X \cap \beta) \in X$ and is unbounded iff $\sup \{\beta + 1 : \beta \in X\} = \alpha$. Since these are both first-order conditions, "$D = MC$" is a $\Diamond_2^1$ condition.

THEOREM 5.10. *For any* $\bigvee_2^1$ *thinning* T *and any* $\Diamond_2^1$ *spectrum* $\mathscr{S}$ *containing only infinite cardinals,* $T\mathscr{S}$ *is a* $\Diamond_2^1$ *spectrum.*

PROOF. By Lemma 5.6 there is an $\bigvee_2^1$ class $\mathscr{C}$ such that a structure $\langle A, R, D\rangle$ is in $\mathscr{C}$ iff there exists a subset C of A such that $\langle A, R, C, D\rangle$ is in the graph of T and

$$\forall x(x \in C \Leftrightarrow |R \upharpoonright x| \in \mathscr{S} \cap |R|).$$

Hence, for any ordinal α and any $S \subseteq \alpha$, $\langle \alpha, \in, S\rangle \in \mathscr{C}$ iff $S = T(\mathscr{S} \cap \alpha) = T\mathscr{S} \cap \alpha$, so that $T\mathscr{S}$ is a $\Diamond_2^1$ spectrum by Lemma 5.6.

It follows from Theorem 5.10 and previous results, for example, that the first fixed point of the $\aleph$ function is $\Diamond_2^1$ characterizable, as is the first fixed point in the enumeration of fixed points of the $\aleph$ function, and so on. In general, as we shall now show, iterating an $\bigvee_2^1$ thinning α times, where α is a $\Diamond_2^1$ characterizable ordinal, still preserves $\Diamond_2^1$ spectra. For any thinning T and any class $\mathscr{F}$ of ordinals, let

$$\begin{aligned} T^\alpha\mathscr{F} &= \mathscr{F} && \text{if } \alpha = 0, \\ &= T\bigcap\{T^\beta\mathscr{F} : \beta < \alpha\} && \text{if } \alpha > 0. \end{aligned}$$

LEMMA 5.11. *For any thinning T, any class $\mathscr{F}$ of ordinals, and any ordinals α, β, if $\alpha \leqq \beta$, then $T^{\alpha}(\mathscr{F} \cap \alpha) = T^{\beta}(\mathscr{F} \cap \alpha)$.*

PROOF. Suppose not, and let α be the least ordinal such that for some $\beta \geqq \alpha$ and some $\gamma < \alpha$, γ is in $T^{\alpha}(\mathscr{F} \cap \alpha)$ but not $T^{\beta}(\mathscr{F} \cap \alpha)$. Then $T^{\gamma+1}(\mathscr{F} \cap \gamma + 1) =$ [since $\gamma \in T^{\alpha}(\mathscr{F} \cap \alpha) \subseteq T^{\gamma+1}(\mathscr{F} \cap \alpha)$] $T^{\gamma+1}(\mathscr{F} \cap \gamma) \cup \{\gamma\} =$ [by the minimality of α] $T^{\gamma}(\mathscr{F} \cap \gamma) \cup \{\gamma\} =$ [since $\gamma \in T^{\alpha}(\mathscr{F} \cap \alpha) \subseteq T^{\gamma}(\mathscr{F} \cap \alpha)$] $T^{\gamma}(\mathscr{F} \cap \gamma + 1)$. Hence

$$T^{\alpha}(\mathscr{F} \cap \gamma + 1) = T^{\gamma}(\mathscr{F} \cap \gamma + 1) = T^{\beta}(\mathscr{F} \cap \gamma + 1),$$

contradicting the choice of γ.

THEOREM 5.12. *For any $\bigvee_2^1$ thinning T and any $\diamondsuit_2^1$ characterizable ordinal α, T^{α} is an $\bigvee_2^1$ thinning.*

PROOF. Again we must show that condition (2) is an $\bigvee_2^1$ condition. But "$D = T^{\alpha}C$" holds iff there is an $S \subseteq A \times A$ such that

(a) $\forall x(|R \restriction x| = 0 \Rightarrow S^{-1}\{x\} = C)$,

(b) $\forall x(|R \restriction x| > 0 \Rightarrow \langle A, R, \bigcap\{S^{-1}\{y\} : \langle y, x\rangle \in R\}, S^{-1}\{x\}\rangle$ is in the graph of T),

and either

(c) $\exists x(|R \restriction x| = \alpha \;\&\; D = S^{-1}\{x\})$, i.e., $\alpha < |R|$ and "$D = T^{\alpha}C$," or

(d) $\neg\, \exists x|R \restriction x| = \alpha \;\&\; \langle A, R, \bigcap\{S^{-1}\{y\} : y \in A\}, D\rangle \in$ graph T, i.e., $\alpha \geqq |R|$ and "$D = T^{|R|}C = T^{\alpha}C$."

For any cardinal κ and any ordinal α, call κ a *fixed point of order* α of the $\aleph$ function iff κ is in $F^{\alpha}\{\lambda : \lambda \geqq \aleph_0\}$, and a *Mahlo cardinal of order* α iff $\kappa \in M^{\alpha}\{\lambda : \lambda$ regular$\}$. The following result is a direct consequence of Corollary 5.5 and Theorems 5.7, 5.10, and 5.12.

COROLLARY 5.13. *For any ordinals α and β, if α and β are $\diamondsuit_2^1$ characterizable (in particular, if $\alpha, \beta < \delta_2$), then the αth fixed point of order β of the $\aleph$ function and the αth Mahlo cardinal of order β (if it exists) are $\diamondsuit_2^1$ characterizable.*

By considering definability of ordinals over universes of at least a given power (cf. Theorem 5.4) one can strengthen Corollary 5.13 to show that, for any α, $\beta < \delta_{(\omega)}$, the αth Mahlo cardinal of order β (if it exists) is $\diamondsuit_2^1$ characterizable provided that the continuum is not extraordinarily large. Even more is possible, since it should be apparent that many results in this section and in the previous sections can be strengthened or generalized considerably if one considers not only second-order definability but also arbitrary higher-order definability. We have avoided such considerations primarily because they complicate the statement and proofs of our results without adding much in the way of ideas or consequences not already present in our study of second-order logic. However, it is probably appropriate at this point to indicate the form such generalizations take.

For any $n < \omega$, an nth *order logic* is that fragment of the simple theory of types which utilizes variables of type at most n, and $\delta_{(\omega)}^n$ is the least ordinal which is not the order type of a well-ordering of ω definable in nth order arithmetic. In the same manner in which Montague [**10**] extended Zykov's theorem to higher types, one can extend Theorem 2.2 to show that, for any nth order characterizable cardinal κ, $(2^{\kappa,n})^+$ is $\diamondsuit_2^1$ characterizable, where

$$2^{\kappa,0} = \kappa \quad \text{and} \quad 2^{\kappa,n+1} = 2^{2^{\kappa,n}}.$$

Among the many possible generalizations of other results are those of Corollaries 5.8 and 5.13 which show that, for any n and any $\alpha, \beta < \delta^n_{(\omega)}$, $\aleph_\alpha$ is $\diamondsuit_2^1$ characterizable if $\beth_n < \aleph_\alpha$, $\beth_\alpha^+$ is $\diamondsuit_2^1$ characterizable, and the αth Mahlo cardinal of order β (if it exists and is larger than $\beth_n$) is $\diamondsuit_2^1$ characterizable.

6. Characterizability and the model theory of set theory. We have already seen in § 2 that the $\diamondsuit_2^1$ characterizability of a particular cardinal (namely, $2^{\aleph_0}$) is both consistent and independent relative to Zermelo-Fraenkel set theory. What can be said about the size of the continuum if $2^{\aleph_0}$ is not $\wedge_2^1$ characterizable? Kunen's result [**8**] shows that $2^{\aleph_0}$ may still be less than $\aleph_\Omega$, while Corollary 5.8 shows that it must be at least $\aleph_{\delta_2}$. Since δ_2 is an $\wedge_2^1$ characterizable ordinal by Theorem 4.12, an examination of the proof of Theorem 5.4 shows that $\aleph_{\delta_2}$ is $\wedge_2^1$ characterizable, and hence $2^{\aleph_0}$ must be greater than $\aleph_{\delta_2}$ if it is not $\wedge_2^1$ characterizable. In fact, the results of § 4 and § 5 can be extended to show that if $2^{\aleph_0}$ is not $\wedge_2^1$ characterizable, then not only is it not among the first δ_2 members of any $\diamondsuit_2^1$ spectrum, but also it is not among the first σ_2 members of any $\wedge_2^1$ spectrum, where $\sigma_2 > \delta_2$ is the least ordinal which is not the order type of a Σ_2^1 well-ordering of a subset of ω. Thus if the power of the continuum does not satisfy a fairly weak describability condition, then it is already excluded from being equal to a large number of cardinals.

The answers to questions (I) and (II) raised in § 2 are also influenced by the model theory of set theory. We have seen that the smallest cardinal which is not $\diamondsuit_2^1$ characterizable lies between $\aleph_{\delta_2}$ and $\aleph_\Omega$, and that if the continuum hypothesis holds then this cardinal is at least as large as $\aleph_{\delta_{(\omega)}}$. How large can this cardinal be? Scott and Kunen observed in a conversation with the author that given any countable standard model $\mathfrak{M}$ of Zermelo-Fraenkel set theory (ZF) and any ordinal $\alpha < \Omega^{\mathfrak{M}}$ there is a countable standard model $\mathfrak{N}$ of ZF with the same ordinals as $\mathfrak{M}$ such that cardinals are absolute from $\mathfrak{M}$ to $\mathfrak{N}$ and, for any $\beta < \alpha$, $\aleph_\beta$ is $\diamondsuit_2^1$ characterizable in $\mathfrak{N}$ (the "trick" is to have $2^{\aleph_n} = \aleph_{f(n)}$ in $\mathfrak{N}$, where f is an increasing function in $\mathfrak{M}$ such that $\{\langle x, y\rangle : \exists n 2^x 3^y = f(n+1) - f(n)\}$ is a well-ordering of ω with order type α). Thus the answer seems to be "as large as possible." How small can the cardinal be? We do not know if it is consistent relative to ZF that $\aleph_{\delta_2}$ is not $\vee_2^1$ characterizable.

Turning now to question (II), it appears that we must answer it in the same fashion. For given any countable standard model $\mathfrak{M}$ of ZF and any cardinal κ in $\mathfrak{M}$, there is a countable standard model $\mathfrak{N}$ of ZF with the same ordinals as $\mathfrak{M}$ such that cardinals are absolute from $\mathfrak{M}$ to $\mathfrak{N}$ and $(2^{\aleph_0})^{\mathfrak{N}} > \kappa$ (cf. Cohen [**4**, § 8]); thus the supremum of the cardinals which are $\diamondsuit_2^1$ characterizable in $\mathfrak{N}$ is greater than κ. It follows that an upper bound to the countable class of $\diamondsuit_2^1$ characterizable cardinals is not definable in any fashion which is absolute for models of ZF with the same cardinals; e.g., it is not arithmetical in the aleph function, that is, definable in a first-order language with variables ranging over the ordinals and with constants denoting the aleph function and the operations of ordinal addition and multiplication.

The answer to question (II) is affected by still other considerations. Suppose that we start with a model of ZF in which there is no inaccessible cardinal and extend it to one containing an inaccessible. Then we have changed the supremum of the class of $\Diamond^1_2$ characterizable cardinals by Corollary 5.8(d). If we add cardinals whose existence cannot be proved from the existence of an inaccessible cardinal (e.g., Mahlo cardinals), then we may raise the supremum still further. Even though Mahlo cardinals are the largest "large cardinals" which we can show to be $\Diamond^1_2$ characterizable, we can add larger cardinals yet and still affect the supremum, as the following examples show.

Hanf and Scott [**5**] call a cardinal κ $\bigwedge^1_n$ *describable* iff there is an $\bigwedge^1_n$ class $\mathscr{C}$ and an $R \subseteq \kappa \times \kappa$ such that $\langle \kappa, R\rangle \in \mathscr{C}$ but $\langle \lambda, R \cap \lambda \times \lambda\rangle \notin \mathscr{C}$ for all $\lambda < \kappa$; they call κ $\bigwedge^1_n$ *indescribable* iff it is inaccessible and not $\bigwedge^1_n$ describable. They show that any $\bigwedge^1_1$ indescribable cardinal is weakly compact, and that the existence of $\bigwedge^1_{n+1}$ indescribable cardinals does not follow from the existence of $\bigwedge^1_n$ indescribable cardinals for any n. The smallest $\bigwedge^1_n$ indescribable cardinal is not $\bigvee^1_{n+1}$ characterizable since a straightforward coding argument shows that any $\bigvee^1_{n+1}$ characterizable cardinal is $\bigwedge^1_n$ describable. However, by the techniques of § 3, the class of $\bigwedge^1_n$ indescribable cardinals is an $\bigwedge^1_{n+1}$ spectrum, and so the smallest $\bigwedge^1_n$ indescribable cardinal κ, if it exists, is at least second-order characterizable as the smallest member of a second-order spectrum; hence $(2^\kappa)^+$ is $\Diamond^1_2$ characterizable, so that $\{\lambda : \lambda \leqq 2^\kappa\}$ is a $\Diamond^1_2$ spectrum by Lemma 5.3 and κ is $\bigwedge^1_{n+1}$ characterizable as the unique $\bigwedge^1_n$ describable cardinal in $\{\lambda : \lambda \leqq 2^\kappa\}$.

An alternative proof of the $\bigwedge^1_2$ characterizability of the first $\bigwedge^1_1$ indescribable cardinal κ is as follows. For any set A and any n, let $[A]^n = \{B : B \subseteq A \;\&\; \mathrm{card}(B) = n\}$. Silver [**15**] showed that κ is the smallest cardinal λ such that $\lambda \to (\lambda)^2$, i.e., such that for every function f from $[\lambda]^2$ into $\{0, 1\}$ there is a subset A of λ such that $\mathrm{card}(A) = \lambda$ and f is constant on $[A]^2$; from this definition, it is apparent that $\{\lambda : \lambda \to (\lambda)^2\}$ is an $\bigwedge^1_2$ spectrum, and that κ is $\bigwedge^1_2$ characterizable as above. The same method shows that the class of Ramsey cardinals, i.e., cardinals κ such that $\kappa \to (\kappa)^{<\omega}$, is also an $\bigwedge^1_2$ spectrum, and that the smallest Ramsey cardinal, if it exists, is $\bigwedge^1_2$ but not $\bigvee^1_2$ characterizable.

It is even possible to add cardinals which are not second-order characterizable at all (e.g., the first measurable cardinal) to a model of set theory and still affect the supremum of the class of $\Diamond^1_2$ characterizable cardinals, since, as noted at the end of § 5, $(2^{2^\kappa})^+$ is $\Diamond^1_2$ characterizable for any third-order characterizable cardinal κ. Is there a cardinal so large that adding it to a model of ZF does not affect the supremum of the class of $\Diamond^1_2$ characterizable cardinals? This question is admittedly rather vague, though we would like to know whether, in view of the above results, there is any precise sense in which an "absolute" upper bound can be placed on the class of $\Diamond^1_2$ characterizable cardinals.

In conclusion, we note that the richness of the class of $\Diamond^1_2$ characterizable cardinals suggests that efforts towards a better understanding of higher-order model theory might begin most profitably with an attempt to isolate natural subclasses of $\Diamond^1_2$ and to analyze the structure of these classes.

References

1. J. W. Addison, *Hierarchies and the axiom of constructability*, Summaries of talks presented at the Summer Institute of Symbolic Logic in 1957 at Cornell University, second edition, Communications Research Division, Institute for Defense Analyses, 1960, pp. 355–362.

2. G. Asser, *Das Repräsentantenproblem im Prädikatenkalkül der ersten Stufe mit Identität*, Z. Math. Logik Grundlagen Math. **1** (1955), 252–263. MR **17**, 1038.

3. J. H. Bennett, *On spectra*, Doctoral Dissertation, Princeton University, Princeton, N.J., 1962.

4. P. J. Cohen, *Set theory and the continuum hypothesis*, Benjamin, New York, 1966. MR **38** #999.

5. W. P. Hanf and D. S. Scott, *Classifying inaccessible cardinals*, Notices Amer. Math. Soc. **8** (1961), 445. Abstract #61T-240.

6. S. C. Kleene, *Hierarchies of number theoretic predicates*, Bull. Amer. Math. Soc. **61** (1955), 193–213. MR **17**, 4.

7. ———, *Quantification of number-theoretic functions*, Compositio Math. **14** (1959), 23–40. MR **21** #2586.

8. K. Kunen, *Indescribability and the continuum*, Axiomatic Set Theory, Proc. Sympos. Pure Math., vol. 13, part I, Amer. Math. Soc., Providence, R.I., 1971, pp. 199–203. MR **44** #63.

9. A. Lévy, *The sizes of the indescribable cardinals*, Axiomatic Set Theory, Proc. Sympos. Pure Math., vol. 13, part I, Amer. Math. Soc., Providence, R.I., 1971, pp. 205–218. MR **43** #7321.

10. R. M. Montague, *Reductions of higher-order logic*, The Theory of Models (Proc. 1963 Internat. Sympos., Berkeley), edited by J. W. Addison, L. Henkin, and A. Tarski, North-Holland, Amsterdam, 1965, pp. 251–264. MR **34** #4114.

11. Y. N. Moschovakis, *Uniformization in a playful universe*, Bull. Amer. Math. Soc. **77** (1971), 731–736.

12. H. Scholz, *Ein ungelöstes Problem in der symbolischen Logik*, J. Symbolic Logic **17** (1952), 160.

13. D. S. Scott, *Axiomatizing set theory*, Lecture notes from the 1967 UCLA Summer Institute on Axiomatic Set Theory.

14. J. R. Shoenfield, *Mathematical logic*, Addison-Wesley, Reading, Mass., 1967.

15. J. H. Silver, *Some applications of model theory in set theory*, Doctoral Dissertation, University of California, Berkeley, Calif., 1966.

16. Y. Suzuki, *A complete classification of the Δ^1_2 functions*, Bull. Amer. Math. Soc. **70** (1964), 246–253. MR **28** #2042.

17. A. A. Zykov, *The spectrum problem in the extended predicate calculus*, Izv. Akad. Nauk SSSR Ser. Mat. **17** (1953), 63–67; English transl., Amer. Math. Soc. Transl. (2) **3** (1956), 1–14. MR **14**, 936; **17**, 1038.

Dartmouth College

Proceedings of Symposia in Pure Mathematics
Volume 13, Part II, 1974

THE CONSISTENCY OF PARTIAL SET THEORY WITHOUT EXTENSIONALITY

P. C. GILMORE

1. Introduction. Consider sets defined over a universe D. The characteristic function f_S of a set S is defined as

$$f_S(x) = 1 \quad \text{if } x \in S,$$
$$= 0 \quad \text{if } x \notin S,$$

where $x \notin S$ means that x is in the complement of S. A set S is a *total set* if f_S is total on D and is a *partial set* if f_S is partial.

The simple theory of types is an example of a theory of partial sets when D is taken to be the union of the type universes. Zermelo-Fraenkel set theory is also an example of a theory of partial sets, since a complement of a set is only defined relative to another set and not relative to the entire universe.

It is natural to attempt a theory of partial sets in which both $\in$ and $\notin$ are primitive. In this paper such a theory PST′ is described in § 2, and in § 3 a model is constructed for it by transfinite induction up to an unspecified ordinal of the second number class. However, the theory is lacking any extensionality axiom or rule. Indeed, it will be shown in § 4 that this axiom and rule are both inconsistent with PST′.

The results of this paper first announced informally in [**3**] supersede the results of [**2**] and questions raised there have now been settled. The theory PST* described in [**2**] and the theory PST described in [**3**] are both proper subtheories of PST′, and their consistency follows from the consistency of PST′. On the other hand, the suggestion made in [**2**] that the extensionality rule might be consistent with PST* is disproved by the counterexample provided in § 4 below.

2. The theory PST′. In addition to $\in$ and $\notin$ the theory PST′ has $=$ as a primitive predicate and is formalized within first-order classical logic with identity. The

---

AMS (MOS) subject classifications (1970). Primary 02K10; Secondary 02C20.

usual definition of $\notin$, namely $(x)(y)\,(x \notin y \equiv \sim x \in y)$, is logically equivalent with the conjunction of

$$(x)(y)\,(x \in y \vee x \notin y), \tag{2.1}$$

$$(x)(y) \sim (x \in y \,\&\, x \notin y). \tag{2.2}$$

But when $\notin$ is interpreted as the complement membership relation, (2.1) is only appropriate for a theory of total sets. Thus, only (2.2) is accepted as an axiom for PST′.

For reasons that will become apparent the theory PST′ has terms which are intended to denote properties or partial sets in any interpretation of PST′. For any variable x and any positive formulae P and Q, $\{x:P, Q\}$ is a term of PST′, where a positive formula, following Lyndon [**4**], is one in which only conjunction, disjunction, and existential and universal quantification are used. The free variables of such a term are the free variables of P and Q other than x. The intended interpretation of these terms is specified by the comprehension axiom schemes which they satisfy:

$$(v)[(v \in \{x:P(x), Q(x)\} \vee (P(v) \,\&\, Q(v))) \equiv P(v)], \tag{2.3}$$

$$(v)[(v \notin \{x:P(x), Q(x)\} \vee (P(v) \,\&\, Q(v))) \equiv Q(v)], \tag{2.4}$$

where x is any variable, $P(x)$ and $Q(x)$ are any positive formulae, and v is any variable free to replace x in $P(x)$ and $Q(x)$.

The given comprehension axiom schemes complete the description of the theory PST′; it is the theory formalized within classical quantificational logic with identity with axiom (2.2) and axiom schemes (2.3) and (2.4). To provide motivation for (2.3) and (2.4), however, we will state an equivalent formulation of these axiom schemes. The scheme (2.3) is equivalent with the conjunction of

$$(v)[v \in \{x:P(x), Q(x)\} \supset P(v)], \tag{2.5}$$

$$(v)[(P(v) \,\&\, \sim Q(v)) \supset v \in \{x:P(x), Q(x)\}], \tag{2.6}$$

while the scheme (2.4) is equivalent with the conjunction of

$$(v)[v \notin \{x:P(x), Q(x)\} \supset Q(v)], \tag{2.7}$$

$$(v)[(\sim P(v) \,\&\, Q(v)) \supset v \notin \{x:P(x), Q(x)\}]. \tag{2.8}$$

These equivalences are easily established by elementary theorems of propositional logic. The schemes (2.5) and (2.7) provide necessary conditions for the relations $\in$ and $\notin$ respectively to hold while (2.6) and (2.8) provide sufficient conditions. Neither the necessary conditions nor the sufficient conditions can be consistently taken to be necessary and sufficient conditions. For if $\supset$ in (2.5) and (2.7) would be replaced by $\equiv$ then the term $\{x:x = x, x = x\}$ would result in a contradiction of (2.2), while if $\supset$ in (2.6) or (2.8) would be replaced by $\equiv$ the term $\{x:x = x, x \in x\}$ or $\{x:x \in x, x = x\}$ respectively would result in a contradiction. For a particular subset of terms, however, the necessary conditions of (2.5) and (2.7) can be taken to be also sufficient.

For any terms s and t, $s \notin t$ is a positive counterpart to $\sim s \in t$ and $s \in t$ is a positive counterpart to $\sim s \notin t$. Further if we define

(2.9) $$s \neq t \quad \text{for} \quad (\exists z)(s \in z \,\&\, t \notin z),$$

where z does not occur free in s or t, then $s \neq t$ is a positive counterpart to $\sim s = t$. More generally for any formula P of PST′ a positive counterpart P^* can be found as follows. First express P only in terms of negation, conjunction, disjunction and existential and universal quantification, with all negations applied only to atomic formulae. Then replace any part $\sim s \in t$ of P by $s \notin t$, any part $\sim s \notin t$ by $s \in t$, and any part $\sim s = t$ by $s \neq t$. It is then a simple matter to prove

(2.10) $$(P^* \supset P)$$

is a consequence of (2.2) in quantificational logic with identity. For any formula $P(x)$ of PST′, therefore, the term $\{x:(P(x))^*, (\sim P(x))^*\}$ must satisfy

(2.11) $$(v)[v \in \{x:(P(x))^*, (\sim P(x))^*\} \equiv (P(v))^*],$$

(2.12) $$(v)[v \notin \{x:(P(x))^*, (\sim P(x))^*\} \equiv (\sim P(v))^*].$$

The schemes (2.11) and (2.12) are analogous to the comprehension axiom scheme of an ordinary set theory. But such simple schemes are not available for general terms of PST′, since no limitation has been placed on the pairs of positive formulae P and Q for which terms $\{x:P, Q\}$ are permitted. The extra disjunct $(P(v) \,\&\, Q(v))$ in the schemes (2.3) and (2.4) is only removable at the price of introducing such a limitation.

3. The consistency of PST′. No attempt has been made to justify the restriction that P and Q in the terms $\{x:P, Q\}$ of PST′ must be positive except to suggest that if one is willing to interpret any formula P as P^*, then the restriction does not limit set formation. However, the restriction has an important technical advantage which allows a persistence lemma for positive sentences proved in Lyndon [**4**] to be exploited in a consistency proof similar to the one given in [**1**]. We will state Lyndon's persistence lemma in a form suitable for application to the consistency of PST′.

The model to be provided for PST′ has as its domain the set D of all constant terms for the theory PST′; that is, the set of all terms $\{x:P, Q\}$ for which x is the only free variable occurring in either P or Q. To define a structure on D in which the formulae of PST′ can be interpreted it is necessary to

(a) assign to each primitive term $r(u_1, \ldots, u_k)$ of PST′ with k free variables $u_1, \ldots, u_k$ a k-argument function with arguments and values from D;

(b) define the set M of valid atomic sentences; that is, the set of sentences $a = b$, $a \in b$ and $a \notin b$, where a and b are in D, which are to be "true."

The primitive terms of PST′ are among the terms $\{x: P, Q\}$, but we will satisfy (a) without actually determining which terms of PST′ are primitive. For associated with any term $\{x:P(u_1,\ldots,u_k), Q(u_1,\ldots,u_k)\}$ of PST′, for which $u_1,\ldots,u_k$ are the only free variables, is the function which for constant terms $a_1,\ldots,a_k$ of D takes as value the constant term $\{x:P(a_1,\ldots,a_k), Q(a_1,\ldots,a_k)\}$ of D. This obvious interpretation of the terms of PST′ over the domain D will always be used. Therefore, a structure on D in which the formulae of PST′ can be interpreted can be identified with the set M of valid atomic sentences for the structure.

Lyndon's persistence lemma can now be stated.

LEMMA 1. *Let M and M' be two structures with domain D for which $M \subseteq M'$. Then any positive sentence valid in M is also valid in M'.*

The lemma can be proved by a simple induction argument on the number of uses made of conjunction, disjunction and universal and existential quantification in constructing a positive sentence from atomic formulae. This argument is given in [**2**], [**3**], and in [**4**].

The model M constructed for PST′ is the limit of an ascending chain of structures $M_0 \subseteq M_1 \subseteq \cdots \subseteq M_\mu \subseteq$ all with domain D. To obtain the model M it will be necessary later to specialize the structure M_0 to be the set of all identity sentences $a = a$ for a in D. For the present, however, to give the arguments in a more general form, M_0 may be any set of identity sentences $a = b$, for a and b in D. Assuming M_μ defined for some ordinal number μ, $M_{\mu+1}$ is the union of M_μ and the set of all atomic sentences $a \in \{x:P(x), Q(x)\}$ and $a \notin \{x:Q(x), P(x)\}$ for which $(P(a) \,\&\, \sim Q(a))$ is valid in M_μ. For a limit ordinal μ, $M_\mu = \bigcup\{M_\nu : 0 \leqq \nu < \mu\}$.

LEMMA 2. *For any ordinal number μ,* (2.2) *and all instances of* (2.5) *and* (2.7) *are valid in M_μ. Further, if μ is not zero or a limit ordinal and $a \in \{x:P(x), Q(x)\}$ or $a \notin \{x:Q(x), P(x)\}$ is valid in M_μ, then $P(a)$ is valid $M_{\mu-1}$.*

PROOF. The lemma is immediately true for $\mu = 0$. Assume therefore that the result is true for all ordinals ν, $0 \leqq \nu < \mu$, and let $a \in \{x:P(x), Q(x)\}$ or $a \notin \{x:Q(x), P(x)\}$ be valid in M_μ. Let ν be the least ordinal for which one of these atomic sentences is valid in M_ν. ν is necessarily not a limit ordinal and $(P(a) \,\&\, \sim Q(a))$ is valid in $M_{\nu-1}$. Since $M_{\nu-1} \subseteq M_\mu$, it follows by Lemma 1 that $P(a)$ is valid in M_μ. Thus, all instances of (2.5) and (2.7) are valid in M_μ. If μ is not a limit ordinal, then since $\nu \leqq \mu$, it follows that $\nu - 1 \leqq \mu - 1$, and hence $P(a)$ is necessarily valid in $M_{\mu-1}$ also. To complete the proof of the lemma it is only necessary to establish the validity of (2.2) in M_μ. But if (2.2) were not valid in M_μ, then for some constant terms a and $\{x:P(x), Q(x)\}$ of D both $a \in \{x:P(x), Q(x)\}$ and $a \notin \{x:P(x), Q(x)\}$ would be valid in M_μ. μ cannot be a limit ordinal and therefore by the second half of the lemma already established for M_μ, both $P(a)$ and $Q(a)$ would be valid in $M_{\mu-1}$. Further, at least one of $a \in \{x:P(x), Q(x)\}$ and $a \notin \{x:P(x), Q(x)\}$ is not valid in $M_{\mu-1}$ so that necessarily $(P(a) \,\&\, \sim Q(a)) \vee (\sim P(a) \,\&\, Q(a))$ is valid in $M_{\mu-1}$ contradicting the validity of $P(a)$ and $Q(a)$.

Should $M_\mu = M_{\mu+1}$ for some μ, then all instances of (2.6) and (2.8) are also valid in M_μ for then whenever $(P(a) \,\&\, \sim Q(a))$ or $(\sim P(a) \,\&\, Q(a))$ is valid in M_μ so also is $a \in \{x:P(x), Q(x)\}$ or $a \notin \{x:P(x), Q(x)\}$ respectively. It does not follow that

M_μ is a model of PST′, since the validity of the axioms for identity has not been established. However, the following theorem is immediate, since the axioms for identity are valid in all the structures M_μ obtained from the given M_0.

THEOREM 1. *Assume that M_0 consists of all identity sentences $a = a$ for a in D. Then should $M_\mu = M_{\mu+1}$ for some μ, M_μ is a model of PST′.*

The consistency of PST′ follows from

COROLLARY. *For some ordinal number λ of the second number class, M_λ is a model of PST′.*

The increasing chain of structures $M_0 \subseteq M_1 \subseteq \cdots \subseteq M_\mu \subseteq \cdots$ is an increasing chain of subsets of the denumerable set of all atomic sentences of PST′. Further, if $M_\nu = M_{\nu+1}$, then $M_\nu = M_\mu$ for all ordinals μ, $\nu \leqq \mu$. Consequently, the set of ordinals μ for which $M_\mu \neq M_{\mu+1}$ is denumerable. But the set of all ordinals of the second number class is not denumerable and hence for some λ of this class $M_\lambda = M_{\lambda+1}$.

4. Identity in PST′. Extensional identity is defined in PST′ as

(4.1) $$s =_e t \quad \text{for} \quad (u)[(u \in s \equiv u \in t)\,\&\,(u \notin s \equiv u \notin t)],$$

where u is a variable not occurring free in s or t. The extensionality axiom then can be expressed as

(4.2) $$(x)(y)(x =_e y \supset x = y).$$

We will prove that (4.2) is inconsistent with PST′.

Consider the following abbreviated terms:

$$(s \cap t) \quad \text{for} \quad \{x : (x \in s \,\&\, x \in t), (x \notin s \vee x \notin t)\},$$

where x does not occur free in s or t;

$$\Lambda^* \quad \text{for} \quad \{u : u \neq u, u \neq u\};$$

$$R' \quad \text{for} \quad \{u : u \in u, u \notin u\};$$

$$A(s) \quad \text{for} \quad \{x : (x = R' \,\&\, s \in x), (x \neq R' \vee s \notin x)\},$$

where x does not occur free in s;

$$T \quad \text{for} \quad \{u : (A(u) \cap \Lambda^*) = A(u), (A(u) \cap \Lambda^*) \neq A(u)\}.$$

Note that each of the terms, with the exception of Λ^*, is of the form $\{x : P, (\sim P)^*\}$ with P positive, and therefore must satisfy (2.11) and (2.12).

THEOREM 2. *Both*

(4.3) $$\sim(A(T) \cap \Lambda^*) = A(T),$$

(4.4) $$(A(T) \cap \Lambda^*) =_e A(T)$$

are provable in PST′.

PROOF. Note first that

$$\sim s \neq s \tag{4.5}$$

is provable in PST′ for any term s, being a consequence of (2.2) and (2.9). A consequence of (4.5) is that Λ^* also satisfies (2.11) and (2.12), since it satisfies (2.3) and (2.4).

(4.3) is proved by showing that $T \neq T$ is a consequence of $(A(T) \cap \Lambda^*) = A(T)$ from which (4.3) follows by (4.5).

$(A(T) \cap \Lambda^*) = A(T) \supset$

$T \in T$	(2.11) for T,
$T \in R'$	(2.11) for R',
$R' = R' \mathbin{\&} T \in R'$	identity,
$R' \in A(T)$	(2.11) for $A(T)$,
$R' \in A(T) \mathbin{\&} (A(T) \cap \Lambda^*) = A(T)$	repeat hypothesis,
$R' \in (A(T) \cap \Lambda^*)$	identity,
$(R' \in A(T) \mathbin{\&} R' \in \Lambda^*)$	(2.11) for $(A(T) \cap \Lambda^*)$,
$R' \in \Lambda^*$	logic,
$R' \neq R'$	(2.11) for Λ^*.

Now we turn to the proof of (4.4). Because of (4.5) and (2.11) for $(A(T) \cap \Lambda^*)$ and Λ^*, to prove (4.4) it is sufficient to prove

$$(u)(u \in A(T) \supset u \in \Lambda^*). \tag{4.6}$$

The proof of (4.6) follows:

$u \in A(T) \supset$

$(u = R' \mathbin{\&} T \in u)$	(2.11) for $A(T)$,
$T \in R'$	identity,
$T \in T$	(2.11) for R',
$(A(T) \cap \Lambda^*) = A(T)$	(2.11) for T,
$(u \in A(T) \mathbin{\&} (A(T) \cap \Lambda^*) = A(T))$	repeat hypothesis,
$u \in (A(T) \cap \Lambda^*)$	identity,
$(u \in A(T) \mathbin{\&} u \in \Lambda^*)$	(2.11) for $(A(T) \cap \Lambda^*)$,
$u \in \Lambda^*$.	

Since both (4.3) and (4.4) are provable in PST′, it follows that the terms $(A(T) \cap \Lambda^*)$ and $A(T)$ contradict (4.2).

Since terms are available in PST′, a natural substitute for (4.2) would be the rule of deduction:

(4.7) When $s =_e t$ is a theorem then so is $s = t$.

But (4.3) and (4.4) contradict this rule also.

REFERENCES

1. F. B. Fitch, *An extension of basic logic*, J. Symbolic Logic **13** (1948), 95–106. MR **9**, 559.

2. P. C. Gilmore, *The consistency of a positive set theory*, IBM Research Report, RC-1754, Jan. 24, 1967.

3. ———, *Partial set theory*, Lecture Notes prepared in connection with the Summer Institute on Axiomatic Set theory, U.C.L.A., Los Angeles, July 10–August 4, 1967, pp. III-E-1 to 14.

4. Roger C. Lyndon, *Properties preserved under homomorphism*, Pacific J. Math. **9** (1959), 143–154. MR **21** #7157.

IBM, T. J. WATSON RESEARCH CENTER

Proceedings of Symposia in Pure Mathematics
Volume 13, Part II, 1974

ON THE EXISTENCE OF CERTAIN COFINAL SUBSETS OF ${}^{\omega}\omega$

STEPHEN H. HECHLER*

1. Introduction and statement of main theorem. In this paper we shall study the partial order $\Omega = \langle {}^{\omega}\omega, \prec \rangle$ where ${}^{\omega}\omega$ is the set of all functions from ω into ω and $\prec$ is the partial ordering defined by

$$f \prec g \leftrightarrow \exists n \, \forall m > n[f(m) < g(m)].$$

We shall be especially interested in those subsets of ${}^{\omega}\omega$ which are cofinal in Ω (a set $S \subseteq {}^{\omega}\omega$ is said to be cofinal in Ω iff $\forall f \in {}^{\omega}\omega \; \exists g \in S(f \prec g)$). We shall call such subsets *scales*, and if the order type of a scale is that of some ordinal α, then we shall call the set an *α-scale*.[1] Using the techniques of Cohen ([**1**], [**2**]) and Solovay and Tennenbaum [**6**], we shall construct models of Zermelo-Fraenkel set theory (ZFC) with many different types of scales and thus show that the structure of Ω is not determined by these axioms.

It was known as early as 1907 ([**3**], [**4**]) that the continuum hypothesis implies the existence of an $\aleph_1$-scale, and in fact in those of Cohen's original models where the continuum hypothesis fails there are no $\aleph_1$-scales. However, in Solovay's models [**5**] with random reals the set of constructible members of ${}^{\omega}\omega$ is itself a scale.[2] But this set has cardinality $\aleph_1$ and, as we shall see in § 8, the existence of a scale of

AMS (MOS) subject classifications (1970). Primary 02K05.

* This paper has been referred to and has appeared in preprint form under the title *On the set theoretical independence of the existence of certain kinds of cofinal subsets of ${}^{\omega}\omega$ under a natural partial ordering*. The greater portion of it has also appeared as part of the author's doctoral dissertation submitted to the University of California at Berkeley in 1967. The author wishes to express his sincere appreciation to Professor Robert M. Solovay for his help and encouragement without which this work could not have been done.

[1] The existence of a scale of cardinality equal to that of the continuum and well-ordered under $\prec$ is referred to as Proposition C_{76} in [**4**, p. 145].

[2] Private communication from R. M. Solovay.

cardinality $\aleph_1$ implies the existence of an $\aleph_1$-scale. Thus, in such models $\aleph_1$-scales exist even if the continuum hypothesis fails. We shall show that it is consistent to assume the existence of $\aleph_2$, $\aleph_3$, etc. scales, and in fact we shall show that any "reasonable" partial order may be consistently assumed to be order-isomorphic to a scale.

We begin by adopting certain conventions. For any function f we shall denote the domain of f by $dm(f)$. (Later we shall use "domain" and "dm" in other ways, but the exact meaning will always be clear from the context.) For any two sets A and B we shall denote the power set of A by $pw(A)$, the cardinality of A by $cd(A)$, the set of functions from A into B by ${}^AB$, and the cardinality of ${}^AB$ by B^A. When it is not clear from the context, we shall use the notation $(cd(A))_{\mathcal{M}}$, $({}^AB)_{\mathcal{M}}$, etc. to indicate that the operations are to be taken with respect to some model $\mathcal{M}$. We shall use the convention that whenever we speak of a set as being countable, we shall mean that it is either finite or of cardinality $\aleph_0$.

Also, let $\mathcal{M}$ be any model of ZFC and let $\mathcal{N}$ be a model of ZFC extending $\mathcal{M}$ which contains no ordinals not in $\mathcal{M}$. Suppose further, that each cardinal of $\mathcal{N}$ is a cardinal in $\mathcal{M}$ and that any two ordinals are cofinal with respect to $\mathcal{N}$ iff they are cofinal with respect to $\mathcal{M}$; in this case we shall call $\mathcal{N}$ a *regular* extension of $\mathcal{M}$. Finally, if $\mathcal{N}$ is a regular extension of $\mathcal{M}$, and if for each set A in $\mathcal{M}$ we have $(2^A)_{\mathcal{M}} = (2^A)_{\mathcal{N}}$, then we shall call $\mathcal{N}$ a *normal* extension of $\mathcal{M}$. We can now state our main theorem.

1.1. THEOREM. *Let $\mathcal{M} = \langle M, \in \rangle$ be any countable standard model of ZFC, and let $\mathcal{A} = \langle A, <_A \rangle$ be any nonreflexive partial order in $\mathcal{M}$ such that $(cd(A))_{\mathcal{M}} \leqq (2^{\aleph_0})_{\mathcal{M}}$ and such that every countable (with respect to $\mathcal{M}$) subset of A has an upper bound. Then there is a normal extension $\mathcal{N}$ of $\mathcal{M}$ in which $\mathcal{A}$ is order-isomorphic to a scale.*

We note that the restriction mentioned on $\langle A, <_A \rangle$ is not only sufficient but necessary as well, for if $\{f_i : i \in \omega\}$ is any countable subset of ${}^\omega\omega$, then the function $f \in {}^\omega\omega$ defined by $f(n) = 1 + \sum_{i=0}^{n} f_i(n)$ is an upper bound. Thus any countable subset of a scale must have an upper bound and, therefore, an upper bound in the scale.

2. Introduction of the language $\mathcal{L}$. To construct $\mathcal{N}$ we use a modification of the methods of Cohen and Tennenbaum; i.e., we use forcing, but we think intuitively of adding for each member of A not only a new member of ${}^\omega\omega$, but a new model extending those already added for smaller (with respect to $<_A$) members of A.

Ideally, we would like this new model to contain a function $f \in {}^\omega\omega$ which is larger than every $y \in {}^\omega\omega$ contained in any model below the new one. Unfortunately, in order to carry out induction proofs it appears to be necessary to have a well-founded, cofinal subset of these models such that while the above condition holds between those in the subset, only a much weaker condition holds between models not in the subset. We shall discuss this further in § 7.

We begin by choosing a well-founded set $B \subseteq A$ which is cofinal in A. The existence of such a set B follows from the axiom of choice (which we shall assume throughout this paper). Let F be a 1-1 function from some ordinal α onto A, and let G be the function from α into A defined by

$$\begin{aligned} G(\beta) &= F(\beta) \quad \forall \gamma \in \beta(\neg F(\beta) <_A G(\gamma)), \\ &= F(0) \quad \text{otherwise.} \end{aligned}$$

Then we may let B be the range of G. For purposes of notation it turns out to be convenient to adjoin to $\mathscr{A}$ a first and last element. As we may assume without loss of generality that $0 \notin A$, we shall define a new partial order $\mathscr{A}^* = \langle A^*, <_{A^*}\rangle$ where $A^* = A \cup \{0, A\}$ and $<_{A^*} = \{\langle a,b\rangle : a <_A b \vee a = 0 \vee b = A\}$. Similarly, we define $\mathscr{B}^* = \langle B^*, <_{B^*}\rangle$ by letting $B^* = B \cup \{A, 0\}$ and $<_{B^*} = <_{A^*} \restriction \mathscr{B}^*$. Where the meaning is clear from the context, we shall use $<$ for $<_A$, $<_B$, $<_{A^*}$, or $<_{B^*}$.

Using this convention, we assign to each member of A^* an ordinal, which we call its *height* (*ht*), as follows:
(1) $ht(b) = \bigcup\{ht(c) + 1 : c < b \wedge c \in B^*\}$ for $b \in B^*$,
(2) $ht(a) = \mu\beta[\exists b \in B(a < b \wedge ht(b) = \beta)]$ otherwise.
Also, we define the *closure* (*cl*) of a subset $S \subseteq A^*$ by

$$cl(S) = \{a \in A^* : \exists s \in S(a \leqq s)\}.$$

In standard constructions using Cohen's methods, a language is defined in such a way that each formula and term is assigned an ordinal and then forcing is defined by transfinite induction. In this paper this will not be sufficient because we will have, intuitively speaking, many different extensions of $\mathscr{M}$ rather than just one, and, therefore, we will have to, in a sense, know not only the "rank" but the "model" as well. Also, since we are not assuming constructibility in $\mathscr{M}$, we must have individual set constants for each member of $\mathscr{M}$. Thus we define our language $\mathscr{L}$.

2.1. (1) Symbols.
(a) *Variables*: $v_0, v_1, \ldots, v_n, \ldots, n \in \omega$.
(b) *Punctuation symbols*: "(","), and ",".
(c) *Individual constants*: For each $c \in M$, $\hat{c}$ is an individual constant.[3]
(d) *Logical symbols*: $\neg$, $\wedge$, $\bigwedge$, and for each ordinal $\alpha \in M$ and each $a \in A^*$ the quantifier $\bigwedge_\alpha^a$ and the set abstractor $\{\,:_\alpha^a\}$.
(e) *Predicate symbols*: The binary predicate symbol $\in$, and for each $a \in A^* - \{0\}$, a ternary predicate symbol A^a.

(2) Limited formulas, abstraction terms, and formulas.
(a) If t, u, and v, are abstraction terms, individual constants, or variables, then $t \in u$ and $A^a(t, u, v)$ are limited formulas.
(b) If φ and ψ are limited formulas, then so are $\neg\varphi$, $(\varphi \wedge \psi)$, and $\bigwedge_\alpha^a v_i \varphi$.

[3] However, if c is denoted by some terms, e.g., $\langle m, n\rangle$, then we shall denote the corresponding term in $\mathscr{L}$ by $\langle m, n\rangle\hat{}$.

(c) If φ is a limited formula with one free variable v_i, then $\{v_i :_\alpha^b \varphi\}$ is an abstraction term if for every quantifier $\bigwedge_\beta^a$, set abstractor $\{ :_\gamma^a \}$, predicate symbol A^a, or individual constant $\hat{c}$, it is the case that $\beta \leqq \alpha$, $\gamma < \alpha$, $a \leqq b$, and the set rank of $\hat{c}$ is less than α.

(d) Nothing is a limited formula or abstraction term except as required by (a), (b), and (c).

(e) Every limited formula is a formula.

(f) If φ and ψ are formulas then so are $(\varphi \wedge \psi)$, $\neg\varphi$, and $\bigwedge v_i \varphi$.

(g) Nothing is a formula except as required by (e) and (f).

In addition to the usual conventions regarding the dropping and adding of parentheses and the usual abbreviations $\vee$, $\bigvee$, $\bigvee_\alpha^a$, etc., we shall also use $\bigwedge_\alpha$, $\bigvee_\alpha$, and $\{ :_\alpha \}$ as abbreviations for $\bigwedge_\alpha^A$, $\bigvee_\alpha^A$, and $\{ :_\alpha^A \}$.

Intuitively, $\bigwedge_\alpha^a$ and $\{x :_\alpha^a \varphi\}$ should be thought of as ranging over all terms created at levels up to and *including* a, and up to but *not including* α. The symbol $A^a(t, u, v)$ should be thought of as saying that t, u, and v denote (in $\mathcal{N}$) sets b, c, and d such that $b \in A$, $b \leqq_A a$, and f_b (the function onto which b will map in $\mathcal{N}$ in the construction of the scale) takes on the value d at c. The superscript is needed because even though b, c, and d may be in $\mathcal{M}$, the question of whether or not $A^a(b, c, d)$ holds will, in general, not be determined until the model corresponding to b is constructed. The superscript a, then, will allow us to postpone consideration of such sentences until we construct the model corresponding to a.

In order to facilitate inductive definitions and proofs, we assign to each closed term and limited formula various ordinals and sets depending only on its structure.

2.2. (1) For closed terms we define *domain* (*dm*), *rank* (*rk*), *height* (*ht*), and *model* (*md*) by:

(a) For $c \in M$ we let $rk(\hat{c})$ = set rank of c and $ht(\hat{c}) = md(\hat{c}) = dm(\hat{c}) = 0$.

(b) For $t = \{v_i :_\alpha^a \varphi\}$ we set $rk(t) = \alpha$, $md(t) = a$, $ht(t) = ht(a)$, and $dm(t) = cl(\{a\})$.

(2) For limited formulas φ we define *domain* (*dm*), *height* (*ht*), *depth* (*dp*), *rank* (*rk*), *type* (*tp*), and *index* (*in*), by:

(a) $dm(\varphi) = cl(\{a : A^a, \bigwedge_\alpha^a$, or $\{ :_\alpha^a \}$ appears in $\varphi\})$.

(b) $dp(\varphi)$ is defined by induction.

(i) $dp(t \in u) = dp(A^a(t, u, v)) = 0$.

(ii) $dp(\theta \wedge \psi) = 1 + \max(dp(\theta), dp(\psi))$.

(iii) $dp(\neg\psi) = dp(\bigwedge_\alpha^a v_i \psi) = 1 + dp(\psi)$.

(3) $ht(\varphi) = \max\{ht(a) : a \in dm(\varphi)\}$.

(4) $rk(\varphi) = \mu\alpha$ (for every quantifier $\bigwedge_\beta^a$ or closed term t appearing in φ it is the case that $\beta \leqq \alpha$ and $rk(t) < \alpha$).

(5) $tp(\varphi) = 0$ if every instance of A^a in φ is contained within an abstraction symbol, every ordinal appearing in φ attached to a quantifier is less than $rk(\varphi)$, and for every subformula $t \in u$ appearing in φ there is an ordinal β such that $rk(t) < \beta < rk(\varphi)$. Otherwise $tp(\varphi) = 1$.

(6) $in(\varphi) = \langle ht(\varphi), rk(\varphi), tp(\varphi), dp(\varphi)\rangle$.

For any closed term t such that $rk(t) = \check{\alpha}$ and $md(t) = a$ we abbreviate $\bigwedge^a_\alpha$, $\bigvee^a_\alpha$, and $\{ :^a_\alpha \}$ by $\bigwedge^t$, $\bigvee^t$, and $\{ :^t \}$ respectively. Thus, in particular, $\bigwedge^{\hat{\omega}}$ is an abbreviation for $\bigwedge^0_\omega$. We also introduce abbreviations for several rather complicated sentences and terms which will appear frequently in the remainder of this paper.

2.3. (1) $t \cong u$ is $\bigwedge^t x(x \in t \to x \in u) \wedge \bigwedge^u x(x \in u \to x \in t)$.

(2) $t =^a_\alpha u$ is $\bigwedge^a_\alpha x(x \in t \leftrightarrow x \in u)$.

(3) $t =^v u$ is $t =^{md(v)}_{rk(v)} u$.

(4) $t =^a_\alpha \langle u, v\rangle$ is $\bigwedge^a_\alpha x[x \in t \leftrightarrow [\bigwedge^a_\alpha y(y \in x \leftrightarrow y =^a_\alpha u)$
$\vee \bigwedge^a_\alpha y(y \in x \leftrightarrow y =^a_\alpha u \vee y =^a_\alpha v)]]$.

(5) $t =^w \langle u, v\rangle$ is $t =^{md(w)}_{rk(w)} \langle u, v\rangle$.

(6) $t \in {}^\omega\omega$ is $\bigwedge^t x \bigvee^{\hat{\omega}} y \bigvee^{\hat{\omega}} z[x \in t \to (x =^t \langle y, z\rangle \wedge y \in \hat{\omega} \wedge z \in \hat{\omega})]$
$\wedge \bigwedge^t x \bigwedge^t y \bigwedge^{\hat{\omega}} j \bigwedge^{\hat{\omega}} k \bigwedge^{\hat{\omega}} m[x \in t \wedge y \in t \wedge x =^t \langle j, k\rangle$
$\wedge y =^t \langle j, m\rangle \to k =^{\hat{\omega}} m] \wedge \bigwedge^{\hat{\omega}} m \bigvee^{\hat{\omega}} n \bigvee^{\hat{\omega}} x[m \in \hat{\omega} \to x \in t \wedge x =^{\hat{\omega}} \langle m, n\rangle]$.

(7) f_a is $\{v :^a_{rk(A)} \bigvee^{\hat{\omega}} x \bigvee^{\hat{\omega}} y[v =^a_{rk(A)} \langle x, y\rangle \wedge A^a(\hat{a}, x, y)]\}$.

While (1) and (2) will eventually represent the same relation, equality, both are convenient at this point. The first more accurately represents what we mean by equality and is, therefore, usually more convenient when we are comparing two fixed terms. However, there is often no way of expressing $\cong$ in a limited formula when quantifiers are present, e.g., "$\bigwedge^a_\alpha v_i(v_i = t)$" cannot be expressed as $\bigwedge^a_\alpha v_i(v_i \cong t)$ because we have no symbol $\bigwedge^{v_i}$ in $\mathscr{L}$. In such cases we must make do with $\bigwedge^a_\alpha v_i(v_i =^a_\alpha t)$ or $\bigwedge^a_\alpha v_i(v_i =^t t)$, and we must be able to see from its form that the formula we choose actually does express the proper relation. A typical example is (6) which will hold in $\mathscr{N}$ iff the set in $\mathscr{N}$ denoted by t turns out to belong to $({}^\omega\omega)_{\mathscr{N}}$.

3. Definition of conditions and forcing. In [1] and similar constructions, a set of conditions is first given and then forcing is defined inductively. However, in the type of extension we are constructing, the question of whether or not some entity is a condition may depend on whether or not one part of it is forced to have certain properties by another part. Therefore, we must define condition and forcing simultaneously using transfinite induction. We begin by singling out a family of objects which have the proper form to be conditions; from these we will choose those which will actually be conditions.

We first define a *function pair* to be an ordered pair of the form $\langle t, k\rangle$ where t is any closed term and $k \in \omega$. For any $a \in A$ we say a set S of function pairs is *suitable for a* iff S is finite and for each pair $\langle t, k\rangle \in S$, we have $md(t) < a$.

A *potential condition p* is a function from a finite subset of A into a set of ordered pairs such that if we denote $p(a)$ by $\langle p_a, p^*_a\rangle$ (which we shall do henceforth), then $p_a \in {}^n\omega$ for some $n \in \omega$ and p^*_a is a set of function pairs suitable for a. For any potential condition p and any set d, closed term t, or limited sentence φ we denote $p \restriction d$, $p \restriction dm(t)$, and $p \restriction dm(\varphi)$, by p^d, p^t, and p^φ respectively. We define the *height* of a potential condition p by

$$ht(p) = \max(\{ht(a) : a \in dm(p)\}),$$

and if p and q are any two potential conditions such that

$$a \in dm(p) \rightarrow p_a \subseteq q_a \wedge p_a^* \subseteq q_a^*,$$

we say that $p \prec q$ (q *extends* p). Finally, if $p \prec q$ and, for some set d, closed term t, or limited sentence φ, we have $dm(q) \subseteq d$, $dm(q) \subseteq dm(t)$ or $dm(q) \subseteq dm(\varphi)$, then we say that $p \prec_d q$ (q *d-extends* p), $p \prec_t q$, or $p \prec_\varphi q$, respectively.

We now begin our inductive definition of *condition* and *forcing* ($\Vdash$). The induction will be on the height of potential conditions and the index (under the lexicographical order) of formulas.

3.1. (1) Height $= 0$.

(a) The empty potential condition 0 is a condition.

(b) If φ has height 0, then $0 \Vdash \varphi$ iff $\Vdash_{\mathcal{M}} \varphi$ (where $\bigwedge_\alpha^0$ is interpreted as ranging over sets of rank less than α).

We now assume we have defined condition and forcing for all potential conditions and sentences of height less than α.

(2) A potential condition p of height α is a condition iff for each $a \in dm(p)$ and each function pair $\langle t, k\rangle \in p_a^*$ we have either:

(a) $ht(t) < ht(a)$ and:

(i) $p^t \Vdash (t \in {}^\omega\omega \wedge \bigwedge^t x(x \neq^t t))$, and

(ii) $\forall m > k[m \in dm(p_a) \rightarrow \exists n(n < p_a(m) \wedge p^t \Vdash \langle m, n\rangle {}^\frown \in t)]$; or

(b) $ht(t) = ht(a)$, t is f_b for some $b < a$, and $\forall m > k[m \in dm(p_a) \rightarrow p_b(m) < p_a(m)]$.

(3) If p is any condition of height α, and φ is any sentence of height less than α, then we set $p \Vdash \varphi$ iff $p^\varphi \Vdash \varphi$. Thus we need only define the relation forcing between conditions of height at most α and limited sentences of height α. We do this by induction on rank, and for sentences of the same rank, by induction on depth and type. Thus we may assume that we have already defined forcing by conditions p such that $ht(p) \leqq \alpha$ for all limited sentences φ such that $ht(\varphi) \leqq \alpha \wedge rk(\varphi) < \beta$.

(4) Let p be a condition of height at most α, and let φ be a limited sentence of height α and rank β. Then we set $p \Vdash \varphi$ iff:

(a) φ is $\psi \wedge \theta$, and both $p \Vdash \psi$ and $p \Vdash \theta$;

(b) φ is $\bigwedge_\gamma^a v_i \psi(v_i)$, and for every term t such that $md(t) \leqq a$ and $rk(t) < \gamma$ we have $p \Vdash \psi(t)$;

(c) φ is $\neg\psi$ and there is no condition q such that $p^\varphi \prec_\varphi q$ and $q \Vdash \psi$.

(d) φ is $t \in \hat{c}$, and for every condition q which φ-extends p^φ, there is a set $b \in c$ and a condition r such that $q \prec_\varphi r$ and $r \Vdash t \cong \hat{b}$.

(e) φ is $t \in \{v_i :_\gamma^a \psi(v_i)\}$, and for every condition q which φ-extends p^φ there is a condition r and a term u such that $q \prec r$, $md(u) \leqq a$, $rk(u) < \gamma$, $r \Vdash \psi(u)$, and $r \Vdash u \cong t$;

(f) φ is $A^b(t, u, v)$, and for every condition q which φ-extends p^φ there is a set $a \leqq b$, a pair of integers m and n, and a condition r such that $q \prec_\varphi r$, $r_a(m) = n$, and $r \Vdash (t \cong \hat{a} \wedge u \cong \hat{m} \wedge v \cong \hat{n})$.

It should be noted that by using universal quantifiers as primitives we are approximating the formulation of forcing often referred to as weak forcing (due to Solovay) rather than the formulation in [**1**]. The following, while obvious in single extensions, requires some proof in our case.

3.2. LEMMA. *If φ is any limited sentence and p is any condition, then:*

(1) *For any $d \subseteq A^*$ such that d is closed ($d = cl(d)$), p^d is a condition.*

(2) *If q is any condition which extends p, then $p \Vdash \varphi \rightarrow q \Vdash \varphi$.*

(3) $p \Vdash \varphi \leftrightarrow p^{\varphi} \Vdash \varphi$.

(4) *If d is a closed set and q is a condition which d-extends p^d, then $(p - p^d) \cup q$ is a condition.*

(5) *In the definition of forcing the phrases "φ-extension" and "φ-extension of p^{φ}" may be replaced by "extension" and "extension of p."*

(6) *p forces $\neg\varphi$ iff no extension of p forces φ.*

(7) *p forces φ iff no extension of p forces $\neg\varphi$.*

PROOF. (1) follows from the definition of condition. For (2) we use induction on the index of φ beginning with the observation that if $ht(\varphi) = 0$ then $p \Vdash \varphi \leftrightarrow 0 \Vdash \varphi \leftrightarrow q \Vdash \varphi$ and then checking the induction clauses, noting that a φ-extension of q^{φ} is also a φ-extension of p^{φ}. One direction of (3) follows directly from (2); the other follows easily from the definition of forcing. (4) follows from (2) and the definition of condition; (5) follows from (2), (3), and (4); and (6) follows from (5). (7), the property that distinguishes this as weak forcing, is proven by induction using the definitions of forcing and condition as modified by (5). □

For sentences of $\mathscr{L}$ which are not limited, we define forcing in the usual manner by induction on depth with the added condition that $p \Vdash \bigwedge^{a}_{\alpha} v_i \varphi(v_i)$ iff for every term t we have $rk(t) < \alpha \wedge md(t) \leqq a \rightarrow p \Vdash \varphi(t)$.

4. Definability in $\mathscr{M}$ and other properties of forcing. Having defined forcing and condition outside of $\mathscr{M}$, we must now show that for limited statements forcing and condition are definable in $\mathscr{M}$. We assume therefore that we have encoded $\mathscr{L}$ in $\mathscr{M}$ in some reasonable way and that we have extended this encoding to potential conditions. For every sentence φ and every potential condition p we denote the sets in M (the universe of $\mathscr{M}$) which represent them by $\ulcorner\varphi\urcorner$ and $\ulcorner p\urcorner$. We also extend forcing to M by setting

$$a \Vdash b \leftrightarrow \exists p \exists \varphi (a = \ulcorner p\urcorner \wedge b = \ulcorner\varphi\urcorner \wedge p \Vdash \varphi).$$

Thus we are looking for a formula $\mathscr{F}(x, y)$ in the language of first order set theory with parameters from M (which, for the remainder of this section, we shall refer to as $\mathscr{L}^*$) such that for any $a, b \in M$, we have $\models_{\mathscr{M}} \mathscr{F}(a, b)$ iff for some condition p and some limited sentence φ it is the case that $a = \ulcorner p\urcorner$, $b = \ulcorner\varphi\urcorner$, and $p \Vdash \varphi$. We note that we cannot directly define $\mathscr{F}(x, y)$ by the usual method of transfinite induction because we defined forcing by induction on the well ordering generated by the index of formulas, and it is not true that for every limited sentence φ the set $\{\ulcorner\theta\urcorner : in(\theta) < in(\varphi)\}$ belongs to M.

We begin the construction of $\mathscr{F}$ by introducing some new notation. For each $\alpha < ht(A)$ we let

$$\Pi_\alpha = \{\ulcorner p \urcorner : p \text{ is a condition} \wedge ht(p) \leqq \alpha\},$$
$$\Pi_\alpha^- = \bigcup_{\beta \in \alpha} \Pi_\beta,$$
$$\Phi_\alpha = \{\ulcorner \varphi \urcorner : \varphi \text{ is a limited sentence} \wedge ht(\varphi) \leqq \alpha\}.$$

We then let $\Pi = \bigcup_\alpha \Pi_\alpha$ and $\Phi = \bigcup_\alpha \Phi_\alpha$. While it is obvious that Φ cannot belong to M, we shall show that Π does belong to M. We first note that since height, rank, model, etc. can be defined in $\mathscr{M}$, there must be formulas $\mathscr{L}(x)$ and $\mathscr{P}(x)$ in $\mathscr{L}^*$ such that for every $a \in M$ we have

$$\models_{\mathscr{M}} \mathscr{L}(a) \leftrightarrow a \in \Phi$$

and

$$\models_{\mathscr{M}} \mathscr{P}(a) \quad \text{iff} \quad \exists p\, [p \text{ is a potential condition and } \ulcorner p \urcorner = a].$$

Now, from the standard arguments, it is clear that there is a formula $\mathscr{E}(x, y, z)$ in $\mathscr{L}^*$ which has the property that for any ordinal α and any set $P \in M$ if $\Pi_\alpha \subseteq P \subseteq \Pi$ and $y \in \Phi_\alpha$, then

$$\models_{\mathscr{M}} \mathscr{E}(P, x, y) \leftrightarrow x \in P \wedge x \Vdash y.$$

This follows from the fact that if the set of conditions is given, forcing may be defined by induction on rank rather than on index using the definition of forcing as modified by 3.2(5). Thus, if we show that $\Pi \in M$, then $\mathscr{E}(\Pi, x, y)$ will be a formula of $\mathscr{L}^*$ which defines forcing in $\mathscr{M}$.

4.1. LEMMA. *The set Π belongs to M.*

PROOF. We shall use transfinite induction to prove that there exists in $\mathscr{M}$ a function $F: ht(A) \to \{\Pi_\alpha\}$ such that $F(\alpha) = \Pi_\alpha$. Then Π will be the union of the range of F. Since $F(0) = \{0\}$, it is sufficient to first find a formula $\mathscr{G}(x, y, z) \in \mathscr{L}^*$ such that

$$\models_{\mathscr{M}} \mathscr{G}(\Pi_\alpha^-, \alpha, a) \quad \text{iff} \quad a \in \Pi_\alpha,$$

and then to prove that $\Pi_\alpha^- \in M \to \Pi_\alpha \in M$. We begin by noting that we may let $\mathscr{G}(x, y, z)$ be the formula

$$\forall p[p \in z \leftrightarrow \mathscr{P}(p) \wedge ht(p) \leqq y \wedge (p \text{ satisfies 3.1(2) with ``} t \Vdash u\text{'' replaced by } \mathscr{E}(x, t, u))].$$

Now assume that $\Pi_\alpha^- \in M$. We see from the construction of $\mathscr{G}$ that to prove that $\Pi_\alpha \in M$, given that $\Pi_\alpha^- \in M$, it is sufficient to prove that the set

$$U = \{\langle p, t\rangle : p \in \Pi_\alpha^- \wedge ht(t) < \alpha \wedge \mathscr{E}(\Pi_\alpha^-, p, \ulcorner t \in {}^\omega\omega \wedge {}^t v(v \neq {}^t t)\urcorner)\} \in M.$$

However, looking at U and using the fact that both Π_α^- and A^* belong to M, we see that U will belong to M unless for some fixed $p \in \Pi_\alpha^-$ and $a \in \{a \in A^* : ht(a) < \alpha\}$, the set

$$V = \{t : md(t) = a \wedge \mathscr{E}(\Pi_{\alpha^-}^-, p, \ulcorner t \in {}^{\omega}\omega \wedge \bigwedge^t v(v \neq^t t)\urcorner)\} \notin M.$$

Again, it is easy to see that if $V \notin M$, then there is no ordinal β such that $\forall t \in V(rk(t) < \beta)$. Let $\aleph$ be some fixed infinite cardinal (for the remainder of the proof of this lemma all references to cardinality and all computations are to be taken with respect to $\mathscr{M}$) greater than $2^{\Pi_\alpha^-}$. By the above, there is a set $T = \{t_\gamma : \gamma \in \aleph\} \subseteq V$ such that $\delta \in \gamma \in \aleph$ implies $rk(t_\delta) < rk(t_\gamma)$. Now for each $t \in T$ we define

$$K_t = \{\langle q, m, n\rangle : m, n \in \omega \wedge p \prec q \wedge \mathscr{E}(\Pi_\alpha^-, q, \ulcorner\langle m, n\rangle^\frown \in t\urcorner)\},$$
$$K = \{K_t : t \in T\}, \qquad P = \Pi_\alpha^- \times \omega \times \omega.$$

Thus, for each $t \in T$ we have $K_t \subseteq P$, so, since $T \in M$ implies $K \in M$, we have $K \subseteq pw(P)$. Hence

$$cd(K) \leqq cd(pw(P)) = 2^P < \aleph = cd(T).$$

Therefore there exist $t, u \in T$ such that $K_t = K_u$. As no two members of T have the same rank we may assume $rk(t) < rk(u)$. But $u \in V$ implies $p \Vdash \bigwedge^u v(v \neq^u u)$, so $p \Vdash t \neq {}^u u$. Thus there is an extension q of p such that $ht(q) < \alpha$, and there are ordinals $m, n \in \omega$ such that $q \Vdash \langle m, n\rangle^\frown \in t$ and $q \Vdash \langle m, n\rangle^\frown \notin u$. Hence $\langle q, m, n\rangle \in K_t - K_u$ which is a contradiction. $\square$

Finally, using $\mathscr{F}(x, y)$ for the atomic cases we can show by induction that for any formula $\psi(v_i, \ldots, v_n) \in \mathscr{L}$ there is a formula $\varphi_\psi(p, x_i, \ldots, x_n) \in \mathscr{L}^*$ such that $p \Vdash \psi(t_i, \ldots, t_n)$ iff $\vDash_{\mathscr{M}} \varphi_\psi(p, \text{“}t_1\text{”}, \ldots, \text{“}t_n\text{”})$. Thus our construction, in spite of many superficial differences, has the essential properties of the original Cohen construction.

Another property common to many Cohen constructions is the so-called combinatorial lemma or countable chain condition. In the following, all computations are again to be taken with respect to $\mathscr{M}$, and although we shall speak of sets of conditions, etc. rather than subsets of Π, it should be clear that all our proofs could be carried out within $\mathscr{M}$.

We begin by defining two potential conditions p and q to be *weakly incompatible* iff for some $b \in (dm(p) \cap dm(q))$ we have $p_b \neq q_b$. We define the *size* (sz) of a potential condition to be the number of elements in its domain. The usefulness of these notions follows from

4.2. LEMMA. *If P is a set of pairwise weakly incompatible potential conditions each of size n, then P is countable.*

PROOF. Induction on n. For any given $a \in A$ the set of possible values for p_a is countable, so the lemma holds for $n = 1$. Now assume it holds for all $k < m$, and assume that there is a set P containing only potential conditions of size m for which it fails. Let p be some fixed member of P, and let $D = \{a_1, a_2, \ldots, a_m\}$ be its domain. Since every member of $P - \{p\}$ is weakly imcompatible with p, every member of $P - \{p\}$ contains a member of D in its domain. Thus there is a fixed $a \in D$ such that the set $Q = \{q \in P : a \in dm(q)\}$ is uncountable. However, for any potential condition p there are only a countable number of possible values for p_a. Therefore, there is a function f such that the set $R = \{q \in Q : p_a = f\}$ is uncountable. Then the set $\{p \restriction (A - \{a\}) : p \in R\}$ is an uncountable set of pairwise weakly-incompatible potential conditions each of size $m - 1$, which contradicts our induction hypothesis. □

We now adopt the standard definition of compatibility (two conditions p and q are *compatible* iff there is a condition r which extends them both) and prove

4.3. COMBINATORIAL LEMMA. *Every set of pairwise incompatible conditions is countable.*

PROOF. We first note that if two conditions p and q are incompatible then they are also weakly incompatible for, if not, the condition r defined by

$$\begin{aligned} r(b) &= p(b) && \text{if } b \in dm(p) - dm(q), \\ &= q(b) && \text{if } b \in dm(q) - dm(p), \\ &= \langle p_b, p_b^* \cup q_b^* \rangle && \text{if } b \in dm(p) \cap dm(q), \end{aligned}$$

would extend them both. Now let P be any set of pairwise incompatible conditions. For each $m \in \omega$ let $P_m = \{p \in P : sz(p) = m\}$. Then, by the above for each $m \in \omega$, P_m is a set of pairwise weakly incompatible potential conditions and is, therefore, by the previous lemma, countable. Hence $P = \bigcup_{m \in \omega} P_m$ is a countable union of countable sets and is, therefore, itself countable. □

Finally, we shall need to know that if we have a condition p, then for any $a \in A$ we can extend p in such a way that the domain of p_a will be increased, and, on its additional domain, p_a will be arbitrarily large. Thus we prove

4.4. LEMMA. *For any condition p, any $a \in A$, and any $m \in \omega$, if $a \notin dm(p)$ or $m \notin dm(p_a)$, then for any $n \in \omega$ there exists a condition $q \succ p$ such that $q_a(m) > n$ and $b \nleqq a \rightarrow q(b) = p(b)$.*

PROOF. It is clear that if $a \notin dm(p)$ and f is any member of ${}^{m+1}\omega$ such that $f(m) > n$, then we can set $q = p \cup \{\langle a, \langle f, 0 \rangle \rangle\}$.

If, however, $a \in dm(p)$, the situation is much more complicated. Thus we shall consider a, m and n, as fixed and shall look at all conditions p such that $a \in dm(p)$ and $m \notin dm(p_a)$. For the purposes of this lemma we define a function pair $\langle t, k \rangle$ to be *pathological* iff $ht(t) < ht(a)$, $\langle t, k \rangle \in \bigcup \{p_b^* : b \leqq a \wedge ht(b) = ht(a)\}$, and for some $k < i \leqq m$ there is no $j \in \omega$ such that $p^t \Vdash \langle i, j \rangle^\frown \in t$. We define a condition to be *pathological* if it contains at least one pathological function pair. We first prove the lemma for nonpathological conditions using induction on the number of elements in the finite set

$$S_p = \{b : b < a \wedge ht(b) = ht(a) \wedge \exists k \exists c \in dm(p) [\langle f_b, k \rangle \in p_c^* \wedge m \notin dm(p_b)]\}.$$

If S is empty, it is clear that a condition q can be found satisfying the conditions of the lemma. Now assume that the lemma is true for any condition r such that S_r has n elements, and let p be a condition such that S_p has $n + 1$ elements. Let b be a minimal element of S_p. Then again it is clear (because p is nonpathological) that there is a function $f \in {}^{m+1}\omega$ such that $p_b \subseteq f$, $q = (p - \{p(b)\}) \cup \{\langle b, \langle f, p_b^* \rangle\rangle\}$ is a nonpathological condition, and S_q has n elements. Hence the lemma holds for q and, therefore, for p.

To prove the lemma for pathological conditions we again use induction, this time on the number of pathological function pairs in p. We begin by assuming the lemma is true for conditions with fewer than n pathological function pairs and letting p be a condition with n such pairs. Let $\langle t, k \rangle \in p_b^*$ be a pathological function pair. Since $\langle t, k \rangle \in p_b^*$, we know that $p^t \Vdash t \in {}^{\omega}\omega$. Thus there is a condition r which t-extends p^t such that $i \leqq m \to \exists j \in \omega(r \Vdash \langle i, j \rangle\char`^ \in t)$. Now let $s = (p - p^t) \cup r$. Since $ht(t) < \alpha$ and $md(t) < a$, we have $(b \nleqq a \vee ht(b) = \alpha) \to p(b) = s(b)$. Thus s has no pathological pairs which were not in p and, therefore, has fewer than n such pairs. Hence by our induction hypothesis there is a condition $q \succ s \succ p$ such that $b \nleqq a \to q(b) = s(b) = p(b)$ and $q_a(m) > n$. $\square$

5. Construction of $\mathcal{N}$. Having now shown that forcing is definable in $\mathcal{M}$ for all limited sentences and for any fixed unlimited sentence, and that $\Pi \in M$, we shall, for the remainder of this construction unless otherwise specified, identify conditions and formulas with their codes and perform all operations and computation within $\mathcal{M}$. Thus we shall assume that height, domain, etc. are also defined in $\mathcal{M}$, and whenever we talk about some q extending a condition p, we shall assume, unless explicitly stated otherwise, that q is also a condition.

We shall not use the definition of complete sequence found in [**1**] but rather one due to Solovay. We define a set P of conditions to be *dense over a condition p* iff

$$\forall q \succ p \, \exists r \in P(r \succ q).$$

A set of conditions will be said to be *dense* if it is dense over O. We define a sequence $\{p_i\}_{i \in \omega}$ of conditions to be *complete* if $i \prec j \to p_i \prec p_j$ and for every dense set of conditions P there is some $i \in \omega$ such that $p_i \in P$. It is well known that the countability of M guarantees the existence of complete sequences, so at this point we choose a complete sequence $\mathscr{C}$ which will remain fixed for the remainder of the proof of Theorem 1.1. We note that $\mathscr{C}$ is not required to be (and cannot be) in M. The usefulness of this definition of complete sequence follows from

5.1. LEMMA. *If P is any set of conditions which is dense over a condition $p \in \mathscr{C}$, then there is an extension q of p which belongs to $\mathscr{C} \cap P$.*

PROOF. Let $Q = \{q : \neg \exists r(p \prec r \wedge q \prec r)\}$ and let $R = \{q \in P : p \prec q\}$. Then $Q \cup R$ is dense, for if r is any condition then either r is incompatible with p in which case r itself is an extension of r which belongs to $Q \subseteq R \cup Q$, or r is compatible with p in which case there is a condition s which extends both p and r and thus a condition $q \in R \subseteq R \cup Q$ which extends s and therefore r. But if $R \cup Q$ is dense, there is a condition

$$q \in (R \cup Q) \cap \mathscr{C} = (R \cap \mathscr{C}) \cup (Q \cap \mathscr{C}).$$

However, q cannot belong to both Q and $\mathscr{C}$ because if q and p both belong to $\mathscr{C}$, they must be compatible. Hence $q \in R \cap \mathscr{C}$ and thus is a member of $\mathscr{C}$ which belongs to P and extends p. □

We now define a set

$$\mathbf{A} = \{\langle a, m, n\rangle : \exists p \in \mathscr{C}(p_a(m) = n)\},$$

and we define $\mathscr{N}$ to be $\mathscr{M}(\mathbf{A})$. Interpreting $\mathscr{L}$ in $\mathscr{N}$ inductively, using the notation $\boldsymbol{t}$ for the set in $\mathscr{N}$ denoted by t, and $^{\boldsymbol{\omega}}\boldsymbol{\omega}$ for $({}^{\omega}\omega)_{\mathscr{N}}$ we set

5.2. (1) $\hat{\boldsymbol{a}} = a$,

(2) $\{v :^{a}_{\alpha} \varphi(v)\} = \{\boldsymbol{t} : md(t) \leqq a \wedge rk(t) < \alpha \wedge \models_{\mathscr{N}} \varphi(t)\}$,

(3) $\models_{\mathscr{N}} t \in u$ iff $\boldsymbol{t} \in \boldsymbol{u}$,

(4) $\models_{\mathscr{N}} A^a(t, u, v)$ iff $\langle \boldsymbol{t}, \boldsymbol{u}, \boldsymbol{v}\rangle \in \mathbf{A} \wedge \boldsymbol{t} \leqq a$,

(5) $\models_{\mathscr{N}} \bigwedge^{a}_{\alpha} x\varphi(x)$ iff $rk(t) < \alpha \wedge md(t) \leqq \alpha \to \models_{\mathscr{N}} \varphi(t)$,

(6) $\models_{\mathscr{N}} \bigwedge x\varphi(x)$ iff for every term t it is the case that $\models_{\mathscr{N}} \varphi(t)$.

Now again using induction one may prove

5.3. TRUTH LEMMA. *If φ is any sentence, and t, u, and v are any closed terms, then*

(1) $\models_{\mathscr{N}} \varphi$ *iff* $\exists p \in \mathscr{C}(p \Vdash \varphi)$;

(2) $\boldsymbol{t} = \boldsymbol{u}$ *iff* $\exists p \in \mathscr{C}(p \Vdash t \cong u)$;

(3) $\boldsymbol{t} \in \boldsymbol{u}$ *iff there exists a closed term w such that $rk(w) < rk(u)$, $md(w) \leqq md(u)$, $\boldsymbol{w} = \boldsymbol{t}$, and $\boldsymbol{w} \in \boldsymbol{u}$;*

(4) *if $md(t), md(u), md(v) \leqq a$ and $rk(t), rk(u), rk(v) \leqq \alpha$, then*

(a) $\boldsymbol{t} = \boldsymbol{u}$ *iff* $\exists p \in \mathscr{C}(p \Vdash t =^{a}_{\alpha} u)$, *and*

(b) $\boldsymbol{t} = \langle \boldsymbol{u}, \boldsymbol{v}\rangle$ *iff* $\exists p \in \mathscr{C}(p \Vdash t =^{a}_{\alpha} \langle u, v\rangle)$;

(5) $\boldsymbol{t} \in {}^{\boldsymbol{\omega}}\boldsymbol{\omega}$ *iff* $\exists p \in \mathscr{C}(p \Vdash t \in {}^{\omega}\omega)$;

(6) $\boldsymbol{f_b} = \{\langle m, n\rangle : \exists p \in \mathscr{C}(p \Vdash A^b(\hat{b}, \hat{m}, \hat{n}))\}$. □

5.4. LEMMA. *$\mathscr{N}$ is a normal extension of $\mathscr{M}$.*

PROOF. It is a well-known consequence of Lemmas 4.3 and 5.3 that $\mathscr{N}$ is a regular extension of $\mathscr{M}$. However, determining cardinality in $\mathscr{N}$ is more difficult. It will be necessary to find a smaller set of conditions $\Pi^* \subset \Pi$ which we shall be able to use in place of Π. We shall assume, for the remainder of the proof of this lemma, that all computations (pw, ${}^{B}A$, A^B, etc.) are to be taken with respect to $\mathscr{M}$.

We begin by defining two terms t and u to be **-equivalent* ($t =^* u$) iff

$$md(t) = md(u) \wedge \forall p(p \Vdash t \in {}^{\omega}\omega \leftrightarrow p \Vdash u \in {}^{\omega}\omega) \wedge \forall p(p \Vdash t \in {}^{\omega}\omega \to p \Vdash t \cong u).$$

We note that * equivalence is an equivalence relation, and for each term t we define t^* to be some fixed member of minimal rank of the equivalence class generated by t (where we assume, of course, that $t =^* u \leftrightarrow t^* = u^*$).

For any condition p we define p^* to be p with all terms t appearing in function pairs replaced by t^*. We note that it follows easily, by induction, that if p is any condition, then p^* is also a condition and, furthermore, if φ is any limited sentence, then $p \Vdash \varphi$ iff $p^* \Vdash \varphi$. Now for each $\alpha < ht(A)$ let

$$\Pi^*_\alpha = \{p^* : ht(p) \leqq \alpha\}, \quad T_\alpha = \{t^* : ht(t) \leqq \alpha \wedge \exists p(p \Vdash t \in {}^{\omega}\omega)\},$$
$$\Pi^* = \bigcup\{\Pi_\alpha : \alpha < ht(A)\}, \; T = \{T_\alpha : \alpha < ht(A)\}.$$

In order to prove that $\Pi^* \in M$ and $cd(\Pi^*) = 2^{\aleph_0}$, we prove by induction that for each α we have $\Pi^* \in M$, $T_\alpha \in M$, $cd(T_\alpha) \leqq 2^{\aleph_0}$, and $cd(\Pi^*_\alpha) \leqq 2^{\aleph_0}$. Thus we assume that for each $\beta < \alpha$ we have $\Pi^*_\beta \in M$, $T_\beta \in M$, $cd(\Pi^*_\beta) \leqq 2^{\aleph_0}$, and $cd(T_\beta) \leqq 2^{\aleph_0}$. For each term t such that $ht(t) \leqq \alpha$ and each $m, n \in \omega$, we let

$$R_t^{mn} = \{p \in \Pi^*_\alpha : p \Vdash (t \in {}^{\omega}\omega \wedge \langle m, n\rangle\hat{} \in t)\},$$

and we let S_t^{mn} be any maximal pairwise-incompatible subset of R_t^{mn}. Finally, we let

$$K_t = \{\langle p, m, n\rangle : p \in S_t^{mn}\}.$$

We note that by Lemma 4.3 each S_t^{mn} is a countable subset of $\Pi^*_a \times \omega \times \omega$. Thus the set $K = \{K_t : ht(t) = \alpha\}$ is a subset of the countable (in $\mathscr{M}$) subsets of $\Pi^*_\alpha \times \omega \times \omega$ and therefore belongs to M and has cardinality no greater than $(2^{\aleph_0 \cdot \aleph_0})_{\mathscr{M}} = (2^{\aleph_0})_{\mathscr{M}}$. Now let t^* and u^* be any two distinct members of T_α. We wish to show that $K_{t^*} \neq K_{u^*}$. We begin by noting that since t^* and u^* are distinct and of height less than or equal to α, there must be a condition $p \in \Pi^*_\alpha$ which either forces $t^* \in {}^{\omega}\omega \wedge u^* \in {}^{\omega}\omega$ but does not force $t^* \cong u^*$, or which forces, say $t^* \in {}^{\omega}\omega$, but not $u^* \in {}^{\omega}\omega$. In either case there must exist (since we are using weak forcing) $m, n \in \omega$, and a condition $q \in \Pi^*_\alpha$ such that $q \Vdash (t^* \in {}^{\omega}\omega \wedge \langle m, n\rangle\hat{} \in t^*)$, and either $q \Vdash u^* \notin {}^{\omega}\omega$ or $q \Vdash \langle m, n\rangle\hat{} \notin u^*$. Thus there is some condition r such that $\langle r, m, n\rangle \in K_{t^*}$ and r is compatible with q. But if $\langle r, m, n\rangle \in K_{u^*}$ we would have $r \Vdash u^* \in {}^{\omega}\omega$ and $r \Vdash \langle m, n\rangle\hat{} \in u^*$ and thus r could not be compatible with q. Hence $K_{t^*} \neq K_{u^*}$. Thus $cd(T_\alpha) \leqq cd(K) \leqq (2^{\aleph_0})_{\mathscr{M}}$.

Now look at any cardinal $\aleph_\alpha$. For each closed term t and each $\beta \in \aleph_\alpha$ we can find a set R_t^β which is a maximal set of pairwise-incompatible conditions with respect to the requirement that $R_t^\beta \in \Pi^*$ and $p \in R_t^\beta \to p \Vdash \hat{\beta} \in t$. Let $S_t^\beta = R_t^\beta \times \{\beta\}$, $K_t = \bigcup\{S_t^\beta : \beta \in \aleph_2\}$, and $K = \{K_t : t \in \mathscr{L}\}$. Because each R_t^β is countable (by Lemma 4.3), we have $cd(K_t) \leqq \aleph_\alpha$ for each $K_t \in K$. Thus if, for any set A, we define $pw_\alpha(A)$ to be the set $\{B \subseteq A : cd(B) \leqq \aleph_\alpha\}$, then we see that, for each closed term t, $K_t \in pw_\alpha(\Pi^* \times \aleph_\alpha)$. Therefore, we have

$$cd(K) \leqq cd(pw_\alpha(\Pi^* \times \aleph_\alpha)) = (\Pi^* \times \aleph_\alpha)^{\aleph_\alpha} = \max(2^{\aleph_0 \cdot \aleph_\alpha}, \aleph_\alpha^{\aleph_\alpha}) = 2^{\aleph_\alpha}.$$

But, using an argument analogous to that used in the first part of this proof, we note that if t and u are two distinct subsets of $\aleph_\alpha$ in $\mathscr{N}$, then $K_t \neq K_u$. Thus $[cd(pw(\aleph_\alpha))]_{\mathscr{M}} = [cd(pw(\aleph_\alpha))]_{\mathscr{N}}$. $\square$

6. The embedding of $\mathscr{A}$. Let $S = \{f_a : a \in A\}$. Then since $S \subseteq {}^{\omega}\omega$ we may speak of the partial order $\mathscr{S} = \langle S, \prec\rangle$.

6.1. LEMMA. *$\mathscr{A} = \langle A, <_A\rangle$ is order-isomorphic to $\mathscr{S}$ under the mapping $a \to f_a$.*

PROOF. First choose any $a, b \in A$ such that $a < b$. To prove that $\boldsymbol{f_a} \prec \boldsymbol{f_b}$ we need only show that the set of conditions which forces $f_a \prec f_b$ is dense. But let p be any condition. Then we can define a new condition q by letting

$$\begin{aligned} q(c) &= p(c) && \text{if } c \neq b, \\ &= \langle b, \langle \{\langle 0, 0\rangle\}, \{\langle f_a, 0\rangle\}\rangle\rangle && \text{if } c = b \text{ and } b \notin dm(p), \\ &= \langle b, \langle p_b, p_b^* \cup \{\langle f_a, dm(p_b)\rangle\}\rangle\rangle && \text{otherwise.} \end{aligned}$$

It is clear that $p \prec q$ and $q \Vdash f_a \prec f_b$.

Now assume that for some $a, b \in A$ we have $\boldsymbol{f_a} \prec \boldsymbol{f_b}$ but $a \nless b$. Then there must be some condition $p \in \mathscr{C}$ and some $k \in \omega$ such that $p \Vdash \forall m > k(f_a(m) < f_b(m))$. If we let $n = \max(dm(p_a), k + 1)$, then it is clear that p must force $f_a(n) < f_b(n)$. But by Lemma 4.4 there is a condition $q \succ p$ such that $n \in dm(q_b)$ and $q(a) = p(a)$. Now again using Lemma 4.4 there is a condition $r \succ q$ such that $r_a(n) > r_b(n)$. Hence $r \Vdash (\boldsymbol{f_a}(n))\hat{} > (\boldsymbol{f_b}(n))\hat{}$. But $r \succ p$, so $r \Vdash (\boldsymbol{f_a}(n))\hat{} < (\boldsymbol{f_b}(n))\hat{}$ which is a contradiction. □

It should be noted that we have not yet made use of the fact that every countable subset of A is bounded. Thus if a did not satisfy this condition, we could still construct $\mathscr{N}$ and embed $\mathscr{A}$ in Ω. However, we do need the condition in order to show that $\mathscr{S}$, the resulting image, is a scale. We shall do this by showing that if $f \in {}^\omega\omega$ then there is some term t such that $f = \boldsymbol{t}$ and $md(t) \in A$. (It is absolutely crucial here that $md(t) \neq A$.) We begin by defining a condition to be *consistent* iff for each $a \in dm(p)$ we have $p_a \subseteq \boldsymbol{f_a}$, and for every function pair $\langle t, k\rangle \in p_a^*$ we have $\boldsymbol{t} \in {}^\omega\omega$ and $n > k \to \boldsymbol{t}(n) < \boldsymbol{f_a}(n)$. Thus a condition is consistent iff the "information" it contains is correct.

6.2. LEMMA. *A condition p is consistent iff there exists a condition $u \in \mathscr{C}$ such that $p \prec u$.*

PROOF. Let p be consistent. To prove the existence of u we use induction on $sz(p)$. Assume the lemma is true for all conditions of size less than n and let p be a condition of size n. We begin by choosing some maximal element $a \in dm(p)$ and using our induction hypothesis to obtain a condition $q \in \mathscr{C}$ such that $q \succ p \upharpoonright (dm(p) - \{a\})$. Now we note that

$$R = \{r : \langle t, k\rangle \in p_a^* \to \exists n_k \geqq k(\langle t, n_k\rangle \in r_a^*)\}$$

is clearly dense over q, so let r be any condition such that $q \prec r$ and $r \in \mathscr{C} \cap R$. Since $\mathscr{C}$ is complete, and all abstraction terms in function pairs belonging to p_a^* are forced by q to be functions from ω into ω, there is a condition $s \in \mathscr{C}$ such that $s \succ r$, $dm(p_a) \subseteq dm(s_a)$, and for each function pair $\langle t, k\rangle \in p_a^*$ it is the case that $k < m \leqq n_k$ implies $s^t \Vdash \langle m, \boldsymbol{t}(n)\rangle\hat{} \in t$. Furthermore, since $s \in \mathscr{C}$ and p is consistent, it is clear from the truth lemma that $p_a \subseteq s_a$, and for any condition $u \succ s$, the function

$$u \upharpoonright (A - a) \cup \{\langle u_a, u_a^* \cup p_a^*\rangle\}$$

is a condition. Thus the set of conditions $U = \{u : p_a^* \subseteq u_a^*\}$ is dense over s. Hence there is a condition $u \succ s$ such that $u \in U \cap \mathscr{C}$. But since $u \in U$ and $s \prec u$ it is clear that $p \prec u$. If, on the other hand, $p \prec u$ for some $u \in \mathscr{C}$, then it follows from the truth lemma that p is consistent. $\square$

We next note that consistency is definable in $\mathscr{N}$ and, more importantly, for each $a \in A$ there is a limited formula $\mathscr{C}_a$ such that both $md(\mathscr{C}_a) = a$, and for each condition p if $b \in dm(p) \to b \leqq a$, then $\models_{\mathscr{N}} \mathscr{C}_a(p)$ iff p is consistent. This follows from the fact that the consistency of such a condition depends only on the functions corresponding to elements of A which are less than or equal to a.

6.3. LEMMA. *For every $f \in {}^{\omega}\omega$ there exists a term t such that $f = \boldsymbol{t}$ and $md(t) \in A$.*

PROOF. From the construction of $\mathscr{N}$ it is clear that we can choose some term u such that $\boldsymbol{u} = f$. If $md(u) \in A$ we are done. Otherwise, let $P_{mn} = \{p : p \Vdash \langle m, n\rangle\hat{} \in u\}$. Let Q_{mn} be any maximal pairwise-incompatible subset of P_{mn} and let $R_{mn} = \{\langle p, m, n\rangle : p \in Q_{mn}\}$. Finally, let $Q = \bigcup\{Q_{mn} : m, n \in \omega\}$ and $R = \bigcup\{R_{mn} : m, n \in \omega\}$. By Lemma 4.3 each Q_{mn} is countable; hence so are Q and R. Thus the set $D = \{a : \exists p \in Q(a \in dm(p))\}$ is a countable subset of A and, therefore, admits an upper bound $a \in A$. Now let t be the term

$$\{x :_{\alpha}^{a} \bigwedge^{\hat{\Pi}} p \bigwedge^{\hat{\omega}} m \bigwedge^{\hat{\omega}} n \bigvee^{\hat{R}} y(y =^{\hat{R}} \langle p, m, n\rangle \wedge y \in \hat{R} \wedge \mathscr{C}_a(p) \wedge x =_{\alpha}^{a} \langle m, n\rangle)\}$$

where $\alpha = rk(\mathscr{C}_a) \cup rk(\{\Pi\})$.

To prove $\boldsymbol{t} = \boldsymbol{u}$, we first show that $\boldsymbol{t} \subseteq \boldsymbol{u}$. It is clear from its construction that $\boldsymbol{t} \subseteq \omega \times \omega$, so choose any pair $\langle m, n\rangle \in \boldsymbol{t}$. Again from the construction, there must be some condition p such that $\langle p, m, n\rangle \in R$ and $\models_{\mathscr{N}} \mathscr{C}_a(p)$. Hence p is consistent, and by Lemma 6.2 there exists a condition $q \in \mathscr{C}$ such that $p \prec q$. But

$$\langle p, m, n\rangle \in R_{mn} \to p \Vdash \langle m, n\rangle\hat{} \in u.$$

Now using the fact that $p \prec q \wedge q \in \mathscr{C}$, we have $q \Vdash \langle m, n\rangle\hat{} \in u$ and thus $\langle m, n\rangle \in \boldsymbol{u}$.

To show that $\boldsymbol{u} \leqq \boldsymbol{t}$ we choose any pair $\langle m, n\rangle \in \boldsymbol{u}$. Then there must exist a condition $p \in \mathscr{C}$ such that $p \Vdash \langle m, n\rangle\hat{} \in u$. We now show that the set $Q_{mn}^* = \{q : \exists r \in Q_{mn}(r \prec q)\}$ is dense over p. Let r be any extension of p. Then by the maximality of Q_{mn} there is a condition $q \in Q_{mn}$ which is compatible with r. Thus there is a condition which both extends r and, because it extends q, belongs to Q_{mn}^*. But Q_{mn}^* dense over p implies the existence of a condition $s \succ p$ such that $s \in Q_{mn}^* \cap \mathscr{C}$. Finally, $s \in Q_{mn}^* \cap \mathscr{C}$ implies that there exists a consistent (because it is extended by s) condition $q \in Q_{mn}$. Thus $\langle q, m, n\rangle \in R$ and $\models_{\mathscr{N}} \mathscr{C}_a(q)$, so $\langle m, n\rangle \in \boldsymbol{t}$. $\square$

Now look at any $f \in {}^{\omega}\omega$. By Lemma 6.3 there is term t such that $\boldsymbol{t} = f$ and $md(t) \in A$. Let T be the set of terms $\{u : \boldsymbol{u} = f \wedge md(u) \leqq md(t)\}$. Let v be any member of T which has minimal rank (with respect to T), and let b be any member of B such that $md(v) < b$. It is clear from the truth lemma and from our choice of v that there exists a condition $p \in \mathscr{C}$ such that $p \Vdash (v \in {}^{\omega}\omega \wedge \bigwedge^{v} x(x \neq^{v} v))$. Thus the set

$$P = \{q : p \prec q \wedge \exists k(\langle v, k\rangle \in q_b^*)\}$$

is dense over p, so there must exist a condition $q \in P \cap \mathscr{C}$. Clearly $q \Vdash v \prec f_b$ so, again using the truth lemma, we have $f = \boldsymbol{v} \prec \boldsymbol{f_b}$. Thus $\mathscr{S}$ is a scale, and the proof of Theorem 1.1 is complete. □

7. Strengthening Theorem 1.1. We note that the scale $\mathscr{S}$ which we obtain may have certain additional properties not mentioned in our theorem. For example if $a < b$ and $a, b \in B$, then for any $\boldsymbol{f} \in {}^{\omega}\boldsymbol{\omega}$ constructible (in $\mathscr{N}$) from $\{\boldsymbol{f_c} : c \leqq a\}$ we have $\boldsymbol{f} \prec \boldsymbol{f_b}$. On the other hand, it is not necessarily true that for any $b \in B$ and any $\boldsymbol{f}$ constructible from $\{\boldsymbol{f_c} : a < b \wedge a \in B\}$ we have $\boldsymbol{f} \prec \boldsymbol{f_b}$. This is because terms appearing in function pairs belonging to p_b^* have quantifiers bound by fixed members $a \in A$ such that $a < b \wedge ht(a) < ht(b)$ rather than quantifiers ranging over the entire set $\{a \in B : a < b\}$. Thus, only if b has a unique predecessor in B, can we say that $\boldsymbol{f_b}$ is larger (under $\prec$) than anything constructible from $\{a \in B : a < b\}$. However, if we had constructed A^* by adding not two new elements, but rather one for each member of B, and in the ordering of A^* placed each such point just below the point in B it was associated with, then the construction of $\mathscr{N}$ could have been carried out on the enlarged set B^*. In this construction every member of B would have had a unique predecessor in B^*.

Thus, if we define a scale $\mathscr{S}$ to be *strong* iff $f \in \mathscr{S}$ implies f is larger (under $\prec$) than any function $g \in {}^{\omega}\omega$ which is constructible from $\{h \in \mathscr{S} : h \prec f\}$, then we see that had we enlarged B to B^* as just described, the scale $\{\boldsymbol{f_b} : b \in B^*\}$ would have been strong.

The situation with respect to A is not as clear. If A is well-founded then we may set $B = A$ and, using the above, construct $\mathscr{N}$ in such a way that the resulting scale $\{\boldsymbol{f_a} : a \in A\}$ is strong. If F is not well-founded, however, we are required by the procedures in our proof to choose some well-founded subset B on which we can define height and thus use as a reference for induction proofs. Because of the nature of these induction proofs, we see that we are severely limited as to which terms of height exactly equal to $ht(a)$ we can use in function pairs appearing in p_a^*. In § 3 we required such terms to be of the form f_b, but the restriction is stronger than necessary. What the proof does require is that such a term depend on only a predetermined (i.e., determined by its form) finite number of values of a finite number of the f_b in question. Thus terms equivalent to $f_b + f_c$, $f_b \circ f_c$, $t \circ f_b$ where $ht(t) < ht(a)$, etc. would all be suitable, as would any term belonging to a properly defined set of terms closed under these operations. To actually carry out the proof of Theorem 1.1 while allowing such terms in p_a^*, it would be necessary to use somewhat more complicated notation and the proof of Lemma 4.4 would become even more tedious. Unfortunately, the resulting scale would still not be strong. It would, however, have properties such as $f, g, h \in \mathscr{S} \rightarrow (f \prec h \wedge g \prec h \rightarrow f \circ g \prec h)$, etc.

At this time we do not know if Theorem 1.1 is still true if we require that the scale in question be strong. In fact we do not even know whether or not it is consistent to assume the existence of a set $\mathscr{F} = \{f_i \in {}^{\omega}\omega : i \in \omega\}$ such that, for each $i \in \omega$, f_i is larger than any $g \in {}^{\omega}\omega$ which is constructible from $\{f_j \in \mathscr{F} : j > i\}$.

8. Consequences. We now ask what kind of scales may be consistently assumed to exist and what relationships may hold between various scales in a given model. We begin with

8.1. THEOREM. *Let $\mathcal{M}$ be a countable standard model of ZFC, and let $\lambda \leqq \kappa$ be any two cardinals in $\mathcal{M}$ with cofinality greater than ω. Then if $(2^\omega)_{\mathcal{M}} \leqq \kappa$, there is a regular extension $\mathcal{N}$ of $\mathcal{M}$ such that $(2^\omega)_{\mathcal{N}} = \kappa$ and $\mathcal{N}$ contains a λ-scale.*

PROOF. Let $\mathcal{M}^*$ be a regular extension of $\mathcal{M}$ obtained by adding κ generic reals to $\mathcal{M}$ in such a way that $(2^\omega)_{\mathcal{M}^*} = \kappa$. Then by Theorem 1.1 there is a normal extension $\mathcal{N}$ of $\mathcal{M}^*$ which satisfies the condition of the theorem. □

It will be useful to look not only at scales but at other unbounded (having no upper bound under $\prec$) subsets of ${}^{\omega}\omega$. For any ordinal κ and any unbounded set $S \subseteq {}^{\omega}\omega$, we shall call S a *κ-chain* if S has order type κ under $\prec$.

8.2. THEOREM. *Let S be a scale of cardinality κ having the property that all of its subchains have cardinality κ. Then there exists a κ-scale $T \subseteq S$.*

PROOF. Let $S = \{f_\alpha : \alpha \in \kappa\}$ (where we do not assume that $\alpha \in \beta \in \kappa \to f_\alpha \prec f_\beta$). For each $\alpha \in \kappa$ let $g_\alpha = f_\beta$ where $\beta = \mu\gamma\,(f_\gamma$ is an upper bound for $\{g_\delta : \delta \in \alpha\} \cup \{f_\alpha\})$. Such an f_β must exist, for otherwise there would be a set $T \subseteq \{g_\beta : \beta \in \alpha\}$ which would be a λ-chain for some $\lambda \in \kappa$. But now the set $G = \{g_\alpha : \alpha \in \kappa\} \subseteq S$ is a κ-scale. □

8.3. COROLLARY. *The existence of a scale of cardinality $\aleph_1$ implies the existence of an $\aleph_1$-scale.*

PROOF. Every unbounded subset of a scale is nondenumerable. □

8.4. COROLLARY (HAUSDORFF). *The continuum hypothesis implies the existence of an $\aleph_1$-scale.*

PROOF. If the continuum hypothesis holds, then ${}^{\omega}\omega$ is itself a scale of cardinality $\aleph_1$, so by Corollary 8.3 there exists an $\aleph_1$-scale. □

In Theorem 8.2 we required that all subchains be of the same cardinality and that this be the cardinality of the scale in question. We now show that both of these conditions are necessary.

8.5. THEOREM. *Let $\mathcal{M}$ be a countable standard model of ZFC, and let κ and λ be two regular cardinals in $\mathcal{M}$ satisfying $\omega < \lambda < \kappa \leqq 2^\omega$ and $\kappa < \aleph_\lambda$. Then there exist normal extensions $\mathcal{N}_1$ and $\mathcal{N}_2$ of $\mathcal{M}$ containing scales S_1 and S_2 respectively which satisfy the following:*

(1) *No subchain of S_1 or S_2 is a scale.*

(2) *S_1 has κ-subchains and λ-subchains but no others.*

(3) *S_2 has only λ-subchains.*

PROOF. We use Theorem 1.1 letting $A = \kappa \times \lambda$ in both cases. To construct $\mathcal{N}_1$ we set $\langle\alpha, \beta\rangle <_A \langle\gamma, \delta\rangle$ iff $\alpha \leqq \gamma \wedge \beta \leqq \delta$, while to construct $\mathcal{N}_2$ we let $\langle\alpha, \beta\rangle <_A \langle\gamma, \delta\rangle$ iff $\alpha \in \gamma \wedge \beta \in \delta$. □

We can also ask if it is possible for a scale to have subchains of infinitely many different cardinalities without having a linear subscale.

8.6. THEOREM. *Let $\mathscr{M}$ be a countable standard model of ZFC, and let* $K = \{\kappa_i : i \in \omega\}$ *be an infinite set of regular cardinals in $\mathscr{M}$ such that* $i \in \omega \rightarrow \aleph_0 < \kappa_i \leqq 2^{\omega} < \aleph_{\kappa_i}$. *Then there is a normal extension $\mathscr{N}$ of $\mathscr{M}$ containing a scale S such that*

(1) *the set* $\{\alpha : \text{there exists a } \lambda\text{-subchain of } S\} = K$,

(2) *no subchain of S is a scale.*

PROOF. Again using Theorem 1.1 we let $A = \prod_{i \in \omega} \kappa_i$ and for $f, g \in A$ we let $f <_A g$ iff $\forall i \in \omega [f(i) \leqq g(i)]$. □

In the proofs of Theorems 8.5 and 8.6 we constructed models in which there existed certain scales with no linear subscales, but we did not claim that there were no linear scales at all in these models. We shall now show that this latter condition is implied by the former, and we shall consider the general question of the relationships between different scales existing in the same model.

8.7. LEMMA. *Let T be a scale of cardinality κ and $S \in {}^{\omega}\omega$ an unbounded set* (*scale*) *of cardinality $\lambda \in \kappa$. Then there exists an unbounded set* (*scale*) *$U \subseteq T$ of cardinality no greater than λ.*

PROOF. Since T is a scale, there is a function $\varphi \in {}^{S}T$ such that $\forall f \in S [f \prec \varphi(f)]$. Let U be the range of φ. □

8.8. THEOREM. *The existence of a κ-scale implies that no unbounded subset of ${}^{\omega}\omega$ has cardinality less than $cf(\kappa)$.* □

8.9. THEOREM. *If for some regular cardinal κ there exists a κ-scale, then every scale contains a κ-scale.*

PROOF. Let S be a κ-scale and T be any scale. We first note that $cd(T) \geqq \kappa$, for otherwise, by Lemma 8.7, there would be a scale $V \subseteq S$ of cardinality less than κ which is clearly impossible. Again by Lemma 8.7 there is a scale $W \subseteq T$ of cardinality κ. Let C be any λ-chain of W. Using Lemma 8.7 once more, we see that S must contain an unbounded subset of cardinality no greater than λ, so $\lambda = \kappa$. Thus for every λ-chain of W we have $\lambda = \kappa$, so, by Theorem 8.2, W contains a κ-subscale. □

8.10. COROLLARY. *In Theorems 8.5 and 8.6 the condition "no subchain of S is a scale" can be replaced by "there are no linear scales in $\mathscr{N}$."* □

8.11. COROLLARY. *If there exist two scales S and T such that S is a κ-scale and T is a λ-scale then $cf(\kappa) = cf(\lambda)$.* □

PROOF. Let $\kappa^* = cf(\kappa)$ *and* $\lambda^* = cf(\lambda)$. Then clearly there exist scales $S^* \subseteq S$ and $T^* \subseteq T$ such that S^* is a κ^*-scale and T^* is a λ^*-scale. Now assume that $\kappa^* \in \lambda^*$. Then by Theorem 8.9 there is a κ^*-scale $U \subseteq T^*$. But this is impossible because the cofinality of a cardinal is always a regular cardinal. □

REFERENCES

1. Paul J. Cohen, *The independence of the continuum hypothesis*. I, II, Proc. Nat. Acad. Sci. U.S.A. **50** (1963), 1143–1148; **51** (1964), 105–110. MR **28** #1118; #2962.

2. ———, *Set theory and the continuum hypothesis*, Benjamin, New York, 1966. MR **38** #999.

3. Felix Hausdorff, *Untersuchen über Ordnungstypen*, Ber. Sächs. Acad. Wiss. **59** (1907), 84–159.

4. Wacław Sierpiński, *Hypothèse du continu*, 2nd ed., Chelsea, New York, 1956. MR **19**, 829.

5. Robert M. Solovay, *A model of set theory in which every set of reals is Lebesgue measurable*, Ann. of Math. (2) **92** (1970), 1–56. MR **42** #64.

6. Robert M. Solovay and Stanley Tennenbaum, *Iterated Cohen extensions and Souslin's problem*, Ann. of Math. (2) **94** (1971), 201–245. MR **45** #3212.

QUEENS COLLEGE OF THE CITY UNIVERSITY OF NEW YORK

Proceedings of Symposia in Pure Mathematics
Volume 13, Part II, 1974

MEASURABLE CARDINALS AND THE GCH

RONALD BJÖRN JENSEN

The forcing method may be used to demonstrate the consistency of the GCH with various assumptions on the existence of large cardinals. In this paper we sketch a proof of the following:

THEOREM. *Let M be a countable transitive model of ZF + AC. Then there exists an extension $N \supset M$ having the same ordinals as M such that the following hold:*

(i) *GCH holds in M.*

(ii) *Every measurable cardinal of M is measurable in N.*

(iii) *Every Ramsey cardinal of M is a Ramsey cardinal in N; moreover, if λ is a limit ordinal and $\alpha \to (\lambda)^{<\omega}$ in M, then $\alpha \to (\lambda)^{<\omega}$ in N.*

By standard methods (cf. Mathias in these PROCEEDINGS) one may convert the proof of this theorem into a syntactical demonstration of various consistency results. For instance if we assume the consistency of ZF + AC together with either of the axioms "there exists a measurable (Ramsey) cardinal," "every closed unbounded class of ordinals contains a measurable (Ramsey) cardinal." Since, as will be apparent from our construction, N can be so chosen that $\aleph_1$ is absolute (i.e., $\aleph_1^M = \aleph_1^N$), it also follows from (iii) that GCH is consistent relative to the assumption $\exists\alpha\ \alpha \to (\aleph_1)^{<\omega}$. All of these results continue to hold if we replace GCH by certain forms of its negation such as the statement that $2^\alpha > \alpha^+$ for arbitrarily large α. It also seems likely that, using McAloon's method, one can replace GCH by the axiom of ordinal definability (this was previously obtainable only on the assumption that some closed unbounded class of ordinals does *not* contain a measurable cardinal).

It is interesting that the compatibility of GCH with the existence of a measurable cardinal may be obtained by diametrically opposed methods, for Silver proved the same result earlier by constructing an inner model of M rather than a Cohen

AMS (MOS) subject classifications (1970). Primary 02K25, 02K35.

extension. As yet, however, no variant of Silver's method has been found to prove the consistency of GCH relative to the stronger assumption that the measurable cardinals are cofinal with the ordinals or the weaker one that a Ramsey cardinal exists. Our proof is more closely related to the earlier work of Lévy and Solovay, who showed that a measurable cardinal κ of M remains measurable in a Cohen extension N if the cardinality of the collection of conditions forcing the extension is of cardinality less than κ. This sufficed to show that, for any $\alpha < \kappa$, an extension could be found in which the continuum hypothesis held up to level α (or, alternatively, fails at arbitrary levels up to α).

After describing the construction of the model N mentioned in our Theorem, we shall confine ourselves to proving (ii). The proof of (iii) is similar in spirit but more complicated. We extend M to N by adding a sequence of maps which "collapse" cardinal numbers without increasing the cardinality of power sets. Thus, if $2^{\aleph_0} > \aleph_1$ in M, we first map $2^{\aleph_0}$ onto $\aleph_1$. Hence $\aleph_2^N = (2^{\aleph_0})^+$. We then map $2^{2^{\aleph_0}}$ into the successor of $2^{\aleph_0}$, etc. In general, having determined that $\aleph_\alpha^N = \gamma^+$, we map 2^γ onto γ^+. In this manner we determine a function φ such that $\aleph_\alpha^N = \aleph_{\varphi(\alpha)}^M$ for all α. φ is easily seen to possess the following properties:

(i) φ is normal and M-definable.

(ii) $\varphi(0) = 0$; $\aleph_{\varphi(\alpha+1)}$ is regular.

(iii) $\varphi(\lambda + 1) = \varphi(\lambda) + 1$ whenever λ is a limit ordinal.

(iv) If $\nu < \varphi(\alpha + 2)$ and $i < \varphi(\alpha + 1)$, then $\aleph_\nu^{\aleph_i} < \aleph_{\varphi(\alpha+2)}$.

From now on, suppose φ to be any function satisfying (i)–(iv). For the sake of brevity we shall write γ_α for $\aleph_{\varphi(\alpha)}$. Set

$$I = \bigcup_\alpha \{i | \varphi(\alpha) < i < \varphi(\alpha + 1)\}.$$

We obtain our new model by adding a sequence $\langle f_i | i \in I\rangle$ of functions such that f_i maps $\{i\} \times \gamma_\alpha$ onto $\aleph_i^M$ whenever $\varphi(\alpha) < i < \varphi(\alpha + 1)$. The collection $\boldsymbol{P}$ of *conditions* consists of certain subfunctions of $\bigcup_i f_i$. We define P to be a condition whenever $P \in M$ and P is a function whose domain is a subset of $\bigcup_i \mathrm{Dom}(f_i)$ such that, for each i, $P \restriction \mathrm{Dom}(f_i)$ is a map into $\aleph_i$ and $P \restriction \bigcup_{\nu < \gamma_{\alpha+2}} \mathrm{Dom}(f_\nu)$ is always of cardinality less than $\gamma_{\alpha+2}$. A modification of the usual proof of Easton's theorem then yields

LEMMA 1. *N is a model of $ZF + AC + GCH$; moreover, $\aleph_\alpha^N = \aleph_{\varphi(\alpha)}^M$ for each ordinal $\alpha \in M$.*

It remains to be shown that the property of measurability is preserved for $\kappa \in M$. We first note that each condition P slits into an *upper part* $(P)^\nu$ and a *lower part* $(P)_\nu$ defined by

$$(P)^\nu = P \restriction \bigcup_{i \geqq \nu} \mathrm{Dom}(f_i), \qquad (P)_\nu = P \restriction \bigcup_{i < \nu} \mathrm{Dom}(f_i).$$

The functions $(\ )^\nu$, $(\ )_\nu$ have the following properties:

(a) $(P)^\nu \cap (P)_\nu = \varnothing$; $(P)^\nu \cup (P)_\nu = P$.

(b) If $\langle P_\nu | \nu < \lambda \rangle$ is a sequence of conditions such that $(P_\nu)^\nu \subset P_\tau$ for $\nu < \tau < \lambda$, then $\bigcup_{\nu<\lambda} (P_\nu)^\nu$ is a condition.

Now let κ be a measurable cardinal of M. If the function φ was defined in the manner sketched at the outset, then κ is a fixed point of φ. By this follows:

(c) $\overline{\{(P)_\nu | P \in \boldsymbol{P}\}} < \kappa$ for $\nu < \kappa$. Our Theorem then follows by

LEMMA 2. *Let N be any Cohen extension forced by the conditions P defined above. Then if κ is measurable in M, it remains measurable in N.*

PROOF. Let D be a normal ultrafilter on the power set of κ in M. We wish to extend D to a nonprincipal κ-complete ultrafilter D^* in N. We define D^* (in N) by

$$X \in D^* \leftrightarrow \forall Y \in M(Y \in D \wedge Y \subset X)$$

for $X \subset \kappa$. We must now show that D^* has the required properties. Clearly, it suffices to prove the following:

(*) If $\lambda < \kappa$ and $\langle X_i | i < \lambda \rangle$ is a sequence such that $\bigcup_{i<\lambda} X_i = \kappa$, then, for some i, $X_i \in D^*$.

Let the sequence $X = \langle X_i | i < \lambda \rangle$ be denoted by a term $\dot{X}$. Since the statement $\bigcup_{i<\lambda} \dot{X}_i \equiv \kappa$ is true in N, it must be forced by a condition P. We assert that P forces the statement $\forall i \dot{X}_i \in D^*$. To prove this it suffices to show that, for each $P' \supset P$, a $Q \supset P'$ and an $i < \lambda$ can be found such that $\{\nu < \kappa | Q \Vdash \nu \in \dot{X}_i\} \in D$. Since $\lambda < \kappa$, we need only show that

$$\bigcup_{i<\lambda} \{\nu < \kappa | Q \Vdash \nu \in \dot{X}_i\} < D.$$

Choose a sequence $P_\nu \supset P'$ $(\nu < \kappa)$ such that

(i) $\forall i < \lambda P_\nu \models \nu \in \dot{X}_i$,

(ii) $P_\nu \supset \bigcup_{\tau<\nu} (P_\tau)^\tau$.

(This is possible by (b).) Now define a sequence Q_ν $(\nu < \kappa)$ by

$$\{\tau | (P_\tau)_\nu = Q_\nu\} \in D.$$

Obviously $(Q_\tau)_\nu = Q_\nu$ for $\nu < \tau$. If we set $A = \{\nu | (P_\nu)_\nu = Q_\nu\}$, we obtain $A \in D$. In order to see this we set $B = \{\tau | (P_\tau)_\nu = Q_\nu\}$ and note that $B_\nu \in D$ and $B_\nu = \bigcap_{i<\nu} B_{i+1}$ for $\nu < \kappa$. By the normality of D we conclude

$$A = \{\nu | \nu \in B\} \in D.$$

Now set $Q = \bigcup_{\nu \in A} P_\nu$. Q is a condition since $Q = \bigcup_{\nu<\kappa} (P_{\alpha_\nu})^\nu$, where $\langle \alpha_\nu | \nu < \kappa \rangle$ is the monotonic enumeration of A. By the definition of Q it follows immediately that

$$A \subset \bigcup_{i<\lambda} \{\nu | Q \Vdash \nu \in \dot{X}_i\},$$

which completes the proof.

Note. The essential combinatorial fact which we have used in the above proof is the following: If R is a tree such that, for each $\nu < \kappa$, the set of branches of order

ν is of cardinality $<\kappa$, and if the field of R has the form $\bigcup_{\nu<\kappa} B_\nu$, where B_ν is a branch of length ν, then there exists a branch B of length κ such that

$$\{\nu < \kappa | B_\nu \subset B\} \in D.$$

The proof of (iii) uses an analogous but more complicated combinatorial principle for Ramsey cardinals. However, this proof is lengthier by a factor of ten.

Proceedings of Symposia in Pure Mathematics
Volume 13, Part II, 1974

THE ORDER EXTENSION PRINCIPLE

A. R. D. MATHIAS

Let O be the sentence "Every set has a total ordering", OE the sentence "Every partial ordering can be extended to a total ordering".

THEOREM. *Let $\mathfrak{M}$ be a countable standard model of $ZF + V = L$. Then $\mathfrak{M}$ can be extended to a countable standard model $\mathfrak{N}$ of $ZF + O + \neg\, OE$.*

COROLLARY (PROOF BY HANDWAVING).

$$\mathrm{Con}\,(ZF) \rightarrow \mathrm{Con}\,(ZF + O + \neg\, OE).$$

PROOF OF THE THEOREM. A theorem of Jońsson [**3**] shows that there is in $\mathfrak{M}$ a partially ordered set $\langle I, R\rangle$ satisfying the following conditions in $\mathfrak{M}$.

P1. Every finite partial ordering can be embedded as a submodel of $\langle I, R\rangle$.

P2. Every isomorphism of finite submodels of $\langle I, R\rangle$ can be extended to an automorphism of $\langle I, R\rangle$.

I specify that R is irreflexive; that is, that for no $i \in I$ is $\langle i, i\rangle \in R$. I write $i \|_R j$ for "neither $i\,R\,j$ nor $j\,R\,i$", and emphasize that if $\langle J, S\rangle$ is a submodel of $\langle I, R\rangle$, and $i \|_S j$ then $i \|_R j$.

Construct in $\mathfrak{M}$ a ramified language $\mathfrak{L}$ with the usual limited quantifiers $\forall_\alpha$, limited comprehension operators E_α, variables, brackets, et cetera. $\mathfrak{L}$ has two predicate symbols $\in$ and $\equiv$, a name $\boldsymbol{u}$ for each $u \in \mathfrak{M}$, an individual constant $\dot{a}_i$ for each $i \in I$ (so that the set $\{\langle \dot{a}_i, i\rangle : i \in I\} \in \mathfrak{M}$), and two further constants $\dot{A}$ and $\dot{<}$. The wffs and limited comprehension terms (henceforth "nouns") of $\mathfrak{L}$ are defined as usual, with the restriction that if $E_\alpha \times \mathfrak{A}$ is a noun and $\dot{A}$ or $\dot{<}$ occurs in $\mathfrak{M}$ then $\alpha > \omega$. For details of the forcing method, the reader is referred to the expository article by Dr. R. B. Jensen [**2**]. In the rest of these notes, ramification indices will be omitted. $s, t, \ldots$ will be used as variables for nouns.

A *condition* is a finite consistent set of wffs of the form $\boldsymbol{n} \in \dot{a}_i$ or $\neg \boldsymbol{n} \in \dot{a}_i$ $(i \in I, n \in \omega)$. $P, P', Q, \ldots$ are variables for conditions.

AMS (MOS) subject classifications (1970). Primary 02H13, 06A05, 06A10.

Define a weak forcing relation $\Vdash$ by

$$P \Vdash \boldsymbol{n} \in \dot{a}_i \leftrightarrow_{\mathrm{Df}} \boldsymbol{n} \in \dot{a}_i \in P;$$

$$P \Vdash s \in \dot{A} \leftrightarrow_{\mathrm{Df}} \forall P' \supseteq P \exists Q \supseteq P' \exists i \in I\, Q \Vdash s \in \dot{a}_i;$$

$$P \Vdash s < t \leftrightarrow_{\mathrm{Df}} \forall P' \supseteq P \exists Q \supseteq P' \exists i, j \in I$$

such that $i\,R\,j$ and $Q \Vdash s \equiv \dot{a}_i \wedge t \equiv \dot{a}_j$ so that, for $i, j \in I$,

$$i\,R\,j \rightarrow O \Vdash \dot{a}_i < \dot{a}_j, \quad i\|_R j \rightarrow O \Vdash \dot{a}_i \| \dot{a}_j \quad \text{and} \quad O \Vdash \dot{a}_i \in \dot{A}.$$

Obtain a complete sequence of conditions and thereby an interpretation Ω of the nouns of the language $\mathfrak{L}$, which defines the model $\mathfrak{N}$.

Write $a_i = \Omega(\dot{a}_i)$, $A = \Omega(\dot{A})$, $< \; = \Omega(\dot{<})$. Then $a_i \subseteq \omega$, all i; $A = \{a_i : i \in I\}$; and $<$ is a partial ordering of A.

LEMMA 1. *$<$ cannot be extended in $\mathfrak{N}$ to a total ordering of A.*

Let $G \in \mathfrak{M}$ be the group of all automorphisms $\in \mathfrak{M}$ of $\langle I, R\rangle$. For $\tau \in G$, define $\tau(P)$, $\tau(\mathfrak{A})$ for conditions P and $\mathfrak{L}$-sentences $\mathfrak{A}$ to be the condition or wff obtained by replacing each occurrence of $\dot{a}_i$ by $\dot{a}_{\tau(i)}$, for each $i \in I$.

S-LEMMA. $P \Vdash \mathfrak{A} \leftrightarrow \tau(P) \Vdash \tau(\mathfrak{A})$.

R-LEMMA. *Suppose $P \Vdash \mathfrak{A}$, and $\boldsymbol{c} = (\dot{a}_1, \ldots, \dot{a}_n)$ is the set of $\dot{a}_i$ occurring in $\mathfrak{A}$. Let $P|\boldsymbol{c}$ be the set of conditions in P in which $\dot{a}_1$ or $\cdots$ or $\dot{a}_n$ occurs. Then $P|\boldsymbol{c} \Vdash \mathfrak{A}$.*

PROOF. Let $\boldsymbol{d}$ be the set of $\dot{a}_i$ occurring in P but not in $\mathfrak{A}$. Write $Q = P \mid \boldsymbol{c}$, and suppose $\sim Q \Vdash \mathfrak{A}$. Then $\exists Q' \supseteq Q(Q' \Vdash \neg\mathfrak{A})$; this Q' mentions (names of) reals in $\boldsymbol{c}$, and also others, say those in the finite set $\boldsymbol{e}$. $\boldsymbol{c} \cap \boldsymbol{d} = \boldsymbol{c} \cap \boldsymbol{e} = O$. By properties P1, P2 of $\langle I, R\rangle$, there is a $\tau \in G$ with $\dot{a} \in \boldsymbol{c} \rightarrow \tau(\dot{a}) = \dot{a}$; $\dot{a} \in \boldsymbol{e} \rightarrow \tau(\dot{a}) \notin \boldsymbol{c} \cup \boldsymbol{d}$. Then $\tau(Q') \Vdash \neg\mathfrak{A}$; $\tau(Q') \cup P$ is a condition but $\tau(Q') \cup P \Vdash \mathfrak{A} \wedge \neg\mathfrak{A}$—a contradiction. (Regard τ as acting on $\{\dot{a}_i : i \in I\}$ by defining $\tau(\dot{a}_i) = \dot{a}_{\tau(i)}$.)

PROOF OF LEMMA 1. Suppose Δ is a noun such that $\Omega(\Delta)$ is a total ordering of A extending $<$. Let $\mathfrak{A}(\Delta)$ be the $\mathfrak{L}$-sentence "Δ is a total ordering of $\dot{A}$ extending $\dot{<}$." Let $\boldsymbol{c}$ be the set of $\dot{a}_i$ occurring in Δ. Pick $\dot{a}_1, \dot{a}_2, \dot{a}_3, \dot{a}_4, \dot{a}_5, \dot{a}_6$ so that $O \Vdash \dot{a}_4 \dot{<} \dot{a}_6$; for $i, j = 1, \ldots, 6$, $i < j$, $\langle i, j\rangle \neq \langle 4, 6\rangle$, $O \Vdash \dot{a}_i \| \dot{a}_j$; for $\dot{a} \in \boldsymbol{c}$, $i = 1, \ldots, 6$, $O \Vdash \dot{a} \| a_i$; and $\mathfrak{N} = \dot{a}_1 \,\Delta\, \dot{a}_2 \wedge \dot{a}_2 \,\Delta\, \dot{a}_3$. Then there is a condition P with $P \Vdash \mathfrak{A}(\Delta) \wedge \dot{a}_1 \,\Delta\, \dot{a}_2 \wedge \dot{a}_2 \,\Delta\, \dot{a}_3$.

By the R-Lemma, we may assume that P mentions only reals in $\boldsymbol{c} \cup \{\dot{a}_1, \dot{a}_2, \dot{a}_3\}$. Set

$$P_1(\boldsymbol{c}, \dot{a}_1, \dot{a}_2) = P|\boldsymbol{c} \cup \{\dot{a}_1, \dot{a}_2\},$$

$$P_2(\boldsymbol{c}, \dot{a}_2, \dot{a}_3) = P|\boldsymbol{c} \cup \{\dot{a}_2, \dot{a}_3\}.$$

Then $P_1(\boldsymbol{c}, \dot{a}_1, \dot{a}_2) \Vdash \mathfrak{A}(\Delta) \wedge \dot{a}_1 \,\Delta\, \dot{a}_2$. There is a $\tau_1 \in G$ with $\dot{a} \in \boldsymbol{c} \rightarrow \tau_1(\dot{a}) = \dot{a}$; $\tau_1(\dot{a}_1) = \dot{a}_6$ and $\tau(\dot{a}_2) = \dot{a}_5$. By the S-Lemma,

$$(*) \qquad Q_1 =_{\mathrm{Df}} P_1(\boldsymbol{c}, \dot{a}_6, \dot{a}_5) \Vdash \mathfrak{A}(\Delta) \wedge \dot{a}_6 \,\Delta\, \dot{a}_5.$$

There is a $\tau_2 \in G$ with $\dot{a} \in c \to \tau_2(\dot{a}) = \dot{a}$, and which sends $\dot{a}_2$ to $\dot{a}_5$ and $\dot{a}_3$ to $\dot{a}_4$. Applying τ_2,

$$Q_2 =_{\mathrm{Df}} P_2(c, \dot{a}_5, \dot{a}_4) \Vdash \mathfrak{A}(\Delta) \wedge \dot{a}_5 \, \Delta \, \dot{a}_4.$$

Now $O \Vdash \dot{a}_4 \dot{<} \dot{a}_6$; $Q_1 \cup Q_2$ is consistent and so a condition. But

$$Q_1 \cup Q_2 \Vdash \mathfrak{A}(\Delta) \wedge \dot{a}_4 < \dot{a}_6 \wedge \dot{a}_6 \, \Delta \, \dot{a}_5 \wedge \dot{a}_5 \, \Delta \, \dot{a}_4,$$

which is a contradiction; Lemma 1 is thus proved.

LEMMA 2. *O is true in $\mathfrak{N}$.*

The argument is an adaptation of that of Mostowski [**4**] which he used to show the truth of O in a model with urelements. The fundamental step is the following:

LEMMA. *Let $t(\boldsymbol{c}, \boldsymbol{d})$, $t'(\boldsymbol{c}, \boldsymbol{e})$ be nouns mentioning only reals in $\boldsymbol{c} \cup \boldsymbol{d}$, $\boldsymbol{c} \cup \boldsymbol{e}$ respectively, where $\boldsymbol{c}, \boldsymbol{d}, \boldsymbol{e}$ are finite disjoint subsets of $\{\dot{a}_i : i \in I\}$. Suppose that $\mathfrak{N} \Vdash t(\boldsymbol{c}, \boldsymbol{d}) \equiv t'(\boldsymbol{c}, \boldsymbol{e})$. Then there is a noun $t''(\boldsymbol{c})$, mentioning only the reals in $\boldsymbol{c}$, with $\mathfrak{N} \Vdash t(\boldsymbol{c}, \boldsymbol{d}) \equiv t''(\boldsymbol{c})$.*

PROOF. Let $P = P(\boldsymbol{c}, \boldsymbol{d}, \boldsymbol{e})$ be a condition in the complete sequence defining $\mathfrak{N}$ such that

$$P \Vdash t(\boldsymbol{c}, \boldsymbol{d}) \equiv t'(\boldsymbol{c}, \boldsymbol{e}).$$

A noun $t''(\boldsymbol{c})$ will be found for which

$$P \Vdash t(\boldsymbol{c}, \boldsymbol{d}) \equiv t''(\boldsymbol{c}).$$

It will be enough to consider the case when $\boldsymbol{e}$ contains the single real $\dot{a}_0$; the general case follows by induction.

Let $\dot{a}_1, \dot{a}_2, \dot{a}_3$ be reals not in $\boldsymbol{c} \cup \boldsymbol{d} \cup \{\dot{a}_0\}$ such that $O \Vdash \dot{a}_1 \dot{<} \dot{a}_0 \wedge \dot{a}_2 \parallel \dot{a}_0 \wedge \dot{a}_0 \dot{<} \dot{a}_3$ and for $i = 1, 2, 3$ there is a $\tau_i \in G$ with $\tau_i(\dot{a}_0) = \dot{a}_i$, and $a \in \boldsymbol{c} \cup \boldsymbol{d} \to \tau(a) = a$. (This last clause will be abbreviated as $\tau_i : \boldsymbol{c}, \boldsymbol{d}, \dot{a}_0 \cong \boldsymbol{c}, \boldsymbol{d}, \dot{a}_i$.)

Write $P = P_0(\boldsymbol{c}) \cup P_1(\boldsymbol{d}) \cup P_2(\dot{a}_0)$. Then applying τ_i $(i = 1, 2, 3)$,

$$P_0(\boldsymbol{c}) \cup P_1(\boldsymbol{d}) \cup P_2(\dot{a}_i) \Vdash t(\boldsymbol{c}, \boldsymbol{d}) \equiv t'(\boldsymbol{c}, \dot{a}_i);$$

$$P \Vdash t(\boldsymbol{c}, \boldsymbol{d}) \equiv t'(\boldsymbol{c}, \dot{a}_0);$$

so $P \cup P_2(\dot{a}_i) \Vdash t(\boldsymbol{c}, \boldsymbol{d}) \equiv t'(\boldsymbol{c}, \dot{a}_0) \equiv t'(\boldsymbol{c}, \dot{a}_i)$, and by the R-Lemma

(*) $$P_0(\boldsymbol{c}) \cup P_2(\dot{a}_0) \cup P_2(\dot{a}_i) \Vdash t'(\boldsymbol{c}, \dot{a}_0) \equiv t'(\boldsymbol{c}, \dot{a}_i).$$

Let $\bar{P}_2(x)$ denote the conjunction of the sentences in $\bar{P}_2(\dot{a}_0)$ with $\dot{a}_0$ replaced throughout by the $\mathfrak{L}$-variable x. Let $D(\boldsymbol{c}, x)$ denote the result of replacing $\dot{a}_0$ by x in the conjunction of the sentences in the diagram of the partial ordering $\boldsymbol{c}, \dot{a}_0$. Set $Q =_{\mathrm{Df}} P_0(\boldsymbol{c}) \cup P_2(\dot{a}_0)$. Note that $Q \subseteq P$. I assert that

(**) $$Q \Vdash \forall x (x \in \dot{A} \wedge \bar{P}_2(x) \wedge D(\boldsymbol{c}, x) : \to t'(\boldsymbol{c}, \dot{a}_0) \equiv t'(\boldsymbol{c}, x)).$$

Let s be a noun. I have to show that if $Q' \supseteq Q$ and $Q \Vdash s \in \dot{A} \wedge \bar{P}_2(s) \wedge D(\boldsymbol{c}, s)$ then there is a $Q'' \supseteq Q'$ with $Q'' \Vdash t'(\boldsymbol{c}, \dot{a}_0) \equiv t'(\boldsymbol{c}, s)$.

Now given such a Q', there is a $Q'' \supseteq Q'$, and a $j \in I$ with $Q'' \Vdash s \equiv \dot{a}_j$. Pick $i = 1, 2$ or 3 so that $\dot{a}_0, \dot{a}_j \cong \dot{a}_0, \dot{a}_i$. There is a $\tau \in G$ so that, for this choice of i, $\tau : \boldsymbol{c}, \dot{a}_i, \dot{a}_0 \cong \boldsymbol{c}, \dot{a}_j, \dot{a}_0$; for $Q' \subseteq Q''$ and so $Q'' \Vdash s \equiv \dot{a}_j \wedge \bar{P}_2(\dot{a}_j) \wedge D(\boldsymbol{c}, \dot{a}_j)$; and therefore $O \Vdash D(\boldsymbol{c}, \dot{a}_j)$ and so $\boldsymbol{c}, \dot{a}_j \cong \boldsymbol{c}, \dot{a}_0$. But $\boldsymbol{c}, \dot{a}_0 \cong \boldsymbol{c}, \dot{a}_i$; by the choice of i, $\boldsymbol{c}, \dot{a}_j, \dot{a}_0 \cong \boldsymbol{c}, \dot{a}_i, \dot{a}_0$, and so such a τ exists. Further, as $Q'' \Vdash \bar{P}_2(\dot{a}_j)$, $P_2(\dot{a}_j) \subseteq Q''$, and so $P_0(\boldsymbol{c}) \cup P_2(\dot{a}_0) \cup P_2(\dot{a}_j) \subseteq Q''$; applying τ to (*),

$$Q'' \Vdash t'(\boldsymbol{c}, \dot{a}_0) \equiv t'(\boldsymbol{c}, \dot{a}_j).$$

As $Q'' \Vdash \dot{a}_j \equiv s$, $Q'' \Vdash t'(\boldsymbol{c}, \dot{a}_0) \equiv t'(\boldsymbol{c}, s)$, as required.

The noun $t''(\boldsymbol{c})$ can now be constructed. Suppose $t'(\boldsymbol{c}, \dot{a}_0)$ is $Ey\mathfrak{A}(\boldsymbol{c}, \dot{a}_0, y)$. Then set

$$t''(\boldsymbol{c}) =_{\mathrm{Df}} Ey\, \exists x(x \in \dot{A} \wedge \bar{P}_2(x) \wedge D(\boldsymbol{c}, x) \wedge \mathfrak{A}(\boldsymbol{c}, x, y)).$$

By (**), $Q \Vdash t'(\boldsymbol{c}, \dot{a}_0) \equiv t''(\boldsymbol{c})$. As $Q \subseteq P$, and everything forced by P is true in $\mathfrak{N}$, the Lemma is proved.

Now set up in $\mathfrak{N}$ a ramified language $\mathfrak{L}^*$ with a name a^* for each $a \in A$ (so that $\{\langle a^*, a\rangle : a \in A\} \in \mathfrak{L}$), names A^*, $<^*$ for A, $<$; names u^* for each $u < \mathfrak{M}$ (possible as $\mathfrak{M} = L_{\mathfrak{L}}$ and is therefore definable in $\mathfrak{N}$) and variables and quantifiers and the rest, as in $\mathfrak{L}$. Define an interpretation Ω^* for nouns of $\mathfrak{L}^*$, by setting

$$\Omega^*(a^*) = a, \quad \text{for each } a \in A;$$

$$\Omega^*(A^*) = A; \quad \Omega^*(<^*) = < \quad \text{and} \quad \Omega^*(u^*) = u;$$

and then extending to all nouns of $\mathfrak{L}^*$ *so that* Ω^* *is* $\mathfrak{N}$*-definable*. There is a clear 1-1 correspondence * between nouns t of $\mathfrak{L}$ and nouns t^* of $\mathfrak{L}^*$ so that $\Omega^*(t^*) = \Omega(t)$ for every t. Now define supp* (x), for $x \in \mathfrak{N}$, as the finite subset (call it $\boldsymbol{d}$) of A of minimal cardinality such that there is a term $t^*(\boldsymbol{d})$ mentioning only the names a^* for $a \in \boldsymbol{d}$ with $\Omega^*(t^*) = x$. By the Lemma, supp* is always well-defined. Further, supp* *is* $\mathfrak{N}$*-definable*. For each finite subset $\boldsymbol{d}$, let $V_{\boldsymbol{d}} = \{x \in \mathfrak{N} : \text{supp}^*(x) = \boldsymbol{d}\}$. Each $V_{\boldsymbol{d}}$ has an $\mathfrak{N}$-definable well-ordering. Well-order the nouns of $\mathfrak{L}^*$ mentioning only the reals in $\boldsymbol{d}$, and use the interpretation Ω^* to induce a well-ordering of $V_{\boldsymbol{d}}$. Each $\boldsymbol{d}$ is a finite subset of A, which is ordered, being a subset of the real line. The set of such $\boldsymbol{d}$ can be ordered by the usual lexicographical method.

This argument shows that more than Lemma 2 holds in $\mathfrak{N}$. Observe that $x \in \mathfrak{N}$ is well-orderable (in $\mathfrak{N}$) iff there is a $\boldsymbol{d}$ with $\forall y(y \in x \rightarrow \text{supp}^*(y) \subseteq \boldsymbol{d})$.

Then there is an $\mathfrak{N}$-definable function which assigns to each set x in $\mathfrak{N}$ an ordering, which is a well-ordering whenever x has a well-ordering in $\mathfrak{N}$. Thus the axiom of choice for families of well-ordered sets holds in $\mathfrak{N}$.

CODA. Professor Scott asked at the end of my oration whether there was a general method for translating Fraenkel-Mostowski independence proofs into Cohen ones. Jech and Sochor have a method for certain classes of existential statements [**1**], and the papers of Pincus ([**5**], [**6**]) contain an extensive discussion of the question. On p. 740 of [**6**], my observation that the statement that the power set of every well-ordered set can be well-ordered, which is equivalent in ZF to the axiom of choice, is true in at least one Fraenkel-Mostowski model in which choice fails is misquoted. Kripke suggested considering instead the statement that every linearly ordered set can be well-ordered, which implies the other, but which avoids the suggestion of a hierarchy contained in the words "power set."

REFERENCES

1. a) T. Jech and A. Sochor, *On θ-model of the set theory*, Bull. Acad. Polon. Sci. Sér. Sci. Math. Astronom. Phys. **14** (1966), 297–303. (Russian) MR **34** #2441.

b) *Applications of the θ-model*, Bull. Acad. Polon. Sci. Sér. Sci. Math. Astronom. Phys. **14** (1966), 351–355. (Russian) MR **37** #3918.

2. R. B. Jensen, *Concrete models of set theory*, Sets, Models and Recursion Theory (Proc. Summer School Math. Logic and Tenth Logic Colloq., Leicester, 1965), North-Holland, Amsterdam, 1967, pp. 44–74. MR **37** #53.

3. B. Jońsson, *Homogeneous universal relational systems*, Math. Scand. **8** (1960), 137–142. MR **23** #A2328.

4. A. Mostowski, *Über die Unabhängigkeit des Auswahlaxioms vom Ordnungsprinzip*, Fund. Math. **32** (1939), 201–252.

5. D. Pincus, *Support structures for the axiom of choice*, J. Symbolic Logic **36** (1971), 28–38. MR **44** #61.

6. ———, *Zermelo-Fraenkel consistency results by Fraenkel-Mostowski methods*, J. Symbolic Logic **37** (1972), 721–743.

PETERHOUSE
CAMBRIDGE, ENGLAND

Proceedings of Symposia in Pure Mathematics
Volume 13, Part II, 1974

"EMBEDDING CLASSICAL TYPE THEORY IN 'INTUITIONISTIC' TYPE THEORY" A CORRECTION

JOHN MYHILL

In the paper of the above title (these PROCEEDINGS, Part I, pp. 267–270) I claimed to extend Gödel's negative interpretation [3] from arithmetic to the finite theory of types. However, the interpretation given is incorrect; specifically we do not see that the translation of the classical comprehension-schema is a theorem of the "intuitionistic" system. The purpose of this note is to correct the error of my earlier paper by giving an interpretation which works; in fact a stronger result will be obtained because we shall succeed in embedding classical type theory *with* extensionality in "intuitionistic" type theory *without* extensionality.

The formalism of the classical theory C contains numerical variables $x_0, y_0, z_0, \ldots$; numerical terms $f(\mathfrak{s}_1, \ldots, \mathfrak{s}_n)$ where $\mathfrak{s}_1, \ldots, \mathfrak{s}_n$ are numerical terms and f a primitive recursive function; and set variables $x_i, y_i, z_i, \ldots$ of type $i > 0$. Atomic formulas are $\mathfrak{s}_1 = \mathfrak{s}_2$ *where* $\mathfrak{s}_1$ *and* $\mathfrak{s}_2$ *are of type* 0, and $\mathfrak{s}_1 \in \mathfrak{s}_2$ where $\mathfrak{s}_1$ and $\mathfrak{s}_2$ are of consecutive ascending types. The other formulas are formed from these by $\forall x_i$, $\wedge$, $\neg$ in the usual way. The logical axioms are the usual ones for the classical many-sorted predicate calculus with identity in type 0 only; the nonlogical axioms are, firstly, the Peano ones with induction for arbitrary formulas and primitive recursive definitions, and, secondly, the axiom-schemata of comprehension:

C1 $$(\exists_C x_{i+1})(\forall y_1)(y \in x \leftrightarrow_C A(y))$$

and extensionality:

C2 $$(\forall x_i)(x \in y_{i+1} \leftrightarrow_C x \in z_{i+1}) \wedge y \in w_{i+2} \rightarrow_C z \in w_{i+2},$$

AMS (MOS) subject classifications (1970). Primary 02C15; Secondary 02D99.

where $A \to_C B$ abbreviates $\neg(A \wedge \neg B)$, $A \leftrightarrow_C B$ abbreviates $(A \to_C B) \wedge (B \to_C A)$, and $(\exists_C x)A(x)$ abbreviates $\neg(\forall x)\neg A(x)$.

The *rules* of C can be taken to be e.g. modus ponens (for $\to_C$), and generalization

$$\frac{A \to_C B(x)}{A \to_C (\forall x)B(x)} \quad (x \text{ not free in } A).$$

The intuitionistic system I has the same atomic formulas; other formulas are built up by $\forall x_i$, $\wedge$, $\neg$ and the intuitionistic connectives $\to_I$, $\vee_I$ and quantifier $(\exists_I x)$. The logical axioms are those of the intuitionistic many-sorted predicate calculus with identity in type 0. The nonlogical axioms are the Peano ones as in C (but written with intuitionistic connectives) and

C1′ $$(\exists_I x_{i+1})(\forall y)(y \in x \leftrightarrow_I A(y)),$$

where $A \leftrightarrow_I B$ abbreviates $(A \to_I B) \wedge (B \to_I A)$.

The rules are modus ponens and generalization (both with $\to_I$) and

$$\frac{A \to_I B}{(\exists x)A \to_I B} \quad (x \text{ not free in } A).$$

Now we define by simultaneous recursion on types a predicate S ("stable") and a relation $=_S$ ("equality modulo stable sets").

$$S(x_0) \equiv x = x,$$

$$x_0 =_S y_0 \equiv x = y,$$

$$S(x_{i+1}) \equiv (\forall y_i)(S(y) \wedge \neg\neg y \in x \to_I y \in x)$$
$$\wedge (\forall y_i z_i)(S(y) \wedge S(z) \wedge y =_S z \wedge y \in x \to_I z \in x),$$

$$x_{i+1} =_S y_{i+1} \equiv (\forall z_i)(S(z) \to_I (z \in x \leftrightarrow_I z \in y)).$$

A formula A of I with free variables $x^1, \ldots, x^n$ is called *stable* if

$$\vdash_I S(x^1) \wedge \cdots \wedge S(x^n) \wedge \neg\neg A \to_I A$$

and

$$\vdash_I S(x^1) \wedge \cdots \wedge S(x^n) \wedge S(y^1) \wedge \cdots \wedge S(y^n) \wedge x_1 =_S y_1$$
$$\wedge \cdots \wedge x_n =_S y_n \wedge A(\mathbf{x}) \to_I A(\mathbf{y}).$$

LEMMA. *If A is a formula of* C *and A_S is obtained from A by restricting all bound variables to* S, *i.e.,* $(\forall x)$ *is replaced by* $(\forall x)(S(x) \to_I)$, *then A_S is stable.*

The proof is by a straightforward induction on the structure of A (cf. [**1**, Lemma I]).

THEOREM. *If $\vdash_C A$ and A has the free variables $x^1, \ldots, x^n$, then*

$$\vdash_I S(x^1) \wedge \cdots \wedge S(x^n) \to_I A_S.$$

In particular, if $\vdash_C A$ *and* A *has no free variables then* $\vdash_I A_S$.

The proof is by induction on the length of the proof of A. The inductive step and most of the basis proceed exactly as in [**1**, p. 37]. The only cases that need be checked are C1 and C2.

Re C1. We must show

$$\vdash_I S(x^1) \wedge \cdots \wedge S(x^n) \to_I [(\exists_C x_{i+1})(\forall y_i)(y \in x \leftrightarrow A(y))]_S,$$

i.e.,

$$\vdash_I S(x^1) \wedge \cdots \wedge S(x^n) \to_I \neg(\forall x_{i+1})(S(x_{i+1}) \to_I \neg(\forall y_i)(S(y) \to_I (y \in x \leftrightarrow_C A_S(y)))).$$

Assume then the hypotheses $S(x^1), \ldots, S(x^n)$; reasoning informally in I we shall establish the conclusion

$$(\exists_I x_{i+1})(S(x) \wedge (\forall y_i)(S(y) \to_I (y \in x \leftrightarrow_I A_S(y))))$$

which is stronger than we need.

By C1′ there is a set, say α, for which always

$$y \in \alpha_{i+1} \leftrightarrow_I (\forall z)(S(z) \wedge z =_S y \to_I A_S(z, \boldsymbol{x}))$$

where we exhibit the parameters $\boldsymbol{x}$ explicitly.

We need $S(\alpha)$, i.e.,

(1) $$S(y) \wedge \neg\neg y \in \alpha \to_I y \in \alpha$$

and

(2) $$S(y) \wedge S(z) \wedge y =_S z \wedge y \in \alpha \to_I z \in \alpha;$$

also

(3) $$S(y) \wedge y \in \alpha \to_I A_S(y, \boldsymbol{x})$$

and

(4) $$S(y) \wedge A_S(y, \boldsymbol{x}) \to_I y \in \alpha.$$

Re (1). Assume the hypothesis. By the definition of α,

$$\neg\neg(\forall z)(S(z) \wedge z =_S y \to_I A_S(z, \boldsymbol{x})).$$

By the predicate calculus

$$S(z) \wedge z =_S y \to_I \neg\neg A_S(z, \boldsymbol{x}).$$

By the hypotheses $S(x^1), \ldots, S(x^n)$ and the Lemma,

$$S(z) \wedge A_S(z, \boldsymbol{x}) \to_I A_S(z, \boldsymbol{x})$$

whence $S(z) \wedge z =_S y \to_I A_S(z, \boldsymbol{x})$ and so $y \in \alpha$. Q.E.D.

Re (2). Assume the hypothesis. Let w satisfy $S(w) \wedge w =_S z$. Then $w =_S y$ and so by the definition of α, $A_S(w, \boldsymbol{x})$. So

$$(\forall w)(\mathrm{S}(w) \wedge w =_{\mathrm{S}} z \rightarrow_{\mathrm{I}} A_{\mathrm{S}}(w, \boldsymbol{x})) \quad \text{and} \quad w \in \alpha.$$

Re (3). This is obvious since $y =_{\mathrm{S}} y$.

Re (4). Assume the hypothesis and $\mathrm{S}(z)$, $z =_{\mathrm{S}} y$; we must prove $A_{\mathrm{S}}(z, \boldsymbol{x})$. But this is clear, since by the Lemma

$$\mathrm{S}(x^1) \wedge \cdots \wedge \mathrm{S}(x^n) \wedge \mathrm{S}(y) \wedge \mathrm{S}(z) \wedge y =_{\mathrm{S}} z \wedge A_{\mathrm{S}}(y, \boldsymbol{x}) \rightarrow_{\mathrm{I}} A_{\mathrm{S}}(z, \boldsymbol{x}).$$

This takes care of C1.

Re C2. We must show

$$\vdash_{\mathrm{I}} \mathrm{S}(y) \wedge \mathrm{S}(z) \wedge \mathrm{S}(w) \rightarrow_{\mathrm{I}} [(\forall x)(\mathrm{S}(x) \rightarrow_{\mathrm{I}} (x \in y \leftrightarrow_{\mathrm{C}} x \in z)) \wedge y \in w \rightarrow_{\mathrm{C}} z \in w].$$

Since $\vdash_{\mathrm{I}} (A \rightarrow_{\mathrm{I}} B) \rightarrow_{\mathrm{I}} (A \rightarrow_{\mathrm{C}} B)$, it is enough to prove

$$\mathrm{S}(y) \wedge \mathrm{S}(z) \wedge \mathrm{S}(w) \wedge (\forall x)(\mathrm{S}(x) \rightarrow_{\mathrm{I}} (x \in y \leftrightarrow_{\mathrm{C}} x \in z)) \wedge y \in w \rightarrow_{\mathrm{I}} z \in w.$$

Assume the hypothesis. We have, for all x,

$$\mathrm{S}x \wedge x \in y \rightarrow_{\mathrm{I}} \neg\neg x \in z \rightarrow_{\mathrm{I}} x \in z$$

(since $\mathrm{S}(z)$). Likewise

$$\mathrm{S}x \wedge x \in z \rightarrow_{\mathrm{I}} x \in y,$$

so

$$(\forall x)(\mathrm{S}x \rightarrow_{\mathrm{I}} (x \in z \leftrightarrow_{\mathrm{I}} x \in y)),$$

i.e., $z =_{\mathrm{S}} y$. But $\mathrm{S}(w)$, whence $y \in w \wedge \mathrm{S}(y) \wedge \mathrm{S}(z) \wedge z =_{\mathrm{S}} y \rightarrow_{\mathrm{I}} z \in w$ and $z \in w$.

Q.E.D.

This completes the proof.

References

1. K. Gödel, *Zur intuitionistischen Arithmetik und Zahlentheorie*, Ergebnisse eines Math. Kolloq., no. 4, 1933, pp. 34–38.

STATE UNIVERSITY OF NEW YORK AT BUFFALO

Proceedings of Symposia in Pure Mathematics
Volume 13, Part II, 1974

REMARKS ON REFLECTION PRINCIPLES, LARGE CARDINALS, AND ELEMENTARY EMBEDDINGS[1]

W. N. REINHARDT

This paper is intended as a brief introduction to recent work directed at the formulation of new axioms of infinity or large cardinal properties. It is intended to be as free from technicalities as possible, and aspires to explain the motivating ideas as clearly and fully as possible, with attention to the problems, puzzles, and intuitions that lie behind them. We include some comments on developments since the 1967 symposium.

A paper to be titled *Strong axioms of infinity and elementary embeddings* is in preparation with R. M. Solovay, and has been owed for some time now. The present paper may serve as a first installment on this debt. (Solovay is not, however, to be blamed for any deficiencies or excesses of this exposition.) The full payment will include a more complete discussion of the relation between large cardinal-embedding conditions and ultrafilter-measure conditions, and also of the known consequences of axioms we have considered.

AMS (MOS) subject classifications (1970). Primary 02—02, 02K35, 02A05; Secondary 02K15, 04A10.

[1] A significant part of this paper was written while the author was visiting the University of Amsterdam. I am especially indebted to Professor M. H. Löb for a number of discussions concerning the problem of existence in mathematics.

I have also benefited from discussions with Professor K. Gödel. However, as to the justification of the axioms attempted here, Gödel feels that it is rather unsatisfactory and that something better could be achieved by a proper analysis of the notion of *structural property* of the concept of set, which would then lead to the reflection principles in the form "any structural property of the concept of set is reflected by some set." What is done here may be viewed as a proposal that the structural properties of properties of sets are those of the form jP for some property of sets P. This, of course, makes the concept of structural property relative to the particular representation of "imaginary situations," and is thus at best an approximation to Gödel's idea. As for extendibility itself, Gödel feels it is mathematically natural, but is about as well justified by the observation that it asserts an analogy between κ and $\lambda > \kappa$, which is plausible in the way that compactness of ω suggests the existence of a strongly compact cardinal.

1. Reflection principles.

1.1. *Paradoxes in logic; Cantor's Ω and reflection principles.* It has been pointed out that the paradoxes of Russell, Burali-Forti never really caused a crisis in mathematics (where one deals only with unproblematic examples of sets), but rather in logic (and general set theory) where one attempts to provide a general and universal framework for mathematics and in particular for arbitrary sets. We now consider such a frame to have been provided for set theory by the clarification of the intuitive idea of the cumulative hierarchy (due chiefly to Zermelo) (i.e., the sets in the series $R_0, R_1, \ldots$ given by $R_{\alpha+1} = \mathscr{P}R_\alpha$ carried into the transfinite for arbitrary ordinals α by the rule $R_\lambda = \bigcup_{\alpha<\lambda} R_\alpha$ in case λ is a limit ordinal; here $\mathscr{P}X = \{t \mid t \subseteq X\}$.) We might say that this reduces all structural questions to questions about "arbitrary subset of" and "arbitrary ordinal." (The apparent restriction on universality arising from sets whose elements are not sets and "irregular" sets is not believed to be significant for the expressive power of the framework. The reason is that *every* set X is supposed to be equivalent to an ordinal, and since we are concerned only with structural questions (i.e., up to isomorphism) it suffices to consider the corresponding ordinal. This is independent of the possible interpretational or epistemic interest of such sets.) The picture provided suffices to set up the basic axioms of set theory. It rather explicitly refrains from a definite division of "ordinals" from other existing objects, and, in fact, does not tell us much about the transfinite sequence of ordinals Ω (which Cantor conceived as "absolutely infinite"). Insofar as we know anything more about this, our knowledge seems to depend on so-called reflection principles. There are concrete examples here which seem quite unproblematic. (Exactly what is included is perhaps debatable, but we would certainly include inaccessible cardinals, Mahlo cardinals, and indescribable cardinals. If our analysis is correct, then measurable, supercompact, and extendable cardinals are to be included, as well as Vopěnka's principle. The latter however requires a further development than that given here.) The difficulties arise in formulating a general principle. For example, analogous to the "ideal" comprehension principle of naive logic, we have the naive reflection principle:

(P1) If Ω has any property P then there is an ordinal $\kappa < \Omega$ which also has the property P.

If we consider the property $P(x) \leftrightarrow x = \Omega$, we see that something more subtle is required. However, this form brings out two aspects of reflection principles clearly. There is (i) some consideration of ("reflection on") the universe of all sets or on Ω and its properties, and (ii) there is claimed to exist an object (here, a set κ) which mirrors ("reflects") this universe. The difficulties lie in assigning suitable meanings to "Ω has P" (which generally seems to involve consideration of objects which are not sets) while preserving in some way the universality of set theory.

1.2. *Examples of reflection arguments.* It may be helpful to give some informal arguments illustrating the use of reflection principles.

The simplest is perhaps: the universe of sets is inaccessible (i.e., satisfies the replacement axiom), *therefore* there is an inaccessible cardinal. This can be elaborated somewhat, as follows. Let θ_ν enumerate the inaccessible cardinals. By the same sort of reasoning, θ_ν is not bounded: the Cantor absolute Ω (all ordinals) is an inaccessible above any proposed bound β, *therefore* there is an inaccessible cardinal above β. Clearly, then, there are Ω inaccessibles below Ω; *therefore* there is an inaccessible κ such that there are κ inaccessibles below it (i.e., $\kappa = \theta_\kappa$). Now suppose that F is a set function $F = \Omega \to \Omega$ such that it is nondecreasing and continuous (i.e., at limits $F\lambda = \bigcup \{F\alpha | \alpha < \lambda\}$). Now it is natural to think of $F\Omega$ as making sense (we can "define" it by the continuity condition) in which case it is clear that $F\Omega = \Omega$. Thus Ω is an inaccessible fixed point of F. If we allow reflection on *this* statement, we get that F has an inaccessible fixed point. Notice that we could have thought of the argument above for κ similarly: $\theta_\Omega = \Omega \to (\exists \kappa < \Omega)\theta_\kappa = \kappa$.

We give one more example. A *tree* is a partial ordering in which each initial segment is well-ordered. The rank of x is the order type of $\{y | y < x\}$. Let T be any tree structure on Ω such that (i) for each $\alpha < \Omega$ the points of T of rank $< \alpha$ form a set (rather than a proper class), and (ii) T has points of arbitrarily high rank. We argue that T must have a branch of length Ω. First choose F_α to be the tree obtained from T by keeping only points of rank $< \alpha$. Now F_α is a set, and, for $\alpha < \beta$, F_β is an end extension of F_α. For limits λ, $F_\lambda = \bigcup\{F_\alpha | \alpha < \lambda\}$. Notice that each F_α has a branch of length α (since $F_{\alpha+1} \sim F_\alpha \neq 0$). Now let us apply F formally to Ω. Then formally we expect F_Ω to have a branch of length Ω. And, indeed, if we suppose F_Ω does *not* have such a branch, and apply reflection, we get a contradiction. Finally, $F_\Omega = \bigcup \{F_\alpha | \alpha < \Omega\} = T$, so T has such a branch.

At this point we can say that our method of making such informal arguments precise is (i) to make precise the sense in which a function $F : \Omega \to \Omega$ can be applied to Ω, and (ii) to make precise the statements about F for which reflection is allowed.

1.3. *An imprecise remark.* It is possible to give a metaphysical motivation for the form of (P1) along Cantorian lines. We present this in the hope that some readers will find it illuminating. According to Cantor, Ω is unlimited.[2] To the extent that our thinking is limited, then, it should be compatible with what we think or understand (about Ω for example) that the same could be thought or understood of some $\kappa < \Omega$. Thus we do not understand (in the requisite sense) the property $P(x) \leftrightarrow x = \Omega$.

A similar doctrine can be formulated about the totality of all things which exist: whatever theory we have about what exists, it should be compatible with our understanding of our theory that the totality of existing things should be a set. This places a restriction on the expressive power of "understandable" languages. (In the case of pure-unapplied-first-order logic, the doctrine is proved by the Löwenheim-Skolem theorem.)

2. Ω from outside. As a first attempt at understanding the mechanics of reflection arguments, let us imagine for a moment that we could get outside of Cantor's

[2] More precisely, Cantor conceived of the *process of formation* of the ordinals as "*absolut* grenzenlos" (letter to Dedekind). Cantor shared with classical Greece a certain distaste for the ἄπειρον (unbounded, indeterminate), under which he classified the potential infinite.

universe $V = R_\Omega$, and think of ("reflect on") V as if it were a set. Then there would be objects (such as Ω, $\Omega + 1$, $R_{\Omega+\Omega}$) of rank $\geqq \Omega$. This of course already violates part of our intuition about Ω, namely, that it is all *possible* ordinals (or well-order types), but aside from this there seems nothing inconsistent in such an imaginative foray. It may be compared to the "virtual displacements" which play a role in the extremum principles of physics. Formally, it is very easy to implement (the difficulties come in interpreting the formalism). We add to the language of ZF individual constants V, Ω together with the usual axioms relativized to V. Also we add an axiom asserting that $V = R_\Omega$. (This last assures that for $x \in V$, $\mathscr{P}x$ contains all possible subsets of x.) The question arises: What theory do we assume in dealing with the new objects we have projected? Now Cantor's universe V is intended to comprehend all possibilities as regards well-orderings; so it is natural to suppose that the theory of R_Ω would be applicable in our projected situation. If we allow the complete theory (sets as parameters) we obtain by this analogical procedure the schema

$$(\forall x, y \in V)[\theta^V(x, y) \leftrightarrow \theta(x, y)]. \tag{S2}$$

(Here θ is any $\in$-formula whose free variables are x, y, and θ^V is obtained from θ by relativizing all quantifiers to V. Syntactically, V is an individual constant.) The schema simply asserts that any first-order sentence θ^V of the theory of R_Ω is true in the projected universe.

2.1. *Reflection arguments from the projection schema.* We indicate how (S2) may be used to formalize reflection principle arguments. Now the formula of set theory which expresses "x is inaccessible" can be formally applied to V. In this sense we can, within this formalism, assert that V is inaccessible (which of course is not expressible by a first-order formula of the language of ZF when the quantifiers range over V). Let us add the axiom "V is inaccessible" to our formalism. It is now possible, in those cases where the set function F is definable, to carry out the informal arguments given earlier. Utilizing informal criteria to judge when the passage from "for all definable F" to "for all F" is acceptable, one can in this way obtain all the Mahlo cardinals. (The informal step is in each case analogous to the passage from "V satisfies replacement for definable functions" to "V is inaccessible.") For details, see Lévy [**L1**], [**L2**]. The schema (S2) can be used to provide an elegant axiomatization of set theory. The resulting theory, which we will call ZA (because of the similarity to Ackermann's set theory [**A1**]) is equivalent to ZF + (S2) + $V = R_\Omega$, and its $\in$-part is ZF. (See [**R1**], [**R3**], [**L1**].)

3. The language of ZF vs. language with Ω: a digression on theories, methods of extending theories, and properties of natural models. The reader has probably noticed that the schema (S2) corresponds to the following relation $E(\kappa; \lambda)$ on ordinals.

DEFINITION 3.1. Let κ, λ be ordinals. We write $E(\kappa; \lambda)$, and say that κ is first-order extendible to λ, in case

$$(R_\kappa, \in) \prec (R_\lambda, \in) \;\&\; \kappa < \lambda. \tag{3.1}$$

Why not dispense with formalisms involving nonsets (or "imagined but nonexisting" sets) and discuss properties of sets like the above? This is usually desirable (especially for convenience in precise formulation), but our concern here is with motivation and evidence, which, like it or not, seem to depend on consideration of Ω. Moreover, the schema (S2) is supposed to assert something about correct theories of sets and correct methods of extending such theories, which is not contained for example in the bare assertion that there are κ, λ such that $E(\kappa; \lambda)$. We stop here to discuss this because it helps explain the role that Ω plays in building set theories.

In devising set theories we are interested in

(a) theories T which are correct on the intended interpretation,

(b) procedures $T \to T'$ which always lead from correct theories to correct theories.

Now "correctness of T on the intended interpretation" is supposed to be expressible in the language of (S2); that is what Ω or V is for. Thus, letting $\psi(T, V)$ be a formula expressing "T is true in V," whenever

(a1) $\mathrm{ZA} \vdash \psi(T, V)$,

(b1) $\mathrm{ZA} \vdash \forall T[\psi(T, V) \to \psi(T', V)]$,

(where $T \mapsto T'$ is a function expressible in ZF) then we have corresponding principles of type (a), (b) above. Thus uses of ZA to get either type of assertion are seen to be special cases of

$$[\mathrm{ZA} \vdash \varphi(V)] \to \varphi(V)$$

which is just the "proof theoretic reflection principle" for ZA (written in a sloppy manner, however). Notice that common methods of extending theories are contained in this; for example $T \mapsto T +$ "T is consistent," $T \mapsto T +$ "T has a natural model," etc.[3]

3.1. *Large cardinal properties.* Within the language of ZA we can also say (for a single $\in$-formula φ) what it means for φ to express a large cardinal property. It means that $\varphi(\Omega)$. Now one can obtain principles of types (a), (b) from a large cardinal property and a given correct theory of sets (such as ZF) as follows. If φ is a large cardinal property

(a2) $\mathrm{ZF} \vdash \forall\kappa(\varphi(\kappa) \to \psi(T, R_\kappa))$,

(b2) $\mathrm{ZF} \vdash \forall\kappa(\varphi(\kappa) \to \forall T[\psi(T, R_\kappa) \to \psi(T', R_\kappa)])$,

then we have the corresponding principles of type (a), (b) above. (Again, $\psi(T, X)$ expresses "T is true in X.")

Thus in giving a model theoretic condition (instead of a formal theory) one must specify what is intended as the large cardinal. For example, in (3.1), κ is

[3] The connection between such reflection principles and set theoretic reflection principles does not seem to me to be merely verbal. Here one considers what is true, and this is mirrored by what is provable. There is also the element of ostensible reflexivity: one's considerations are turned back upon themselves (e.g. one tries to be conscious of the formal system one is using). It seems, however, that they always fall back to something less than themselves. This element also occurs in set theoretic reflection principles, but the ostensible reflexivity is more ontological: we reflect on the (mathematical) existence of that which we consider, as we consider mathematical existence.

intended as the large cardinal. Since in (S2) we consider not only statements of the form $\psi(T, V)$ (T is correct) but also $\varphi(V)$ (φ is a large cardinal property), we get from a condition such as (3.1) (and a correct theory such as ZF) versions of (a), (b) for large cardinal properties. Let $\theta(\kappa, \lambda)$ formalize (3.1). If

(a3) $\mathrm{ZF} \vdash \theta(\kappa, \lambda) \rightarrow \varphi_0^{R_\lambda}(\kappa)$,

(b3) $\mathrm{ZF} \vdash \theta(\kappa, \lambda) \wedge \varphi^{R_\lambda}(\kappa) \rightarrow \bar{\varphi}^{R_\lambda}(\kappa)$,

then φ_0 is a large cardinal property and $\varphi \mapsto \bar{\varphi}$ is an operation which preserves large cardinal properties.

3.2. *Formal consistency*. One further point can be made here. The theory ZA (ZF + (S2) + $V = R_\Omega$) is provably consistent relative to ZF (see Lévy [**L1**]). But the proof of this does not appear to provide an interpretation of ZA which gives Ω the role which we are suggesing here. In fact, the formal consistency of ZA merely asserts the existence of models $\mathfrak{A}$, $\mathfrak{B}$ such that $\mathfrak{A} \prec \mathfrak{B}$, etc., and is thus not as strong as $\exists\kappa, \lambda E(\kappa, \lambda)$, which we have already observed does not capture the role of Ω.[4]

4. Bernays' reflection principle using classes. The rather formalistic approach via (S2) which we have taken to reflection principles has the advantage that (ostensibly at least) we are never required to accept the existence of objects other than sets, or of relations other than $\in$. Everything else is merely a formal device. The appropriateness of the device must be judged informally. In its defense it may be pointed out that, assuming we want to think of V at all, the analogical procedure (think about it as you would think about anything else) seems to have an edge until either (a) it runs into trouble or proves fruitless or (b) some informal considerations suggest an alternative to classical logic or classical set theory which has some chance of proving fruitful in this context.

One disadvantage is that we would like a better account of the passage from "all definable F" to "all F." Allowing proper classes to exist would seem to be a fair price to pay for this, especially since proper classes seem needed anyway just to *express* the properties we are concerned with (such as inaccessibility). We therefore consider means of expressing reflection principles in the language of class-set theory (GB or KM). (Since we now have objects which are not sets, the universality of set theory is threatened. We shall return to this point.)

[4] Some further remarks may help clarify the situation. If N is a model of GB, then $(V^N, \in^N)$, or V^N for short, is a model of ZF. Similarly for ZA. If C is a class of models of the language for ZF, write $T \vdash_C \sigma$ iff $\forall N(N \models T \,\&\, V^N \in C \rightarrow V^N \models \sigma)$. Since *any* model M of ZF can be extended to a model N of GB having the same sets, for all C we have $\mathrm{ZF} \vdash_C \sigma$ iff $\mathrm{GB} \vdash_C \sigma$, and we may regard ZF and GB as essentially the same theory of sets. The new objects of GB can be regarded as a mere *facon de parler*. (Of course, one need not so regard them, and indeed Gödel did not regard them this way in his monograph introducing L; but this is beside the point.) Although any model M of ZF gives rise via Lévy's proof to a model N^* of ZA, N^* is not in general an extension of M having the same sets (M and $(V^*, \in^*)$ need not be isomorphic). Thus V does not capture the notion of set of the original interpretation M and thus does not play the role we desire. In fact, the equivalence $\mathrm{ZF} \vdash_C \sigma$ iff $\mathrm{ZA} \vdash_C \sigma$ fails if C is the class of natural models R_α (ZA is the stronger theory here). Nevertheless, Lévy's proof shows that $\mathrm{ZF} \vdash \sigma$ iff $\mathrm{ZA} \vdash \sigma^V$ iff $\mathrm{ZA} \vdash \sigma$, so that (in a weaker sense than for GB) ZA and ZF are essentially the same theory, and in this weaker sense the "imaginary" objects can be regarded as a *facon de parler*.

Let us suppose that $\kappa < \Omega$ is going to reflect Ω. Then sets correspond to elements of R_κ, and proper classes to subsets of R_κ. Thus if we wish to allow proper classes $F \subseteq R_\Omega$ as parameters in reflectable statements, it is quite natural to suppose that what is true of F will be true of $F \cap R_\kappa$ in the reflecting universe. This suggests the schema

$$\text{(S3)} \qquad \theta(F, G) \to (\exists\kappa < \Omega)\theta^{R_{\kappa+1}}(F \cap R_\kappa, G \cap R_\kappa)$$

(since statements θ are really of the form $\theta^{R_{\Omega+1}}$ in class-set theory). Note that this is close to the form of (P1).

4.1. *An interpretation eliminating imaginary sets in favor of sets.* Using (S3) one can immediately give an interpretation of (S2) which is closer to (P1). Let F be (or code) the Skolem functions for R_Ω. Then there is a θ which says R_Ω is closed under the Skolem functions. Thus choosing κ for this θ, R_κ is closed also, i.e.,

$$\text{(4.1)} \qquad R_\kappa \prec R_\Omega.$$

(Evidently one can conjoin other properties of Ω with θ before using (S3) to get κ. Those familiar with KM will note that it will suffice to get the unembellished (4.1).) Now take the quantifiers of (S2) to range over R_Ω, and take V to be R_κ. Note that in this interpretation no quantifiers go outside R_Ω. However, the definition or proof of existence of κ appears to require going beyond R_Ω.

4.2. *Eliminating imaginary sets in favor of classes.* Still another interpretation of (S2) exists on the basis of (S3). We can show, in effect, that there is a transitive "set" M such that

$$\text{(4.2)} \qquad (R_\Omega, \in) \prec (M, \in).$$

Actually we can only get M up to isomorphism; the exact statement is: There exists a well-founded relation $E \subseteq R_\Omega$ (say with field M), a point $\bar{\Omega}$ in M, and an isomorphism j such that

$$(R_\Omega, \in) \cong_j (R_{\bar{\Omega}}^M, E) \prec (M, E).$$

The proof (by contradiction) is easy. If it fails, choose an inaccessible κ reflecting the statement that it fails, and such that $R_\kappa \prec R_\Omega$. By the downward Löwenheim-Skolem theorem, there is a transitive set M of cardinality κ such that $R_\kappa \prec M$, a contradiction.

We remark that these interpretations are in a sense nonstandard: In the first, the "large cardinal" κ may not have all large cardinal properties it should; in the second the properties of Ω may not relativize properly to M. However, in either case the interpretation is adequate for expressing first-order properties of R_Ω.

4.3. *Reflection arguments using Bernays' schema.* Using (S3), it is possible to carry out all the informal arguments given earlier (and even prove the replacement axiom). For details see Bernays [**B1**], where (S3) is introduced as the basis of a very elegant axiomatization of class-set theory. (By taking care in the relativization of equality, Bernays even avoids the necessity of relativizing the parameters F to $F \cap R_\kappa$.)

5. Imaginary sets and classes; the set-class distinction. We may regard Bernays' schema as a reflection principle for second-order logic over set theory. There are obvious generalizations to higher orders; these are the so-called indescribability conditions. (Bernays' schema asserts that Ω is second-order indescribable. The notion of indescribability is discussed in Lévy [**L3**].) Passing to higher types gives stronger axioms. None of these generalizations, however, allows objects of type higher than classes of sets as parameters in the reflected statements. Since it is the presence of the classes as parameters in Bernays' schema which appears to be responsible for its great strength, we would eventually like to continue by formulating a third-order reflection principle which allows third-order objects as parameters. It does not appear to be possible to generalize (S3) in this direction very directly. (The reader who does not believe this is welcome to try.) We therefore return to the considerations that led to (S2), and ask whether we can formulate a version of (S2) appropriate to class-set theory.

5.1. *The set-class distinction*; *a projection schema.* In class-set theory we have classes as distinct from sets; but if we conceive of them merely as collections, this looks like a distinction without a difference. In particular, we seem to have nothing not contained already in (S2) plus "V is inaccessible." Moreover, the classes threaten the universality of set theory. (This perhaps is why many mathematicians find ZF far more natural than KM; our idea of set comes from the cumulative hierarchy, so if you are going to add a layer at the top it looks like you just forgot to finish the hierarchy.) A proper class P may however be distinguished from a set x in the following way (if the reader will indulge another counterfactual conditional): If there were more ordinals (or if, as in (S2), Ω were an ordinal), x would have exactly the same members, whereas P would necessarily have new elements. We could say that the extension of x is fixed but that of P depends on what sets exist. Roughly, x *is* its extension, whereas P has more to it than that.[5] Notice that in (S2) the definable properties behave in just this way; the extension of θ is $P = \{x \in V | \theta(x)\}$, whereas in the projected universe it becomes $\{x | \theta(x)\}$. (Thus θ determines a set just in case the two are the same—which agrees with the "size" criterion which is frequently used to motivate the axioms of ZF.) This tells us how to formulate the analogue of (S2) for class-set theory: Write jP for the extension of P in the formally projected universe; then the schema we want is

$$\text{(S4)} \qquad (\forall x, y \in V)(\forall P \subseteq V)[\theta^{\mathscr{P}V}(x, y, P) \leftrightarrow \theta(x, y, jP)].$$

Syntactically j is a unary function symbol. We allow it in the comprehension principle appropriate to class-set theory:

$$\exists z \forall t[t \in z \leftrightarrow \theta \wedge \exists u(t \in u)].$$

Otherwise (S4) is just like (S2). (x, y, P exhaust the free variables of θ; V is a new individual constant; $\theta^{\mathscr{P}V}$ indicates relativization of all quantifiers to "$x \subseteq V$"; we suppose the usual axioms for V and that V has the form R_Ω.)

[5] We could say that P contains the famous three dots or "etc." of mathematics in an essential way.

Note that if $x \in V$, also $x \subseteq V$, so that $x = x \leftrightarrow x = jx$ is an instance of (S4). This assures that j is the identity on V.

In (S2), the original universe V consists of sets; in the formal projection we treat V as a set, thus projecting new sets. In (S4), the original universe consists of sets and classes of sets. In the formal projection we introduce new sets and classes of *these* as well; among these are the jP.

Schema (S4) asserts that sentences of the theory of sets, allowing quantifiers over classes, and both sets and classes as parameters, are true in the projected universe when the class parameter X is given its proper extension jX.

5.2. *Model theoretic version.* The model theoretic condition corresponding to (S4) is $E^1(\kappa; \lambda)$:

DEFINITION 5.1. We write $E^1(\kappa; \lambda)$ and say that κ is second-order extendible to λ in case

$$\kappa < \lambda \text{ and } \exists M, j: R_{\kappa+1} \to M \subseteq R_{\lambda+1} \text{ such that}$$

$$\text{(i) } (M, \in) \prec (R_{\lambda+1}, \in), \tag{5.1}$$

$$\text{(ii) for all } x \in R_\kappa, jx = x.$$

Evidently $j\kappa = \lambda$; we call κ the critical point of the embedding j (since it is the first point moved by j). It is κ which is intended as the large cardinal. In the sequel we write (5.1) in the more abbreviated form

$$j: R_{\kappa+1} \cong \prec R_{\lambda+1},$$

understanding that κ is the critical point (and hence $\kappa < \lambda$).

Note that this condition can also be expressed by saying that the complete theory of the structure $(R_{\kappa+1}, \in, x, X)_{x \in R_\kappa, X \in R_{\kappa+1}}$ has an interpretation $(R_{\lambda+1}, \in, x, X')_{x \in R_\kappa, X \in R_{\kappa+1}}$ in which κ is a set ($\kappa < \lambda$).

5.3. *Universality of set theory: applicability to classes.* In conceiving classes $P \subseteq V$ we must decide whether their extensions are to be treated in a perfectly classical way, or not; whether, for example, the axiom of choice is to hold. If set theory is as universal as we intend, it should be applicable to any collections, including these. We note that (S4) decides strongly in favor of such a classical treatment. The reason for this is that while classes $P \subseteq V$ are treated as having "potential extensions" jP, they (or, if you prefer, their extensions) are simultaneously treated formally as sets. In the formally projected universe, V is a set, so we have in effect identified projected subsets of V with the extensions of our proper classes. Since the theory of V is supposed to apply in the projected universe, $V = R_\Omega$ and $R_{\Omega+1}$ are treated classically. In this way we mitigate the threat to the universality of set theory which is posed by the introduction of proper classes.

5.4. *Virtues of the class-set projection schema.* So far we have introduced (S4) only as a formal device. (We prefer to give it some interpretation. In the sequel we will suggest, for example, interpretations analogous to the interpretations (4.1), (4.2) which Bernays' schema (S3) provides for the first-order case (S2).) As such it has some noteworthy virtues. (i) It works extremely smoothly in formalizing the earlier informal arguments (it is only necessary to read $(jF)\Omega$ for $F\Omega$). (ii) Bernays' schema (S3) follows easily from it. (iii) As we shall see, it is very easy to generalize it to allow higher-order parameters. (iv) It allows us to make an intelligible conceptual distinction between sets and proper classes.

5.5. *References.* The condition (S4) (or its model theoretic version) was first formulated by Silver following a suggestion of Reinhardt. It is closely related to the set theory of Ackermann, to informal ideas of Shoenfield, and to a set theory of Powell (for details, see Jech-Powell [**J1**], Reinhardt [**R1**], and Shoenfield [**S1**, p. 234]). The condition was used by Silver [**S2**] in his proof of the relative consistency of "$2^\kappa \neq \kappa^+$ and κ measurable."

5.6. *Free historical remark.* Formal devices such as j and the idea of analogical predication are hardly new to the history of thought. The author recently stumbled on an early use of a j-like device by the writer known as Pseudo-Dionysius (fifth century(?)); Greek scholars can probably find earlier antecedents. According to the "negative way" of Proclus as followed by Pseudo-Dionysius, created beings are (for example) wise or unwise (as the case may be), but the Creator is neither; he is superwise. According to (S4), of course, ordinals are accessible or not as the case may be, but Ω is j (inaccessible). See Coppleston [**C1**, pp. 51–52; also compare pp. 61–62].

6. Higher type objects (Ω-classes); representing imaginary sets and classes using classes (Ω-classes). We wish to indicate how (S4) may be interpreted. The chief difficulty here is explaining what kind of object the quantifiers range over. Before assuming there are objects satisfying the axioms we would like at least to know what sort of things they are going to be. Of course we can suppose there is a set which is a model, but this defeats the intention that the R_Ω part of the theory really applies to Cantor's universe and that proper classes are really classes of sets. On the other hand, if we introduce new objects (even proper classes) beyond Cantor's universe this violates the universality of the concept of set. We propose to mitigate this sorrow by seeing the universality not in the *extension* of the concept of set but in the applicability of the *theory* of sets. (We have already seen how this works in the case of classes in observing how (S4) treats classes classically.)

6.1. *Ω-classes.* The simplest way to proceed is perhaps the following. We have now introduced proper classes of sets as distinct from sets. (To be sure, the distinction is drawn only by considering "imaginary" sets and classes.) In a similar way we can consider classes of proper classes, classes of these, etc. We call these Ω-classes, since they are built up over Ω. The manner in which proper classes are not "purely extensional" is indicated by (S4). In an analogous manner, the Ω-classes are conceived as "nonextensional"; this will be spelled out below. They may, however, also be considered purely extensionally.[6] Since we regard set theory (the theory of Ω) as *the* theory of extensional objects such as sets, collections, etc.,

[6] That is, without projecting any imaginary universe. The essential nonextensionality appears only when this is done.

we assume this theory applies to Ω-classes. This is expressed by the schema (S2), which we adopt explicitly here.

AXIOM 6.1. *We adopt the schema* (S2) *interpreting* "$\forall x$" *as "for all* Ω*-classes* x," Ω *as the Cantor series of ordinals, and* "V" *as* R_Ω.

Of course, this assumption gives us an enormously rich supply of Ω-classes.

In order to express the distinctive characteristics of Ω-classes we could now introduce imaginary sets and imaginary Ω-classes. However, we wish to avoid as much as possible an unending series of extensions of the types of objects allowed in our theories (especially "imaginary" objects). Therefore we shall use the Ω-classes we already have, making free use of the natural models R_β ($\Omega < \beta$) which are available, to represent imaginary objects and to explain the special axioms for Ω-classes. Thus while our set-class distinction does presuppose the idea of a set-possible set distinction, we do avoid the introduction of possible sets. This still makes liberal use of Ω-classes, so we shall examine how many are needed for certain purposes; but at first we ignore such niceties.

6.2. *The special axiom for* Ω*-classes.* Let $\lambda > \Omega$, so that R_λ is a collection of Ω-classes. We wish to consider a realm $R_{\Omega'}$ of imaginary sets, and a corresponding realm $R_{\lambda'}$ of imaginary Ω-classes, in which Ω is an imaginary set. Moreover we want classes of sets to correspond to classes of imaginary sets, and other Ω-classes x to correspond appropriately to other imaginary classes jx in such a way that the theory is preserved. In the case $\lambda = \Omega + 1$, this preservation is expressed by the satisfaction of (S4); in model theoretic terms for R_λ, $R_{\Omega'}$, $R_{\lambda'}$, this says that there exists $j: R_\lambda \to R_{\lambda'}$, such that

(6a) (i) $\Omega < j\Omega = \Omega' < \lambda'$,

(ii) $\forall x < \Omega, jx = x$,

(iii) if $M = \{jx | x \in R_\lambda\}$, $(M, \in) \prec (R_{\lambda'}, \in)$.

In addition, in the case $\lambda = \Omega + 1$ we automatically have

(6b) (iv) $\lambda < \Omega'$

which, of course, says that the Ω-classes considered are in the realm of imaginary sets. It is (6) which we use to indicate the special character of the Ω-classes.

We put these conditions in a formal definition.

DEFINITION 6.2. Let Ω, λ, Ω', λ' be ordinals, j a function $j: R_\lambda \to R_{\lambda'}$.

(a) We write $j: E_0(\Omega, \lambda; \Omega', \lambda')$ in case $\lambda > \Omega$ and the conditions (6a)(i)–(iii) above are satisfied. We call Ω the critical point of the embedding j. In case the condition (6b)(iv) above is also satisfied, we write $j: E(\Omega, \lambda; \Omega', \lambda')$.

(b) If there is such a j we write simply $E_0(\Omega, \lambda; \Omega', \lambda')$ or $E(\Omega, \lambda; \Omega', \lambda')$ and say that Ω is λ-extendible (to Ω').

(c) We say that Ω is extendible if it is λ-extendible for every $\lambda > \Omega$.

We make explicit our special assumption about Ω-classes.

AXIOM 6.3. *For every* Ω*-class* λ *which is formally an ordinal,* Ω *is* λ*-extendible.*

Several comments are needed about (6). (a) Although it is obvious, we note that it is Ω that is the large cardinal here and not λ. Of course if one thinks of R_λ as all

Ω-classes, or all existing classes, or the like, it seems reasonable to expect that properties which hold of λ will be large cardinal properties; but this does require an additional conceptual step. (b) The condition (i) says that Ω is an imaginary set, or is imagined as a set, (ii) that sets are allowed as parameters without application of j (as in schemas (S2) and (S4)), and (iii) that truth is preserved in the projected universe. The condition (iv) does not seem implicit in the idea of projecting imaginary sets and Ω-classes. However, if we want to treat not only Ω but also the Ω-classes in R_λ as sets, it is clearly necessary that $\lambda \leqq \Omega'$. Thus by taking an Ω' for $\lambda + 1$ we get $\lambda < \lambda + 1 \leqq \Omega'$, so that (iv) is also seen to be a natural condition here.

6.3. *Ω-classes = possible Ω-classes*? Having introduced Ω-classes, we may ask whether it makes sense to consider all possible Ω-classes. Since we already have so much difficulty with all possible *sets*, one may expect at least analogous difficulties here. However, if we do think of λ as representing all possible Ω-classes, it seems fairly clear that $\lambda = \lambda'$. (As λ' cannot be smaller than λ, it has nowhere else to go if λ is all *possible* Ω-classes. In other words λ remains the same in the imaginative projection while Ω changes to Ω'.) This is clearly incompatible with (iv) as we surely must have $\Omega' < \lambda'$ ($\Omega < x < \lambda$ therefore $\Omega' = j\Omega < jx < \lambda'$); but we observed that (iv) did not seem forced by our guiding idea. If we now retain the conditions (6a), we are led to the conclusion that $(R_\lambda, \in)$ has a proper elementary embedding into itself. In 1967 it seemed plausible to the author that such an absolute treatment of λ should be possible, and hence that there should be inaccessible cardinals λ of this sort. Such an axiom was proposed in the author's thesis [**R2**]. This however was a mistake: Kunen has proved (using the axiom of choice) that any such ordinal λ must either be of the form β or $\beta + 1$ where β has cofinality ω [**K1**]. Whether the cofinality condition is necessary and whether the AC is inessential or alternatively the condition provides an interesting way of violating AC (as with the axiom of determinateness) remain open problems.

The above line of thought which suggests a proper elementary embedding of R_λ into itself can be set in a slightly different context. There is a natural tendency to try to keep everything within the realm of sets, and to regard all talk of Ω as actually about an ordinal κ which reflects Ω to a suitable degree. (We of course have opted to allow Ω-classes and to secure the universality of set theory in another way.) If one does this it is natural to consider κ which are λ-extendible for every ordinal λ. But in this context (keeping everything within the realm of sets), the next step, namely κ which are ∞-extendible (meaning one j which works for all λ) is just like setting Ω-classes = imaginary Ω-classes above and yields a proper elementary embedding of the universe of sets into itself. This was the approach of [**R2**], where it was insisted that the contemplation of alternatives to the actual universe of sets (here accomplished by the distinction set-imaginary set, and represented using Ω-classes) be in terms of representations of V which are sets.

6.4. *An alternative motivation.* There is an alternative motivation for elementary embedding conditions which may also suggest considering the condition $V \cong \prec V$ (proper). It is well known (from Scott [**S3**]) that if κ is a measurable cardinal, then there is an elementary embedding j of the universe of regular sets V into a well-founded class of sets M, with κ the first ordinal moved by j. Moreover, as stronger closure conditions are imposed on M (M closed under sequences of length 2^{κ}, $2^{2^{\kappa}}$, etc.), the assertions become arithmetically stronger (actually, κ increases in size). This suggests various elementary embedding conditions including $V \cong \prec V$.

While this line of thought is motivated by the desire for stronger assertions and by a natural generalization, and does motivate some of the conditions we introduced earlier, it is not the original motivation. (Kunen [**K1**] presents it in such a way that the reader might think it was.) Rather, it is very close to the motivation for considering irreducible covers (see § 7). Moreover as this presupposes the idea of a measure, and does not seem to have much direct connection with the idea of reflection or other basic ideas, it seems to have quite different foundational relevance than the approach of §§ 6.2, 6.3.

6.5. *A priori possibilities for Ω-classes.* If R_λ are Ω-classes, and Ω', $R_{\lambda'}$ represent respectively the imaginary ordinals and the imaginary Ω-classes corresponding to R_λ, then there are four a priori possibilities for the order relations among $\Omega, \lambda, \Omega', \lambda'$. Clearly $\Omega < \Omega' < \lambda'$, and $\Omega < \lambda \leqq \lambda'$. Thus the possibilities are given by the placement of $\lambda : \lambda < \Omega'$, $\lambda = \Omega'$, $\Omega' < \lambda < \lambda'$, and $\lambda = \lambda'$. The first and last have been considered above. We note here the four conditions are of increasing strength.

THEOREM 6.4. *Let $\Omega, \lambda_0, \ldots$ be any ordinals. The conditions*

(i) $j : E_0(\Omega, \lambda_0; \Omega', \lambda_0') \,\&\, \lambda_0 = \lambda_0'$,

(ii) $j : E_0(\Omega, \lambda_1; \Omega', \lambda_1') \,\&\, \Omega' < \lambda_1 < \lambda_1'$,

(iii) $j : E_0(\Omega, \lambda_2; \Omega', \lambda_2') \,\&\, \lambda_2 = \Omega'$,

are of decreasing strength:

(a) *if* (i) *holds,* $\&\, \lambda_1 = j\Omega', \lambda_1' = j\lambda_1$, *then* (ii) *holds, and*

(b) *if* (ii) *holds,* $\&\, \lambda_2 = \Omega', \lambda_2' = j\lambda_2$, *then* (iii) *holds.*

Moreover, if (iii) *holds, then in the universe* R_{λ_2}, Ω *is extendible, i.e.,* $\forall \lambda < \lambda_2$, $\exists \Omega', \lambda' < \lambda_2$ *such that*

(iv) $E_0(\Omega, \lambda; \Omega', \lambda')$ *and* $\lambda < \Omega'$.

PROOF. The first part is obvious. For the second, suppose (iii) holds. We note that $R_\Omega \prec R_{\Omega'}$, i.e., that, in $R_{\Omega'} = R_\lambda$,

$$(\forall x \in R_\Omega)(\theta^{R_\Omega}(x) \leftrightarrow \theta(x)),$$

which is a statement of the form $\varphi(\Omega)$. Now by the embedding of R_λ into $R_{\lambda'}$, this means that in $R_{\lambda'}$, $\varphi(j\Omega)$, i.e., that $\varphi(\lambda)$ or

$$(\forall x \in R_\lambda)(\theta^{R_\lambda}(x) \leftrightarrow \theta(x))$$

which means that $R_{\lambda'} \prec R_\lambda$. Since, for any $\beta < \Omega'$, Ω is clearly β-extendible in $R_{\lambda'}$, it must also be so in R_λ as desired.

7. Measures, supercompactness. We now introduce the notion of irreducible cover. This notion arises naturally when one investigates κ-complete ultrafilters

(0-1 measures). It can be regarded as a generalization of the notion of normal measure. (A κ-complete measure μ on κ is said to be normal in case (i) if $\nu < \kappa$, $\mu\{x \in \kappa | \nu < x\} = 1$, (ii) if $f : \kappa \to \kappa$ and $\mu\{x | fx \in x\} = 1$, then there is $\nu \in \kappa$ such that $\mu\{x | fx = \nu\} = 1$. Every measurable cardinal has a normal measure.) A supercompact cardinal is one admitting an irreducible cover of every set. The terminology and this form of the definition are due to Solovay, the notion independently to Solovay and the author. In § 7, Axioms 6.1, 6.3 are not used.

DEFINITION 7.1. (a) $\mu \subseteq \mathscr{P}S$ is called an ultrafilter on S in case for all $X, Y \subseteq S$,

(i) $X \in \mu$ or $S \sim X \in \mu$,

(ii) if $X \in \mu$ and $X \subseteq Y$, then $Y \in \mu$,

(iii) if $X, Y \in \mu$ then $X \cap Y \in \mu$.

(b) An ultrafilter μ is called κ-complete if whenever $\lambda < \kappa$ and for all $\nu < \lambda$, $X_\nu \in \mu$, also $\bigcap\{X_\nu | \nu < \lambda\} \in \mu$.

(c) A κ-complete ultrafilter μ is said to be of degree κ if it is not κ^+ complete.

DEFINITION 7.2. Suppose μ is an ultrafilter of degree κ on S. Then

(a) A function $f : S \to \mathscr{P}A$ is said to be a cover of A (of degree κ) in case

(i) for each $a \in A$, $\{x \in S | a \in fx\} \in \mu$,

(ii) card $fx < \kappa$ for each $x \in S$.

(b) A cover of A is said to be irreducible if in addition

(iii) whenever g is a function on S and $\{x \in S | gx \in fx\} \in \mu$, there is an $a \in A$ such that $\{x \in S | gx = a\} \in \mu$.

DEFINITION 7.3. (a) A cardinal κ is said to be strongly compact in case every set has a cover of degree κ.

(b) A cardinal κ is said to be supercompact in case every set has an irreducible cover of degree κ.

We begin by observing that every cover of A of degree κ determines a cover whose space is $\mathscr{P}_\kappa A = \{H \subseteq A \mid \text{card } H < \kappa\}$ and whose covering function is the identity. Namely, if μ' is the measure on S, define μ on $\mathscr{P}_\kappa A$ by $X \in \mu$ iff $\{x \in S | fx \in X\} \in \mu'$. The degree of μ is also κ, and if μ' is an irreducible cover so is μ.

THEORM 7.3. *If κ is extendible then κ is supercompact.*

PROOF. Let A be given; we will get an irreducible cover for A. Choose $\lambda > \kappa$ so that $A \in R_\lambda$. For convenience we also suppose λ a limit ordinal. By extendibility, choose j, λ' so $j : R_\lambda \cong R_{\lambda'}$ with κ as critical point. We now induce an ultrafilter μ on $\mathscr{P}_\kappa A$ by

$$X \in \mu \quad \text{iff} \quad j''A \in jX$$

where $j''A = \{ja | a \in A\}$. Using the identity function, this provides the required cover of degree κ.

We omit the easy proof that μ is an ultrafilter. To see it is κ-complete, let $X_\nu \in \mu$, $\nu < \lambda < \kappa$. Now let $Y = \{X_\nu | \nu < \lambda\}$. Since $\lambda < \kappa$, $j\lambda = \lambda$, so $jY = \{(jX)_\nu | \nu < \lambda\}$. Similarly $j(X_\nu) = (jX)(j\nu) = (jX)_\nu$, so $jY = \{j(X_\nu) | \nu < \lambda\}$. Evidently $j''A \in jY$ since it is in each $j(X_\nu)$.

To see that μ is a cover, we must see that for each $a \in A$, if $X_a = \{H \in \mathscr{P}_\kappa A | a \in H\}$, then $X_a \in \mu$. Now $jX_a = \{H \in \mathscr{P}_{\kappa'}(jA) | ja \in H\}$, where $\kappa' = j\kappa$. Obviously $ja \in j''A$, so to get $j''A \in jX_a$ it will suffice to see that $j''A \in \mathscr{P}_{\kappa'}(jA)$. Evidently if $a \in A$ then $ja \in jA$, so $j''A \subseteq jA$. Since card $j''A$ = card A and $A \in R_\lambda \in R_{\kappa'}$, we must in fact have $j''A \in \mathscr{P}_{\kappa'}(jA)$. Thus μ is a cover.

To see that μ has degree κ, note that if $H \in \mathscr{P}_\kappa A$, then $\bigcap\{X_a | a \in H\} \in \mu$. Thus if $\alpha < \kappa$, $\{H \mid \text{card } H = \text{card } \alpha\} \notin \mu$, and we have κ sets of measure 0 whose union is the whole space.

It remains to see that the cover is irreducible. Suppose $X = \{x \in \mathscr{P}_\kappa A | gx \in x\} \in \mu$. Then $jX = \{x \in \mathscr{P}_{\kappa'}(jA) | (jg)x \in x\}$, and since $j''A \in jX$, $(jg)(j''A) \in j''A$, i.e., there is $a \in A$ such that $(jg)(j''A) = ja$. But this means that $j''A \in jY$, where $Y = \{x | g(x) = a\}$. Thus we indeed have an irreducible cover.

REMARK 7.4. For those familiar with the linguistic definition of strong compactness, we sketch the proof of the existence of a cover of A from the κ-compactness of $L_{\kappa,\kappa}$. Let $A \in R_\lambda$, λ a limit ordinal. Let T be the complete $L_{\kappa,\kappa}$ theory of $(R_\lambda, \in)$. Add to the language constants c, d and sentences saying "$\nu < c < \kappa$" for each $\nu < \kappa$, and "$a \in d \subseteq A$ & card $d < \kappa$" for $a \in A$. If ζ is the new set of sentences, clearly each subset of ζ of cardinality $<\kappa$ is consistent. Since the sentences include well-foundedness and extensionality, we may take the model of ζ to be a transitive set $(M, \in)$. Putting $jx = x^M$, we have $j:(R_\lambda, \in) \simeq \prec(M, \in)$ with κ as critical point. There is just enough in M so that the argument of the preceding theorem can be carried through to show that A has a cover (use d^M instead of $j''A$ to induce the ultrafilter). Thus compactness is actually characterized by the existence of j, M such that for some $d \in M$, $j''A \subseteq d \subseteq jA$ and $(\text{card } d)^M < j\kappa$. (The argument from a cover to such an M can be given directly using an ultrapower of R_λ for M.) The only additional thing needed for an irreducible cover is that the type "$x \neq a$ & $x \in d$" ($a \in A$) be omitted. Thus it seems that supercompactness adds some kind of type-omitting condition to compactness; how to formulate it precisely is not clear however. Supercompactness can also be characterized by the condition on M: One obtains such an M from an ultrafilter μ on $\mathscr{P}_\kappa A$ by taking the ultrapower R^μ_λ, which is isomorphic to such an M. The embedding is the canonical embedding into ultrapowers, and it turns out that the identity function on $\mathscr{P}_\kappa A$ represents $j''A$.

8. Using fewer Ω-classes. In order to express the idea of projecting imaginary sets and classes we made free use of Ω-classes. Now we try to be more efficient. First there is a rather trivial way we can use fewer Ω-classes: We can weaken our representation of the projected universe of imaginary objects, and use simply a well-founded transitive model M instead of a natural model $R_{\lambda'}$. Let us look at the case $\lambda = \Omega + 1$. If $R_{\Omega+1} \cong \prec R_{\Omega'+1}$ then by the downward Löwenheim-Skolem theorem and the collapsing lemma there is an M of cardinality 2^Ω with $R_{\Omega+1} \subseteq M$ and $R_{\Omega+1} \cong \prec M \cong \prec R_{\Omega'+1}$, and M is therefore a model of (S4) and the accompanying comprehension schema. Now we can suppose the elements of M to be generated by a countable number of Skolem functions, and thus to be represented by terms with sets and classes as parameters, all of which can of course be construed as elements of $R_{\Omega+1}$. It thus seems that in a sense the ontology is limited to classes, while obtaining the benefits of thinking of proper classes as sets (with the ensuing higher type objects). The crucial thing however is the relation of membership among the terms, and a treatment of this requires a further type level. However in this way we can express a significant part of the idea behind (S4) using the universe $R_{\Omega+2}$. This is somewhat analogous to (4.2).

We easily see that there are ordinals $\kappa_0 < \kappa_1 < \Omega$ such that $E^1(\kappa_0;\kappa_1)$ and $E^1(\kappa_1;\Omega)$, which yields an analogue of (4.1): Choose Ω', j so $j:R_{\Omega+2} \cong \prec R_{\Omega'+2}$, Ω critical point. In $R_{\Omega'+2}$ it is expressible, and true, that $E^1(\Omega;\Omega')$. Thus $\exists\kappa E^1(\kappa;\Omega')$ must be true in $R_{\Omega+2}$ also, i.e., $\exists\kappa_0 < \Omega E^1(\kappa_0;\Omega)$. Repeating the argument yields κ_1 with the desired properties. In fact, if μ is the normal measure on κ given by

$$X \in \mu \leftrightarrow j''\kappa = \kappa \in jX,$$

then $\{\kappa \in \Omega | E^1(\kappa;\Omega)\} \in \mu$. Similarly if μ on $\mathscr{P}_\kappa(R_{\Omega+1})$ is induced by $j''(R_{\Omega+1})$, then $\{H \subseteq R_{\Omega+1} | (\exists\kappa \in \Omega) R_{\kappa+1} \cong H \prec R_{\Omega'+1}\} \in \mu$. Similar results hold for higher types of extendibility. Note that κ_0 provides an interpretation of Ω which appears completely adequate for second-order statements.

9. Appraisal. Some comments about what we have tried to accomplish in the introduction of the concept of extendible cardinal (through § 6.2) may be helpful. We have tried to indicate that Cantor's Ω is extendible. It is difficult to describe the epistemological force of this indication; whether for example it should be called a plausibility argument or an informal proof, or just a proposal for a theory abouts sets and properties of sets. Ordinarily a plausibility argument can represent either the understanding of a proof, if you already have the proof, or a heuristic for finding a proof, if you do not yet have one (or a mistake if it is simply wrong). But in the case of an axiom, what does a plausibility argument represent? Perhaps some insight, or structured collection of insights, into the workings of some concepts (somewhat like imaginary experiments). In this case the term proof does not seem inappropriate if the insights are sharp enough. However, such informal proofs do not always carry conviction. Rather they are attempts to put in coherently organized form convictions we have perhaps together with some ideas about the way those convictions arise. While such convictions undoubtedly have an empirical content, we believe it is a gross exaggeration to say, for example, that we accept ZF because experience has not led to a contradiction (neither has NF). Our experience with ZF has perhaps as much to do with the ideas of ZF and the development of our intuition about sets as with the formal manipulations we make using the theory. There is also the element of appealing to the meanings of concepts, and the judgment of naturalness in elaborating and generalizing such meanings, and the recognition of analogies. The process by which we decide that some axioms (such as ZF) are correct, or have correct consequences, is of course not well understood. (It does not seem to me that of itself this casts any doubt on the validity of the decisions. Skepticism applied at the right point always has at least the virtue of preventing hubris, but applied at the wrong point it may not be beneficial.) Part of the process may involve getting experience with the axioms (such as the results Solovay has obtained concerning the GCH using large cardinals, or even the simple observations of §§ 7,8). Nevertheless, there must be a first step in recognizing axioms, and the foregoing introduction attempts to take such a first step, a step which will make the axioms seem worth considering as axioms rather than merely as conjectures or speculations. Thus we have attempted more than simply to motivate the mathematical concept of extendible cardinal. Since success in such informal ventures is not measured by the outcome of a known recursive procedure, the attempt may miss its mark. But the potential gain seems worth the risk; we must always remember that the incompleteness theorems show that large cardinal considerations have a bearing even on arithmetical problems.

REFERENCES

[**A1**] W. Ackermann, *Zur Axiomatik der Mengenlehre*, Math. Ann. **131** (1956), 336–345. MR **18**, 3.

[**B1**] P. Bernays, *Zur Frage der Unendlichkeitsschemata in der axiomatischen Mengenlehre*, Essays on the Foundations of Mathematics, Magnes Press, Hebrew Univ., Jerusalem, 1961, pp. 3–49. MR **30** #1034.

[**C1**] F. Coppleston, *A history of medieval philosophy*, Methuen, London, 1972.

[**H1**] W. Hanf and D. Scott, *Classifying inaccessible cardinals*, Notices Amer. Math. Soc. **8** (1961), 445.

[**J1**] T. Jech and W. Powell, *Standard models of set theory with predication*, Bull. Amer. Math. Soc. **77** (1971), 808–813. MR **43** #6074.

[**K1**] K. Kunen, *Elementary embeddings and infinitary combinatorics*, J. Symbolic Logic **36** (1971), 407–413.

[**L1**] A. Lévy, *On Ackermann's set theory*, J. Symbolic Logic **24** (1959), 154–166. MR **23** #A3667.

[**L2**] ———, *Axiom schemata of strong infinity in axiomatic set theory*, Pacific J. Math. **10** (1960), 223–238. MR **23** #A1522.

[**L3**] ———, *On the sizes of the indescribable cardinals*, Proc. Sympos. Pure Math., vol. 13, Part I, Amer. Math. Soc., Providence, R.I., 1971, pp. 205–218. MR **43** #7321.

[**L4**] A. Lévy and R. L. Vaught, *Principles of partial reflection in the set theories of Zermelo and Ackermann*, Pacific J. Math. **11** (1961), 1045–1062. MR **25** #2951.

[**R1**] W. Reinhardt, *Set existence principles of Shoenfield, Ackermann, and Powell*, Fund. Math. **83** (1974), 12–41.

[**R2**] ———, *Topics in the metamathematics of set theory*, Doctoral Dissertation, University of California, Berkeley, Calif., 1967.

[**R3**] ———, *Ackermann's set theory equals* ZF, Ann. Math. Logic **2** (1970), no. 2, 189–249. MR **43** #42.

[**S1**] J. Shoenfield, *Mathematical logic*, Addison-Wesley, Reading, Mass., 1967. MR **37** #1224.

[**S2**] J. Silver.

[**S3**] D. Scott, *Measurable cardinals and constructible sets*, Bull. Acad. Polon. Sci. Sér. Sci. Math. Astronom. Phys. **9** (1961), 521–524. MR **26** #1263.

INSTITUTE FOR ADVANCED STUDY
UNIVERSITY OF AMSTERDAM

Proceedings of Symposia in Pure Mathematics
Volume 13, Part II, 1974

AXIOMATIZING SET THEORY

DANA SCOTT

As long as an idealistic manner of speaking about abstract objects is popular in mathematics, people will speak about collections of objects, and then collections of collections, and collections of collections of... of collections. In other words, *set theory is inevitable*. Let us not argue here the question of whether set theory is the ultimate foundation for mathematics. Surely it has already been shown to be an adequate foundation for a reasonably large part of mathematics. Furthermore, even though it may not be the last word, it is a simple theory of perfectly clear intent which would have to be interpretable in whatever theory finally would serve as our foundation. It is the purpose of this essay to demonstrate just how inescapable set theory is by re-examining its axioms.

It must be understood from the start that Russell's paradox is *not* to be regarded as a disaster. It and the related paradoxes show that the naive notion of all-inclusive collections is untenable. That *is* an interesting result, no doubt about it. But note that our original intuition of set is based on the idea of having collections of *already* fixed objects. The suggestion of considering all-inclusive collections only came in later by way of formal simplification of language. The suggestion proved to be unfortunate, and so we must return to the primary intuitions. These intuitions can gain an initial precision through formulating the two basic axioms of *extensionality* and *comprehension*, which we now discuss in detail.

Let the variables $a, b, c, a', b', c', a'', \ldots$ range over *sets* and the variables $x, y, z, x', y', z', x'', \ldots$ range over *arbitrary* objects. The symbol $=$ is used for *identity* and $\in$ for *membership*. Whether it is really interesting or profitable to allow for non-sets in the theory is debatable; but let us not exclude them yet. We agree that the condition $x \in y$ should imply that y is a set, a principle that we can formulate in logical symbols thus:

$$\forall x, y[x \in y \rightarrow \exists a[y = a]].$$

AMS (MOS) subject classifications (1970). Primary 02K15.

Of course this must be taken as an axiom; but it is so primitive, so much just a convention of grammar, that we will not even give it a name. The first axioms to be named are the following:

EXTENSIONALITY. $\forall a, b[\forall x[x \in a \leftrightarrow x \in b] \to a = b]$.

COMPREHENSION. $\forall a \exists b \forall x[x \in b \leftrightarrow x \in a \wedge \Phi(x)]$.

The extensionality axiom formalizes our idea that a set is nothing more than a collection of objects: It is uniquely determined by its elements. The comprehension axiom formalizes the idea that once a collection a is fixed, we can then extract from a any arbitrary subcollection b. The extraction process is effected by finding a property $\Phi(x)$ which distinguishes b as the subset of a comprehending all those elements having the property. There is no reason to place any restriction on how the property $\Phi(x)$ is formulated: We believe in the existence of arbitrary subsets. It is a great temptation to erase the condition $x \in a$, thus simplifying the axiom schema; but we all know what happens. It is much more profitable to ask: Where does the a come from?

Zermelo answered the question by giving several construction principles for obtaining new a's from old. Fraenkel and Skolem extended the method, and von Neumann, Bernays and Gödel modified it somewhat. Actually this is a rather sad history—because set theory is made to seem so artificial and formalistic. The naive axioms are contradictory. We block the contradiction and thereby emasculate the theory. Therefore to get anywhere we reinstate a few of the principles we eliminated and hope for the best. Now it would be wrong to accuse any of the above men of holding such a simplistic view of the axiomatic process. Nevertheless it is a very widely held view and one that is easy to fall into when considering only the formal axioms without their intuitive justification. Let us try to see whether there is another path to the same theory more obviously based on the underlying intuition.

The truth is that there is only one satisfactory way of avoiding the paradoxes: namely, the use of some form of the *theory of types*. That was at the basis of both Russell's and Zermelo's intuitions. Indeed the best way to regard Zermelo's theory is as a simplification and extension of Russell's. (We mean Russell's *simple* theory of types, of course.) The simplification was to make the types *cumulative*. Thus mixing of types is easier and annoying repetitions are avoided. Once the later types are allowed to accumulate the earlier ones, we can then easily imagine *extending* the types into the transfinite—just how far we want to go must necessarily be left open. Now Russell made his types *explicit* in his notation and Zermelo left them *implicit*. It is a mistake to leave something so important invisible, because so many people will misunderstand you. What we shall try to do here is to axiomatize the types in as simple a way as possible so that everyone can agree that the idea is natural.

Let us proceed on a very primitive basis. In order to obtain the sets to which the comprehension axiom can be applied, we imagine some way of dividing the sets into levels. There will be earlier levels and later levels. Let the sets up to a certain level be thought of as forming a *partial* universe V which is regarded as a legitimate set. We can be generous and assume that all the nonsets, the set-theoretical atoms, belong to all of the levels. In a *later* universe V' we have not only the elements of V, but also V itself to be used to form subcollections of V'; that is, $V \in V'$ as well as $V \subseteq V'$, where we define

$$\forall x, y[x \subseteq y \leftrightarrow \forall z[z \in x \rightarrow z \in y]].$$

Furthermore not only V, but by the same token all the subcollections of V, should also be *elements* of V'.

Once a set is fixed at one level, all its subsets are fixed at a later level—that is certainly the basic idea of the theory of types. We formalize this idea *not* by introducing type indices, but more simply by identifying a level with the collection of all sets (and nonsets) up to that level. We let variables $V, V', V'', \ldots$ range over these levels—that is, we take the idea of a *type level* (as identified with certain sets) as a primitive notion. The "later than" relation is transcribed simply as $V \in V'$. It need hardly be mentioned that we assume that there is at least one level and that each level is a set—axioms that we do not stop to name. What is important is the idea that a given level is *nothing more than* the accumulation of all the members and subsets of all the *earlier* levels (and all the nonsets, if any there be). In formal terms we have this axiom:

ACCUMULATION. $\forall V' \forall x[x \in V' \leftrightarrow \neg \exists a[x = a] \vee \exists V \in V'[x \in V \vee x \subseteq V]]$.

(By the way, just because we use the *variables* $V, V', V'', \ldots$ we should *not* think of the levels *themselves* as being arranged in an ω-type sequence. In general we will want a transfinite sequence. Also note that $V \in V'$ does *not* imply that V' is the *next* level; it may be a much later level.) The purpose of this axiom is to show how the levels *fit together*.

The question of *how far* the levels go out will be postponed for the moment. Of course, however far they do go, our intention is that they eventually capture everything. As an axiom this idea gives the following:

RESTRICTION. $\forall x \exists V[x \subseteq V]$.

In other words, the *whole universe*, if only it were a set, would behave as the ultimate level in the sense of the previous axiom. (Note that this axiom gives the existence of at least one level. It really should have been formulated with the clause $[x \in V \vee x \subseteq V]$, but we shall show below that $[x \in V \rightarrow x \subseteq V]$.) This will turn out to be nothing more or less than the well-known axiom of foundation, which is generally poorly understood. We feel that in the present context it appears as a quite natural expression of the fact that the sets are restricted to the levels.

Well, except for axioms of extent (infinity), what else do we need to explain the behavior of the levels and their subsets? Do we not have to go on to specify that the levels are ordered—even well-ordered? That the levels are comparable? Even that there is a first level? *No, we do not*: All these essential facts will follow from the above apparently primitive assumptions. This at first surprising result shows how little choice there is in setting up the type hierarchy. Furthermore, we will find that the principles of construction (union and power set) also drop out of these axioms; so that really only the principles of extent are needed beyond what we have. It seems safe to say that it would be hard to reduce the axioms of set theory to less than the above and still retain a simple intuitive basis.

As our first deduction, notice that as an immediate consequence of the accumulation axiom we have

(1) $$V \in V' \to V \subseteq V'.$$

This implies that the "less than" relation among levels is *transitive*. Though it follows from what we will prove below, it is instructive to show that the relation is *irreflexive*:

(2) $$V \notin V.$$

Because suppose the opposite. By the comprehension axiom we can form the set $a = \{x \in V : x \notin x\}$. (We use here the usual notation for set abstraction, which is justified by our axioms.) The set looks familiar, somehow. We have assumed $V \in V$, and by construction $a \subseteq V$. Thus $a \in V$, in view of the accumulation axiom. Now we proceed as with the Russell paradox to derive a contraction. Thus we have put a paradox to work to obtain a useful conclusion which is not at all paradoxical.

To make further progress it is helpful to employ the paradox of the set of all grounded classes just as we employed the Russell paradox. This paradox shows that there is no set consisting of *all* the x such that

$$\forall a[x \in a \to \exists y \in a[y \cap a = 0]].$$

(Here, 0 denotes the empty set and $y \cap a = \{z \in a : z \in y\}$, the use of both of which is justified by the comprehension axiom.) As in the above argument, we introduce the auxiliary set

$$\|V\| = \{x \in V : \forall a[x \in a \to \exists y \in a[y \cap a = 0]]\}$$

and ask about the interesting properties of this well-determined subset. The main fact is

(3) $$V \in V' \to \|V\| \in \|V'\|.$$

Because assume $V \in V'$. Now $\|V\| \subseteq V$, so by the accumulation axiom $\|V\| \in V'$. To establish the desired conclusion we need only show that $\|V\|$ is grounded. Thus, assume $\|V\| \in a$. If $\|V\| \cap a = 0$, we are done. If $\|V\| \cap a \neq 0$, let $x \in \|V\|$ and $x \in a$. By the definition of $\|V\|$, we have in this case again the required $\exists y \in a[y \cap a = 0]$. Thus another paradox has been put to work.

The first reward from the rehabilitation of the grounded classes is this important fact:

(4) $$x \in V \to x \subseteq V.$$

Because suppose $x \in V$ and let

$$a = \{b : \exists V' \in V[x \in V' \wedge b = \|V'\|]\}.$$

Clearly, by (3), a exists and $a \subseteq \|V\|$. If $a = 0$, then $\forall V' \in V[x \notin V']$. Thus by the accumulation axiom we find

$$\neg \exists c[x = c] \vee \exists V' \in V[x \subseteq V'].$$

The first alternative implies $x \subseteq V$ (by our grammatical convention), while the second implies the same in view of (1). Next, if $a \neq 0$, we can find $b \in a$ with $b \cap a = 0$ because $a \subseteq \|V\|$. But then $b = \|V'\|$ where $V' \in V$ and $x \in V'$. In effect we have already eliminated the case in which $\neg \exists c[x = c]$, so by accumulation we have $V'' \in V'$ with $[x \in V'' \vee x \subseteq V'']$. The first alternative is impossible, because it would imply $\|V''\| \in b$ and $\|V''\| \in a$. From the second alternative we obtain $x \subseteq V'' \subseteq V' \subseteq V$, as desired.

The second reward is the deduction of the full principle of well-foundedness (foundation regularity) for the $\in$-relation:

$$\exists x \Phi(x) \rightarrow \exists x[\Phi(x) \wedge \neg \exists y \in x[\Phi(y)]]. \tag{5}$$

Assume $\Phi(z)$; by the restriction axiom, $z \subseteq V$ for some V. As in the last proof, let

$$a = \{b : \exists V' \in V[\exists x \subseteq V'[\Phi(x)] \wedge b = \|V'\|]\}.$$

Again this set exists and $a \subseteq \|V\|$. In case $a = 0$, we easily conclude from (4) by means of the accumulation axiom that $\neg \exists y \in z[\Phi(y)]$. In $a \neq 0$, we choose $b \in a$ with $b \cap a = 0$. Now $b = \|V'\|$ with $x \subseteq V'$ and $\Phi(x)$, for suitable x and V'. From $b \cap a = 0$ we easily conclude as above that $\neg \exists y \in x[\Phi(y)]$ to complete the argument. Note that we can relativize quantifiers and deduce at once from (5) the special case:[1]

$$\exists V \Phi(V) \rightarrow \exists V[\Phi(V) \wedge \neg \exists V' \in V[\Phi(V')]]. \tag{6}$$

As an application of (6) we can now show that the levels are indeed linearly ordered:

$$V \in V' \vee V = V' \vee V' \in V. \tag{7}$$

Thus in conjunction with (1) and (6) we see that the levels are in fact *well-ordered*. To prove (7) we argue by contradiction. Assume the negation of the universal generalization of (7), and then pick V so that

$$\neg \forall V'[V \in V' \vee V = V' \vee V' \in V],$$

but so that no $V'' \in V$ has this property. Next let V' be chosen so that

$$V \notin V' \wedge V \neq V' \wedge V' \notin V,$$

[1] As A. Lévy observed to the author, the proof of (4) is, essentially, a proof of (6). Therefore it would seem better to prove (6) first and deduce (4) and (5) as corollaries. Note that this proof of (6) does not use the axiom of restriction.

but so that no $V'' \in V'$ has this property. After these two applications of (6) we shall obtain a contradiction. Suppose $V'' \in V$. Then $V'' \neq V'$ and by transitivity $V' \notin V''$, because $V' \notin V$. Hence by the choice of V, we must conclude $V'' \in V'$. Conversely, suppose $V'' \in V$. Then again $V'' \neq V$ and $V \notin V''$. This time by the choice of V', we must conclude $V'' \in V$. Thus we have shown

$$\forall V''[V'' \in V \leftrightarrow V'' \in V'].$$

Now look at the axiom of accumulation. We easily deduce from this last biconditional that

$$\forall x[x \in V \leftrightarrow x \in V'],$$

which means $V = V'$, a contradiction. Looking back, we certainly have utilized *all* the axioms thoroughly to reach this point: The rest is smooth sailing.

For example, though the existence of the intersection of two sets follows from comprehension, the existence of the union does not. But now from (7) we note that if a and b are given, then for suitable V, V' we have $a \subseteq V \wedge b \subseteq V'$, and $V \subseteq V' \vee V' \subseteq V$. Suppose $V \subseteq V'$, then

$$a \cup b = \{x \in V' : x \in a \vee x \in b\},$$

and similarly in the other case. Note also by (4) that

$$\bigcup a = \{x \in V : \exists c \in a[x \in c]\}.$$

There is, however, a problem about singletons and power sets.

Inasmuch as we have not assumed any axioms of extent, note that the system is trivially consistent with a model consisting of the empty set alone. Hence, we cannot expect to prove the existence of any power sets unconditionally. Nevertheless the situation is not very complicated; what we can prove is

$$\exists V[a \in V] \leftrightarrow \exists b \forall c[c \in b \leftrightarrow c \subseteq a]. \tag{8}$$

Because, if $a \in V$, then $a \subseteq V' \in V$ for some V'. Therefore, if $c \subseteq a$, then $c \subseteq V' \in V$, and so $c \in V$, too. This means we can take $b = \{c \in V : c \subseteq a\}$ to obtain the desired conclusion. Notice that (8) can be further simplified because in view of the axiom of restriction,

$$\exists V[a \in V] \leftrightarrow \exists y[a \in y].$$

We could call objects x such that $\exists y[x \in y]$ *elements*. Principle (8) tells us that sets that are elements always have power sets. Likewise we could show that elements can always be made into singletons, doubletons, and into *finite* sets in general.

At this point we are free to choose various directions: either moving to the Zermelo-Fraenkel-Skolem theory or to the von Neumann-Bernays-Gödel theory by distinguishing between sets (elements) and classes. The author does not feel that the class theory is particularly useful, but, of course, it is a perfectly good theory. In any case the remarks about the Zermelo theory can always be transposed to the other. The main point is that, whatever is desired, the artificial, "ad hoc" axioms have been completely avoided.

The usual way of axiomatizing the Zermelo-Fraenkel-Skolem theory is to adjoin an axiom of *infinity* and an axiom schema of *replacement*. These two principles can be combined into one seemingly more powerful statement.

REFLECTION. $\exists V \forall x \in V[\Phi(x) \rightarrow \Phi^{(V)}(x)]$.

Here the formula $\Phi(x)$ may contain additional free variables and $\Phi^{(V)}(x)$ represents the result of relativizing all quantifiers in $\Phi(x)$ to elements of V (that is, replace $\exists y$ by $\exists y \in V$, etc.) The principle is called the *reflection axiom* because the *partial* universe V that is asserted to exist *reflects* all the properties $\Phi(x)$ for $x \in V$ that hold relative to the *whole* universe. The author has to admit that the replacement axiom is more elementary and probably more intuitive. But the reflection axiom is so easy to formulate and use, it has all the practical advantages of a good axiom. In any case it is interesting to prove the equivalence with the more primitive axioms, though we will not stop to do so here.

To see why reflection gives infinity we proceed in two steps. First we show

$$\forall x \exists V[x \in V]. \tag{9}$$

Because let Φ be the formula $\exists y[x = y]$ with x as a free variable which will *not* be quantified upon in the application of reflection. We obtain by reflection

$$\exists V[\Phi \rightarrow \Phi^{(V)}],$$

which means $\exists V[\exists y[x = y] \rightarrow \exists y \in V[x = y]]$, or, more simply, $\exists V[x \in V]$, as desired. In the same way we can show next that

$$\exists V[\forall x \in V \exists y \in V[x \in y] \wedge \exists x[x \in V]]. \tag{10}$$

Because from (9) the formula $[\forall x \exists y[x \in y] \wedge \exists x[x = x]]$ is provable, and reflection yields (10). Clearly in the well-ordering of the levels the V of (10) is a nonzero limit point and thus represents an infinite set.

We can strengthen the reflection principle to

$$\forall a \exists V[a \in V \wedge \forall y, y', y'', \ldots \in V[\Phi(y, y', y'', \ldots) \leftrightarrow \Phi^{(V)}(y, y', y'', \ldots)]]. \tag{11}$$

An outline of the proof will have to suffice. In the original reflection axiom we can take the formula to be

$$\forall y, y', y'', \ldots, z[x = \langle y, y', y'', \ldots, z\rangle \rightarrow [\Phi(y, y', y'', \ldots) \leftrightarrow z = 0]]$$
$$\wedge\ \forall y, y', y'', \ldots, z \exists x'[x' = \langle y, y', y'', \ldots, z\rangle] \wedge \exists b[b = a],$$

where $\langle y, y', y'', \ldots, z\rangle$ denotes an unordered tuple. It is best to write out the equation $x = \langle y, y', y'', \ldots, z\rangle$ in the obvious way. From this it is very easy to deduce a statement that clearly implies Fraenkel's replacement schema:

$$\forall a \exists V \forall x \in a[\exists y \Phi(x, y) \rightarrow \exists y \in V \Phi(x, y)]. \tag{12}$$

By the way, it should be pointed out that the levels can be defined in terms of membership and so are theoretically eliminable. In fact one can prove

$$\exists V[a = V] \leftrightarrow \P a \subseteq a \wedge \forall x[\neg \exists c[x = c] \rightarrow x \in 0] \tag{13}$$
$$\wedge\ \forall a' \in a \exists b \in a \forall c \subseteq b[\P c \in a \wedge [\P c \in b \vee a' \subseteq \P c]]$$

where we define

$$\P a = \{x : \exists y \in a[x \subseteq y]\}.$$

After a suitable effort we can finally show that the axioms of accumulation and restriction together are equivalent to this single statement:

$$\forall a \exists b \forall c \subseteq b \exists c'[\forall x[x \in c' \leftrightarrow \exists y \in c[x \subseteq y]] \wedge [c' \in b \vee a \subseteq c']].$$

Deductions from this axiom (which were carried out by Montague and the author) are quite lengthy, so there seems to be no technical or conceptual advantage in reducing the number of primitive notions to the minimum. Of course, in some model-theoretic discussions we may want to know that the levels are definable, so the result has some interest.

Looking now to the future we ask: What should the new axioms of set theory be like? We can of course give axioms of infinity that increase the extent of our sets far beyond the inaccessible cardinalities. This chapter of set theory is far from being closed and probably never will be. As yet none of these axioms have settled the continuum hypothesis, though some of them do contradict Gödel's constructibility axiom. It will be very interesting to see what develops along this direction. Also we have not discussed the axiom of choice. No doubt it will always be desirable despite the technical interest of various independence questions involving it and weaker principles. If only it could be deduced from some more primitive principle!

OXFORD UNIVERSITY

Author Index

Roman numbers refer to pages on which a reference is made to an author or a work of an author. *Italic numbers* refer to pages on which a complete reference to a work by the author is given. **Boldface numbers** indicate the first page of the articles in the book.

Subject Index

LECTURES DELIVERED DURING THE INSTITUTE

John Addison
Current problems in descriptive set theory

Robert Bradford
Undecidability of the theory of Dedekind cardinal addition

C. C. Chang
Sets constructible using $L_{\kappa,\kappa}$

Paul Cohen
Remarks on the foundation of set theory

John E. Doner and Alfred Tarski
Extended arithmetic of ordinal numbers

Paul Erdös
Finite and infinite combinatorial analysis

Paul Erdös
Problems and results on combinatorial set theory

S. Feferman
Predicative set theory

Paul Fjelstad
Set theory as algebra

William Frascella
Tactical configurations for infinite sets

Haim Gaifman
Pushing up the measurable cardinal

Robin Gandy
Recursive functions of sets

Stephen Garland
Second-order cardinal characterizability

R. J. Gauntt
Undefinability of cardinality

P. C. Gilmore
Partial set theory

Petr Hájek
Bibliography of the Prague seminar on foundations of set theory (1962–1967)

James Halpern
The Boolean prime ideal theorem

Stephen Hechler
Generalized scales on number-theoretic functions

Ronald Jensen
Measurable cardinals and the GCH

Carol Karp
Some interconnections between infinitary logic and set theory

H. Jerome Keisler
End extensions of models of set theory

Saul Kripke
On the application of Boolean-valued models to solutions of problems in Boolean algebra

Saul Kripke
Transfinite recursion, constructible sets, and analogues of cardinals

Kenneth Kunen
Indescribability and the continuum

F. William Lawvere
Category-valued higher-order logic

Azriel Lévy
The independence of the axiom of choice, the generalized continuum hypothesis and the axiom of constructibility of Σ_2 statements of set theory

Azriel Lévy
Independence proofs by the Cohen forcing method. I

Azriel Lévy
Independence proofs by the Cohen forcing method. II

Azriel Lévy
The sizes of the indescribable cardinals

Saunders Mac Lane
Categorical alternatives to set theory

R. Mansfield
The solution of one of Ulam's problems concerning analytic rectangles

A. R. D. Mathias
The order-extension principle

Kenneth McAloon
Cohen's method applied to questions about ordinal definability

Yiannis N. Moschovakis
Predicative classes

Jan Mycielski
On the axiom of determinateness

John Myhill
Imbedding classical type theory in 'intuitionistic' type theory

John Myhill and Dana Scott
Ordinal definability

K. Namba
An axiom of strong infinity and the analytic hierarchy of ordinal numbers

Katuzi Ono
On Russell type paradoxes and some related problems

Richard Platek
Eliminating the continuum hypothesis

Lawrence Pozsgay
Liberal intuitionism as a basis for set theory

Karel Prikry
Models constructed using perfect sets

Hilary Putnam
Degrees of unsolvability and constructible sets

William Reinhardt
Conditions on natural models of set theory

Gerald Sacks
Degrees of nonconstructibility

Dana Scott
Axiomatizing set theory

Dana Scott
Lectures on Boolean-valued models for set theory

J. R. Shoenfield
Constructible sets

Jack Silver
On the consistency of the GCH with the existence of a measurable cardinal

Jack Silver
The independence of Kurepa's conjecture and two-cardinal conjectures in model theory

Robert Solovay
Measurable cardinals and the axiom of determinateness

Robert Solovay
Real-valued measurable cardinals

Robert Solovay
Solution of a problem of Fodor and Hajnal

Robert Solovay
If ZF + AC + MC is consistent, then ZF + MC + $2^{\aleph_0}$ *is a real-valued measurable cardinal is consistent. A sketch of the proof*

G. L. Sward
Transfinite sequences of axiom systems for set theory

Gaisi Takeuti
Hypotheses on power set

R. L. Vaught
Axiomatizability by a schema

Martin Zuckerman
Finite versions of the axiom of choice